Cybersecurity and Data Science

Cybersecurity and Data Science

Editor

Krzysztof Szczypiorski

MDPI • Basel • Beijing • Wuhan • Barcelona • Belgrade • Manchester • Tokyo • Cluj • Tianjin

Editor
Krzysztof Szczypiorski
Warsaw University of Technology
Warszawa, Poland

Editorial Office
MDPI
St. Alban-Anlage 66
4052 Basel, Switzerland

This is a reprint of articles from the Special Issue published online in the open access journal *Electronics* (ISSN 2079-9292) (available at: https://www.mdpi.com/journal/electronics/special_issues/Cybersecurity_Data).

For citation purposes, cite each article independently as indicated on the article page online and as indicated below:

LastName, A.A.; LastName, B.B.; LastName, C.C. Article Title. *Journal Name* **Year**, *Volume Number*, Page Range.

ISBN 978-3-0365-6906-2 (Hbk)
ISBN 978-3-0365-6907-9 (PDF)

Cover image courtesy of Krzysztof Szczypiorski.

Contents

About the Editor

Krzysztof Szczypiorski

Krzysztof Szczypiorski received the M.Sc. (Hons.), Ph.D. (Hons.), and D.Sc. (Habilitation) degrees in telecommunications from the Faculty of Electronics and Information Technology (FEIT), Warsaw University of Technology (WUT), Poland, in 1997, 2007, and 2012. He also completed his postgraduate studies in the psychology of motivation at the University of Social Sciences and Humanities, Warsaw, in 2013 and the Master of Business Administration (MBA) from the Warsaw University of Technology Business School, in 2022. He graduated in advanced networking from Budapest Tech (now Óbuda University), Hungary, in 2003, and the Hass School of Business, University of California at Berkeley, in 2013. His research interests include: new methods of phenomenon observation especially in communication and social networks, medicine, and stock exchange, new algorithms in cybersecurity (network steganography and steganalysis, anomaly detection, and fraud management), exploring cyber crimes through digital forensics, and design of advanced intelligence systems. He is currently a Professor of telecommunications with the Institute of Telecommunications (IT), FEIT, WUT, where he is the head and the co-Founder of the Cybersecurity Division. Head and co-founder of Research Centre for Cybersecurity and Data Science at WUT (2020-), head and co-founder B.Sc., M.Sc., MBA Programs in Cybersecurity at WUT (2019-). Since 2015, he has been the co-founder and the R&D Director of Cryptomage S.A. — a cybersecurity company. He is the author or the co-author of over 220 papers, three patent applications (one is granted), and over 80 invited talks.

Preface to "Cybersecurity and Data Science"

The increasing availability of big data, structured and unstructured datasets, raises new challenges in cybersecurity, efficient data processing and knowledge extraction. The field of cybersecurity and data science fuels the data-driven economy, so innovations in this field require strong foundations in mathematics, statistics, machine learning and information security.

This book is a reprint of the Special Issue (SI) on Cybersecurity and Data Science published in Electronics. The SI comprises high-quality papers dealing with the challenging research topics in cybersecurity combined with data science, especially with artificial intelligence and machine learning.

The papers included in this reprint discuss various topics ranging from cyberattacks, steganography, anomaly detection, evaluation of the attacker skills, modelling of the threats, and wireless security evaluation. Given this diversity of topics, I hope that the book will represent a valuable reference for researchers in cybersecurity security and data science.

Krzysztof Szczypiorski
Editor

electronics

Editorial

Cybersecurity and Data Science

Krzysztof Szczypiorski

Institute of Telecommunications, Warsaw University of Technology, 00-661 Warsaw, Poland; krzysztof.szczypiorski@pw.edu.pl

Towards the end of the Cold War in 1985, in reference to the theory of leadership for the first time, in the book 'Leaders: The Strategies For Taking Charge' by Warren Bennis and Burt Nanus [1], a modelled world concept with the acronym VUCA appeared to properly express its volatility, uncertainty, complexity, and ambiguity. The model adopted by the military and business circles spoke to a tragic paradigm of regular, often severe, and confusing changes. The catastrophic world of VUCA, which also fits the view of cyberspace, has become heavily exploited after almost 35 years; hence, the updated approach presented by Jamais Cascio in 2020 is BANI—brittle, anxious, non-linear, and incomprehensible. At first glance, you can treat the BANI world as a VUCA world with a new descriptive language. Still, a deeper look allows you to have, perhaps absurdly, hope that there is a method of "controlling" chaos by paving the way for proactive solutions by creating new roadmaps for the overwhelming world formed in the last few years, mainly due to the COVID-19 pandemic, and now due to hostilities in Eastern Europe.

The world of BANI excellently describes the challenges faced by modern cybersecurity [2]. When faced with existing phenomena, it has no chance to completely protect the world from all the unexpected vulnerabilities and defend against all attacks and their often-unknown consequences. This Special Issue is devoted to promoting the latest cybersecurity and data science research in a world where digital transformation turns data into the new oil. The increasing availability of big data, structured, and unstructured datasets raise new challenges in cybersecurity, efficient data processing, and knowledge extraction. The field of cybersecurity and data science fuels the data-driven economy. Innovations in this field require strong foundations in mathematics, statistics, machine learning, and information security.

The unprecedented increase in data availability in many science and technology fields (e.g., genomic data, data from industrial environments, sensory data of smart cities, and social network data) require new methods and solutions for data processing, information extraction, and decision support. This stimulates the development of new data analysis methods, including those adapted to analysing new data structures and the growing volume of data.

This Special Issue, 'Cybersecurity and Data Science', includes fifteen contributions from reputable researchers from Canada, China, Ecuador, India, Lithuania, Poland, Ukraine, the United Kingdom, and the USA.

In the first article entitled 'Multilayer Detection of Network Steganography', Smolarczyk et al. [3] proposed a new method for steganography detection in network protocols to provide a steganalysis capability for entities with large numbers of devices and connections. The solution was based on a multilayer approach for the selective analysis of derived and aggregated metrics utilising machine learning algorithms.

In the article 'A Wireless Covert Channel Based on Dirty Constellation with Phase Drift', Grzesiak et al. [4] presented a novel method of steganographic transmission based on phase drift in phase-shift keying or quadrature amplitude modulation. The proposed approach was based on the drift correction modulation method previously used in watermarking audio signals.

'Multi-Language Spam/Phishing Classification by Email Body Text: Toward Automated Security Incident Investigation' by Rastenis et al. [5] includes a solution based on

Citation: Szczypiorski, K. Cybersecurity and Data Science. *Electronics* **2022**, *11*, 2309. https://doi.org/10.3390/electronics11152309

Received: 15 July 2022
Accepted: 22 July 2022
Published: 25 July 2022

email message body text-automated classification into spam and phishing emails written in three languages: English, Russian, and Lithuanian. As most public email datasets almost exclusively collect English emails, the authors investigated the suitability of automated dataset translation to adapt it to email classification written in other languages.

In the article entitled 'Discussion on IoT Security Recommendations against the State-of-the-Art Solutions', Chmiel et al. [6] presented an overview of security guidelines for IoT proposed by various organisations and evaluated some of the existing technologies applied to ensure IoT security against these guidelines. The authors gathered recommendations offered by selected government organisations, international associations, and advisory groups. They compiled them into a set of the most common and essential considerations, divided into eight categories.

The topics of threat assessment were studied by Sharma et al. in 'Analysis and Implementation of Threat Agents Profiles in Semi-Automated Manner for a Network Traffic in Real-Time Information Environment' [7]. The authors proposed a semi-automatic information security model, which can deal with situational awareness data, strategies prevailing information security activities, and protocols monitoring specific types of the network next to the real-time information environment.

Krupski et al. [8] presented a survey on data transformation schemes for CNN-based network traffic analysis. The authors showed a consequence of the fact that network traffic data and machine learning data have different structures. They introduced a taxonomy of data transformation schemes and used this categorisation to describe various CNN-based approaches found in the state-of-the-art of network trafficking analysis.

'A Method for Fast Selection of Machine-Learning Classifiers for Spam Filtering' by Rapacz et al. [9] elaborated on how text analysis influences classification—a key part of the spam-filtering process. The authors proposed a multistage meta-algorithm for checking the classifiers' performance.

Bieniasz et al. [10] proposed a new approach to generating datasets for cyber threat research in a multi-node system. Towards this purpose, the proof-of-concept of such a system was implemented and could be used to collect unique datasets with examples of information hiding techniques.

Maksymovych et al. [11] developed a modification to additive Fibonacci generators, the essence of which was to use prime numbers as modules of recurrent equations describing the operation of generators. This modification made it possible to ensure the constancy of the repetition period of the output pseudorandom pulse sequence in the entire range of possible values of the initial settings–keys (called seeds) at specific values of the module.

In the article 'A Hybrid Machine Learning-Based Malware Detection Model for Android Devices', Rodrigo et al. [12] proposed the BrainShield as a hybrid malware detection model trained on the Omnidroid dataset to reduce attacks on Android devices. The simulation results showed that BrainShield improved the accuracy and the precision of well-known malware detection methods.

'Detection of Image Steganography Using Deep Learning and Ensemble Classifiers' by Płachta et al. [13] dealt with the problem of detecting JPEG images that have been steganographically manipulated. The performance of employing various shallow and deep learning algorithms in image steganography detection was analysed. The data, images from the BOSS (Break Our Steganographic System) database, were used with the information hidden using three popular steganographic algorithms.

Korona et al. in 'Comparison of Hash Functions for Network Traffic Acquisition Using a Hardware-Accelerated Probe' [14], addressed the problem of efficient and secure monitoring of computer network traffic. The authors proposed, implemented, and tested a hardware-accelerated implementation of a network probe using the DE5-Net FPGA development platform. They also researched the problem of choosing an optimal hash function to be used in a network probe for addressing network flows in a flow cache.

Andrade et al. in 'An Exploratory Study of Cognitive Sciences Applied to Cybersecurity [15], identified the fundamental concepts related to the application of cognitive sciences in cybersecurity for establishing defence strategies to minimise the impact of cyberattacks. The authors developed an exploratory study based on two stages: a text mining process to identify the main interest areas of research in the cybersecurity field and a valuable review of the papers chosen in a systematic literature review that was carried out using PRISMA methodology.

The machine learning-based implementation of Chinese Ludo, also known as Aeroplan Chess, a trendy board game for several decades, is the main topic of the paper by Han et al. [16]. Unlike most chess programs, which depend on high machine performance, the evaluation function in the proposed implementation was only a linear sum of four-factor values. The other contribution of this research was that the authors innovatively constructed a threat matrix that allows for the quick acquisition of the threat between any two dice from any two positions.

Finally, the paper entitled 'Simulation of Authentication in Information-Processing Electronic Devices Based on Poisson Pulse Sequence Generators' by Maksymovych [17] was devoted to modelling authenticators of information-processing electronic devices by creating a bit template simulator based on a Poisson pulse sequence generator. The developed generator had improved statistical characteristics for the output pulse sequence and expanded capabilities for solving specific practical problems.

I would like to thank all the contributors to this Special Issue, including the authors, reviewers, and the *Electronics* publishing team. I firmly believe the findings presented in this Special Issue will benefit the reading of interested researchers and general audiences.

Funding: This research received no external funding.

Acknowledgments: The editor would like to thank all the contributors to this Special Issue, including the authors and reviewers.

Conflicts of Interest: The authors declare no conflict of interest.

References

1. Bennis, W.; Nanus, B. *Leaders: The Strategies for Taking Charge*; Harper & Row: New York, NY, USA, 1985; 244p.
2. Szczypiorski, K. Cyber(in)security. *Int. J. Electron. Telecommun.* **2020**, *66*, 243–248. [CrossRef]
3. Smolarczyk, M.; Szczypiorski, K.; Pawluk, J. Multilayer Detection of Network Steganography. *Electronics* **2020**, *9*, 2128. [CrossRef]
4. Grzesiak, K.; Piotrowski, Z.; Kelner, J. A Wireless Covert Channel Based on Dirty Constellation with Phase Drift. *Electronics* **2021**, *10*, 647. [CrossRef]
5. Rastenis, J.; Ramanauskaitė, S.; Suzdalev, I.; Tunaitytė, K.; Janulevičius, J.; Čenys, A. Multi-Language Spam/Phishing Classification by Email Body Text: Toward Automated Security Incident Investigation. *Electronics* **2021**, *10*, 668. [CrossRef]
6. Chmiel, M.; Korona, M.; Kozioł, F.; Szczypiorski, K.; Rawski, M. Discussion on IoT Security Recommendations against the State-of-the-Art Solutions. *Electronics* **2021**, *10*, 1814. [CrossRef]
7. Sharma, G.; Vidalis, S.; Menon, C.; Anand, N.; Kumar, S. Analysis and Implementation of Threat Agents Profiles in Semi-Automated Manner for a Network Traffic in Real-Time Information Environment. *Electronics* **2021**, *10*, 1849. [CrossRef]
8. Krupski, J.; Graniszewski, W.; Iwanowski, M. Data Transformation Schemes for CNN-Based Network Traffic Analysis: A Survey. *Electronics* **2021**, *10*, 2042. [CrossRef]
9. Rapacz, S.; Chołda, P.; Natkaniec, M. A Method for Fast Selection of Machine-Learning Classifiers for Spam Filtering. *Electronics* **2021**, *10*, 2083. [CrossRef]
10. Bieniasz, J.; Szczypiorski, K. Dataset Generation for Development of Multi-Node Cyber Threat Detection Systems. *Electronics* **2021**, *10*, 2711. [CrossRef]
11. Maksymovych, V.; Harasymchuk, O.; Karpinski, M.; Shabatura, M.; Jancarczyk, D.; Kajstura, K. A New Approach to the Development of Additive Fibonacci Generators Based on Prime Numbers. *Electronics* **2021**, *10*, 2912. [CrossRef]
12. Rodrigo, C.; Pierre, S.; Beaubrun, R.; El Khoury, F. BrainShield: A Hybrid Machine Learning-Based Malware Detection Model for Android Devices. *Electronics* **2021**, *10*, 2948. [CrossRef]
13. Płachta, M.; Krzemień, M.; Szczypiorski, K.; Janicki, A. Detection of Image Steganography Using Deep Learning and Ensemble Classifiers. *Electronics* **2022**, *11*, 1565. [CrossRef]
14. Korona, M.; Szumełda, P.; Rawski, M.; Janicki, A. Comparison of Hash Functions for Network Traffic Acquisition Using a Hardware-Accelerated Probe. *Electronics* **2022**, *11*, 1688. [CrossRef]
15. Andrade, R.; Fuertes, W.; Cazares, M.; Ortiz-Garcés, I.; Navas, G. An Exploratory Study of Cognitive Sciences Applied to Cybersecurity. *Electronics* **2022**, *11*, 1692. [CrossRef]

16. Han, F.; Zhou, M. Threat Matrix: A Fast Algorithm for Human-Machine Chinese Ludo Gaming. *Electronics* **2022**, *11*, 1699. [CrossRef]
17. Maksymovych, V.; Nyemkova, E.; Justice, C.; Shabatura, M.; Harasymchuk, O.; Lakh, Y.; Rusynko, M. Simulation of Authentication in Information-Processing Electronic Devices Based on Poisson Pulse Sequence Generators. *Electronics* **2022**, *11*, 2039. [CrossRef]

electronics

MDPI

Article

Multilayer Detection of Network Steganography

Milosz Smolarczyk [1], Krzysztof Szczypiorski [1,2,*] and Jakub Pawluk [1]

1 Research & Development Department, Cryptomage SA, 50-130 Wrocław, Poland; Milosz.Smolarczyk@cryptomage.com (M.S.); jakub.pawluk@cryptomage.com (J.P.)

2 Institute of Telecommunications, Warsaw University of Technology, 00-661 Warsaw, Poland

* Correspondence: krzysztof.szczypiorski@pw.edu.pl or ksz@tele.pw.edu.pl

Received: 19 November 2020; Accepted: 10 December 2020; Published: 12 December 2020

Abstract: This paper presents a new method for steganography detection in network protocols. The method is based on a multilayer approach for the selective analysis of derived and aggregated metrics utilizing machine learning algorithms. The main objective is to provide steganalysis capability for networks with large numbers of devices and connections. We discuss considerations for performance analysis and present results. We also describe a means of applying our method for multilayer detection of a popular RSTEG (Retransmission Steganography) technique.

Keywords: steganography; network security; steganography detection; steganalysis; machine learning; big data; IoT; pattern mining

---

1. Introduction

Network steganography has recently gained considerable attention in the scientific community. Many new methods have been developed, and many more will be developed in the near future [1] as new network protocols are constantly being developed. This paper focuses solely on the detection of steganography techniques that operate at the network protocol level.

With the growing number of devices in networks, including IoT, network steganography detection faces new challenges in terms of both accuracy and performance [2]. To be performed effectively, steganography needs to operate:

- In line with analyzed network traffic;
- In near real-time regimes.

If detection is performed off-line or if it causes too much latency, there will be more traffic waiting to be analyzed than can actually be analyzed. Performance optimization is the main focus of the research described here since the main application of network steganography is real-time communication [3,4].

Some of the accurate detection methods tailored for specific network steganography techniques cannot be effectively implemented in real-time regimes because excessive computing and/or memory resources are needed [5]. This makes us question the *overall accuracy* of such methods since they are unable to analyze high-throughput traffic in a multi-host environment.

In this paper, we present a new method to introduce a compromise between detailed packet inspection and optimal detection performance. Our motivation is to provide a generic method that orchestrates network steganography detection in real-time regime, making it possible to implement in multi-host environments that generate high-throughput traffic. As a part of the method, we have presented a steganalysis layer selection method that provides an intelligent selection of steganalysis algorithms, preserving the balance between resource consumption and detection performance. To the authors' best knowledge, this is the first generic network steganography detection method that utilizes a top-down approach for a detection method selection algorithm to ensure optimal computation resource allocation.

2. Related Work

Historically, most network steganography detection methods had been part of research on new steganographic techniques. In recent years, there emerged new detection methods that are not countermeasures for a particular steganographic technique but provide a broader perspective. Based on the literature, we can distinguish two major categories for network steganography detection methods: *technique-specific* and *generic*, as presented in Figure 1.

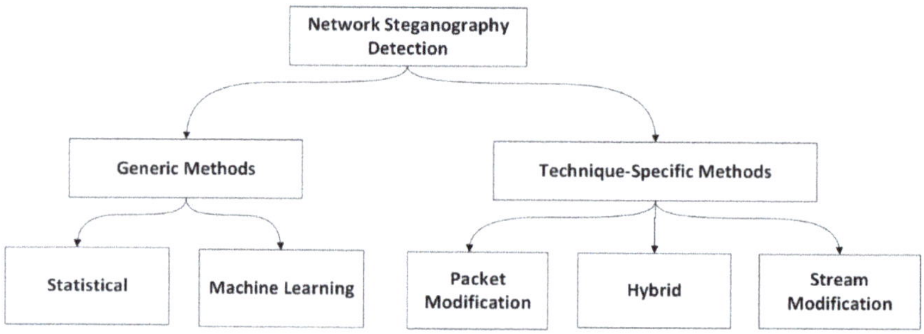

Figure 1. Network steganography detection classification.

The first category: technique-specific, comprises methods proposed as countermeasures for specific steganographic techniques. Methods in this category usually operate on low-level network data, require relatively much computation resources, and are not able to detect other steganographic techniques instead of the one or several for which they are designed.

The second category: generic, comprises methods that are not designed to detect one specific steganographic technique but offer a comprehensive approach to network anomaly detection and categorization of network traffic for potential steganographic utilization. Methods in this category may not provide detailed information on detected suspicious traffic but can label it for further investigation. Most generic methods fall into two subcategories that characterize their approach: statistical or machine learning.

A majority of methods described in the existing literature fall in the first category. Each of those methods is applied to specific steganographic techniques categories, as shown in Figure 2 [6].

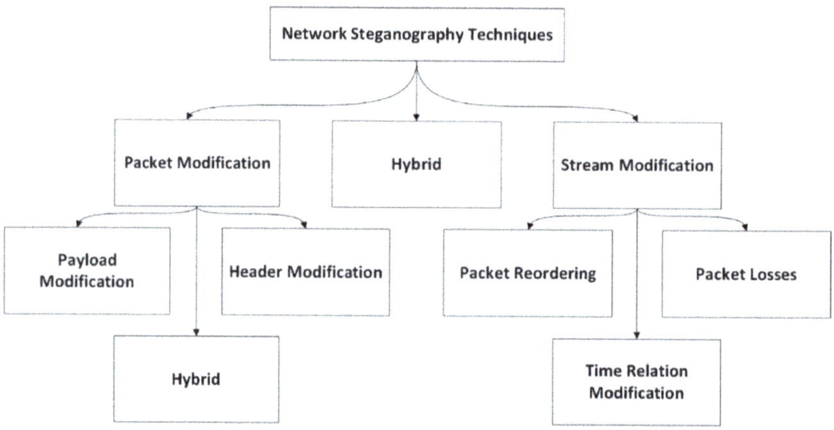

Figure 2. Network steganography classification.

For packet modification techniques, the steganalysis methods presented so far include:

- Header and payload analysis [7], including analysis of Verification Tags values; comparison between values of Maximum Inbound Streams sent by "normal" users (users who do not use steganography) and suspicious users; comparison between values of Stream Sequence Number sent by "normal" users (users who do not use steganography) and suspicious users; checking the value of Payload Stream Identifier; analysis of *a_rwnd* values and sizes of received chunks; analysis of the average number of duplicated chunks; analysis of Shared Key Identifier values; analysis of Padding Data; checking the existence of IP addresses that are sent in these parameters; comparison between values of the Heartbeat Info Parameter sent by a regular user (a user who do not use steganography) and a suspicious user; analysis of RandomNumber; comparison between values of ASCONF-Request Correlation ID sent by a regular user and a suspicious user. The methods presented above are dedicated to all packet modification techniques, including payload modification, header modification, and hybrid techniques.
- Header checksum observation [8], including checksum comparison for retransmitted IEEE 802.11 frames. If the checksum differs for the same payload and header and such observations are frequent, it is likely that a steganographic technique like HICCUPS has been utilized. The method is dedicated to header modification steganographic techniques.
- Observation of selected primary or derived features of header data [9], which includes observation of the least significant bit of the TCP sequence number. The method is dedicated to header modification steganographic techniques.

For stream modification techniques, several detection methods have been described, including:

- A multi-agent approach for observing network traffic time parameters, and intelligent correlation of observed meta-histograms utilizing trained machine learning algorithms [10].
- Analysis of inter-packet delays sequence distribution in multiple dimensions: distribution shape, data variation rule, data statistics. The method proposes an analysis of polarization characteristics, autocorrelation characteristics and clustering characteristics of the above features [11].
- Statistical analysis of selected metrics, header field comparison, and random number analysis [8].

There also exist steganalysis methods designed for hybrid steganographic techniques, including:

- MoveSteg [12], which is a method for detecting an endpoint from which hidden information is transmitted by analyzing a distribution of delay between consecutive packets as well as delay statistical metrics.
- The RSTEG (Retransmission Steganography) detection method [5,13], which is based on outlier detection of selected metrics, such as a retransmission ratio. Detection based on a retransmitted segment payload comparison is also proposed.
- The LACK (Lost Audio PaCKets) detection method [14], which is based on observation and outlier detection of RTP (Real-time Transport Protocol) segment delay.

Some generic methods for steganalysis operating on high-level aggregated metadata have been proposed:

- Data mining and anomaly detection in various metrics for distributed network covert channel detection [15].
- A framework that utilizes a statistical approach for monitoring of selected metrics and anomaly detection in statistical measures, including non-linear chaotic data [2]. The framework analyzes detected outliers and provides a probability of data leakage.
- A deep-learning approach for the detection and classification of covert channels. The method requires a data set comprised of covert communication, which can include a mix of various steganographic techniques [16].
- Detection method based on network traffic visualization [17], in which a fundamental design principle of the anomaly detection approach is the lack of direct, linear time dependencies for the created network traffic visualizations.

In addition, several generic methods for steganalysis have been proposed for steganogram detection in digital media. However, these methods apply for a different range of data-hiding techniques (digital images/media) that are outside the scope of this research. Those methods include:

- A supervised learning-based steganalysis [18], which requires a training phase to learn classification rules to further classify digital data utilizing deep learning algorithms.
- A simple image comparison and its metadata, such as file size, to extract a steganogram [19].
- Utilizing Bayes classifier for observation of peak frequency in audio signals [20],
- Utilizing a sliding window and a convolutional neural network for steganalysis in audio transmission [21].

All told, the existing literature on network steganography detection focuses on countermeasures and methods for the detection of newly described steganography techniques rather than a generic approach, with exceptions described above.

The generic method described in this paper provides a framework for the utilization of various steganalysis methods at once. The method requires the use of other existing network steganography detection methods for optimum effectiveness. The proposed method's main novelty is providing a capability for intelligent selection of best-fit steganalysis methods for analyzed network traffic to maintain optimal resource utilization. While some of the existing methods provide a generic approach to steganography detection, none of those methods provide a unified cooperation model for utilizing other methods.

3. Multilayer Network Steganography Detection

3.1. Method Description

The core concept for our proposed method of network steganography detection is multilayer steganalysis and intelligent detection method selection based on packet classification and optimal resource utilization. We propose a top-down approach for a detection method selection algorithm as it ensures optimal computation resource allocation. In such an approach, we prefer high-layer metrics analysis over methods operating on low-level data (which would require more resources) unless high-level analyzers identify suspicious network traffic.

As shown in Figure 3, the first step is a packet capture (101), which acquires a single network packet from a hardware resource, such as a network card. The next step is feature extraction (102), which is the first stage of building a data model. Extracted features may include protocol headers and other derived data that can be calculated in near real-time. Extracted features serve as an input for metrics aggregation (103) and steganalysis layer selection (104). Metrics aggregation modules provide derived metrics operating on various aggregation layers. The scope of the metrics and calculation algorithms is determined by the steganalysis method(s) for which the method is to be applied. Examples of the metrics aggregation may include aggregated data counters, port utilization, etc. The main assumption for metrics aggregation is that high-layer metrics computation should consume fewer resources and take less time than the computation of low-layer metrics, as shown in Figure 4. We named the lowest-layer metrics "*1st layer* aggregated metrics" and the highest-layer metrics "*N^{th} layer* aggregated metrics."

The calculated metrics and features extracted from each packet serve as input for steganalysis layer selection (104), which determines the optimal steganalysis layer. We discuss the steganalysis layer selection in Section 3.2.

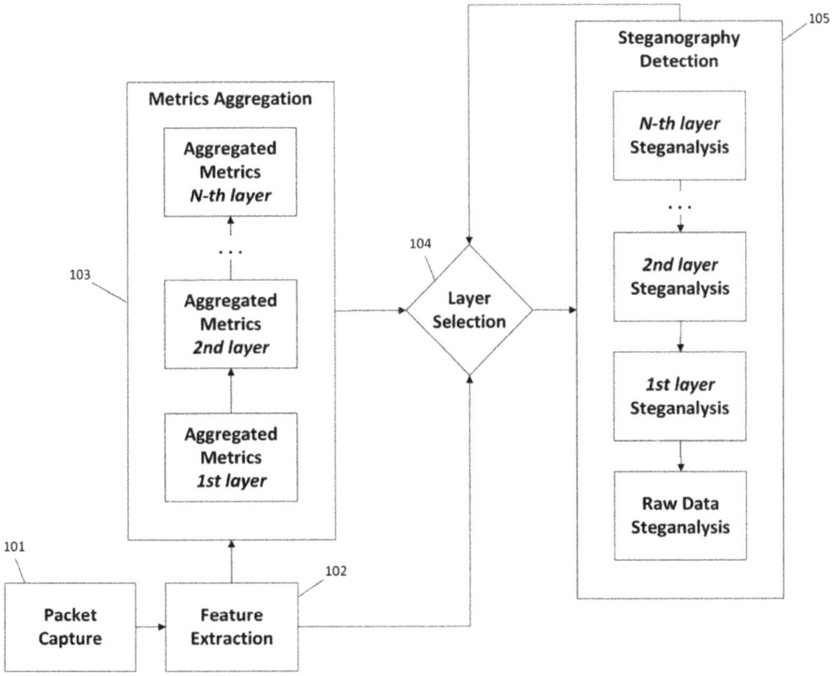

Figure 3. Multilayer detection method description.

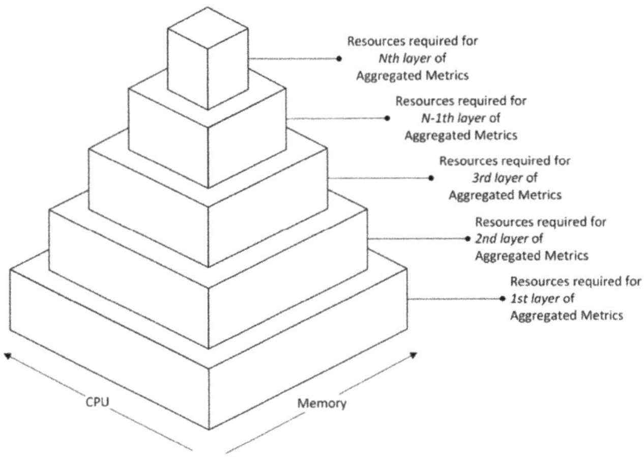

Figure 4. Aggregated Metrics hierarchy.

The Steganography Detection module (105) comprises multiple steganalysis methods. Each steganalysis method is assigned to a specific layer, based on the method's complexity and, in particular, on its resource utilization. Given a maximum of N layers of steganalysis methods, and a function $L(m)$ defining real-time operating resource consumption for each method m belonging to the set of methods M, the following is assumed:

$$\forall m \in M(L(m) < L(m-1)),\ \textit{provided that}\ N \geq m > 1 \quad (1)$$

In other words, steganalysis methods in higher layers require fewer resources to effectively detect network steganography in the real-time regime. Steganography detection methods in each layer may, but do not have to, operate on corresponding aggregated metrics layers.

The result of the performed multilayer steganalysis is provided to the steganography layer selection module to update the classification rules.

3.2. Steganalysis Layer Selection

The performance of our proposed method relies on the accuracy of the steganalysis layer selection algorithm and its parameters. In order to achieve better results, the algorithm should be tailored to fit specific performance requirements and at least the anticipated types of steganography technique. We suggest the following selection method, which should suffice for most applications.

As shown in Figure 5, the steganalysis layer selection method can operate in two modes:

1. Rule learning;
2. Packet classification.

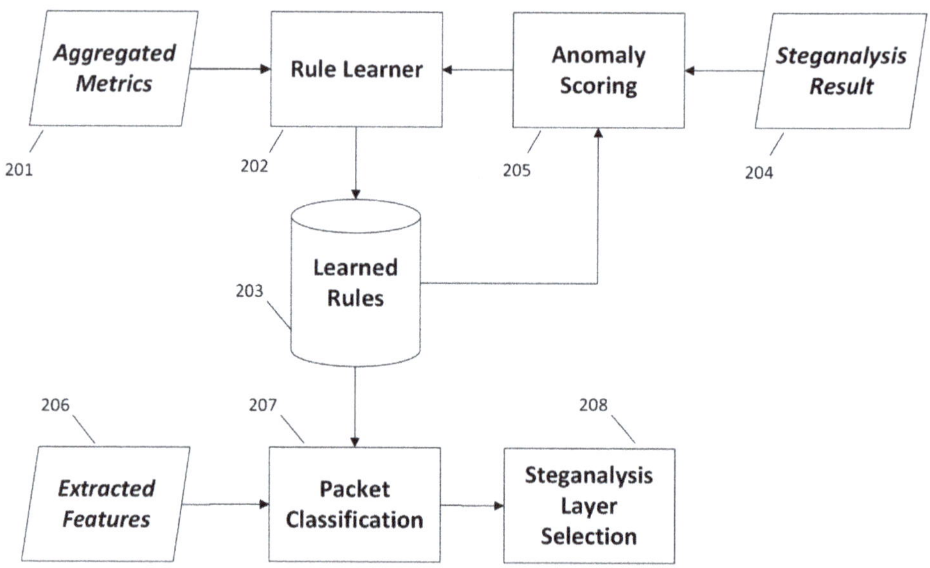

Figure 5. Steganalysis layer selection method.

In the first mode, the method applies various machine learning algorithms for frequent pattern mining, classification, and clustering to the steganalysis result (204) provided by the layered steganalysis module, computed anomaly scoring (205), and aggregated metrics (201). Learned rules are stored in memory (203) for the anomaly scoring module and packet classification.

In the second mode, the layer selection method receives a packet's extracted features (206) to classify the packet (207) for the selection of the optimal steganalysis layer (208). Packet classification (207) operates on previously learned rules and may use various classification methods and metrics, including but not limited to network address classification, network protocol classification, and TCP/UDP port classification.

The selection and application of specific algorithms for frequent pattern mining, classification, and clustering utilized by the rule learner module (202) are beyond the scope of this research work as they are widely discussed in the literature [21,22]. However, we recommend the *k-means clustering* for mining a predefined number of clusters of network devices, the *FP-growth* algorithm for frequent pattern mining, and an optimized SVM (Support-Vector Machine) trainer [23] for classification.

3.3. Applicability

Our proposed method can be applied to optimize the detection of the most known network steganography techniques shown in Figure 2. The spectrum of detected steganographic techniques relies on network steganography detection methods utilized by the presented multilayer detection method. In Table 1, we outline the potential advantages and disadvantages of applying our multilayer network steganography detection method to each group of techniques.

Table 1. Applicability of detection method.

Group of Techniques	Method Applicability
Packet Modification	Network steganography techniques belonging to this group are relatively easy to detect without utilizing significant resources. Applying our proposed method for this group may introduce unnecessary overhead for high-layer steganalysis.
Stream Modification	Detection of network steganography techniques belonging to this group needs significantly more resources to monitor network traffic. Applying our proposed method for this group provides value by optimizing and narrowing the range of detection methods used in the described top-down approach.
Hybrid	Detection of network steganography methods belonging to this group needs at least as many resources as stream modification methods. Applying our proposed method for this group provides value by optimizing and narrowing the range of detection methods used in the described top-down approach.

Based on the above findings, we suggest limiting the use of our method to stream modification and hybrid network steganography detection.

4. Case Study

4.1. Experiment Scope and Methodology

To measure the crucial features of the proposed method, we decided to perform an experiment by applying the method to a chosen network steganographic technique. The main need was to evaluate steganalysis time and its characteristics. To perform accurate measurements, we needed to choose a steganographic technique that has the following features:

- There exists a detection method that compares raw network traffic;
- There exists a detection method that operates on the *1st layer* of aggregated metrics;
- There exists a detection method that operates on the *2nd layer* of aggregated metrics;
- The method preferably operates under the application layer.

The above set of features ensures that the proposed method application is best utilized and operates on at least three layers. In our opinion, applying the proposed method to any steganographic technique satisfying the requirements above should provide performance gains, depending on the chosen steganalysis methods on each layer. Given the requirements, we chose to apply our method to RSTEG (retransmission steganography) [5,13,24]. The application to RSTEG detection provides us a set of steganalysis methods, presented in the literature, that can operate on aggregated metrics as well as raw data.

The main idea of RSTEG is to not acknowledge a successfully received packet in order to intentionally invoke retransmission. The retransmitted packet carries a steganogram instead of user data in the payload field [5]. Although RSTEG is intended for a broad class of protocols that utilize retransmission mechanisms, we chose to conduct the experiment on hidden communication detection in TCP/IP networks.

The objective of our case study is to document the performance of network steganography detection utilizing steganalysis method(s) individually and in the multilayer approach presented in this paper. Various RSTEG steganalysis methods can be implemented using a passive warden [25] in the architecture we describe in Section 4.2. We proposed detection methods and assigned them to particular layers.

We measured packet processing time to determine the effectiveness of the method. We divided the experiment into two parts:

1. Communication capture;
2. Capture analysis.

Processing time was measured between the times the warden started and finished analyzing captured traffic. All measurements were performed on ~100 MB chunks of ~5 GB of captured network traffic on a virtual machine with a single CPU and 2 GB of RAM. Each measurement was repeated 10 times to provide average results.

4.2. RSTEG Steganalysis Methods

The most effective methods for RSTEG communication in TCP/IP networks are based either on payload comparison or anomaly detection in derived stream metrics, i.e.:

1. Comparison of the retransmitted and original payload;
2. Anomaly detection in the number of retransmissions for an individual connection;
3. Anomaly detection in the number of retransmissions for an individual device.

4.2.1. Comparison of the Retransmitted and Original Payload

The method of detection based on a comparison of retransmitted and original payload operates on the assumption that every retransmitted TCP segment should have a similar payload to the original one. Any outliers can be safely assumed to be carrying steganograms.

Processing and memory requirements for this method are excessive [5] and limit the method's application to selected network connections only. Required resources scale with the amount of transmitted data and the number of network connections.

Based on the above description, we assign this method to the "Raw Data Steganalysis" layer.

4.2.2. Anomaly Detection in a Number of Retransmissions for an Individual Connection

Anomaly detection in a number of retransmissions for an individual connection requires the following operations to be performed:

1. Determining whether an individual packet is retransmitted;
2. Determining the TCP segment retransmission ratio for an individual network connection;
3. Outlier detection in the TCP segment retransmission ratio for an individual network connection.

Based on the fact that all of the above steps operate on a packet's extracted features and aggregated metrics, we assign this method to the first layer.

4.2.3. Anomaly DETECTION in a number of Retransmissions for an Individual Device

The method of anomaly detection in a number of retransmissions for an individual device is similar to the method presented above but operates in a broader scope. In this approach, the retransmission ratio for all network device traffic is determined, and outliers are detected.

Based on the fact that this method operates in a higher layer of aggregated metrics, we assign this method to the second layer.

4.3. Architecture

We conducted the experiment utilizing the following architecture for data capture and further investigation.

The architecture presented in Figure 6 comprises two endpoints: Alice (303) and Bob (311), who have established an RSTEG channel and are exchanging steganograms, among other network traffic. Bob's endpoint resides in a local network (310) in which all network traffic goes through the core router (312). The core router sends a copy of all traffic to the passive warden (313). Communication coming from other network devices (314) not necessarily involved in steganographic communication is also analyzed.

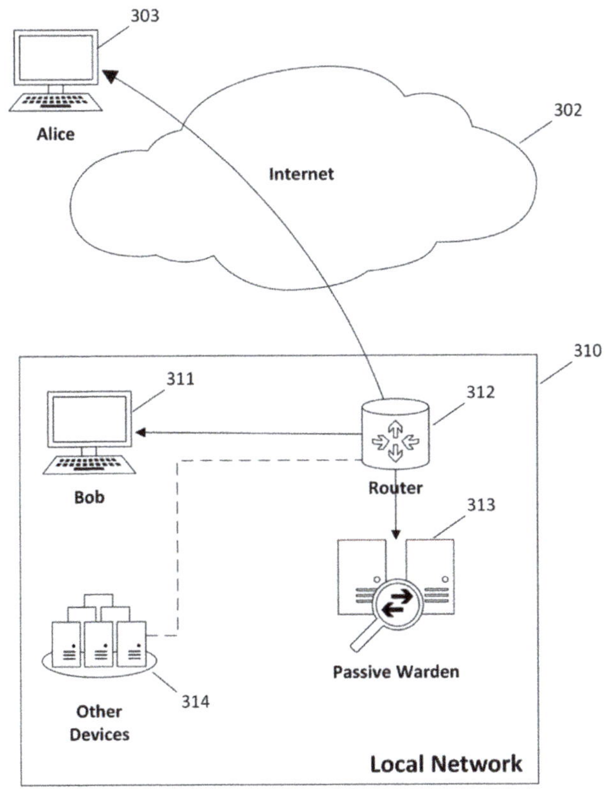

Figure 6. Implementation architecture.

4.4. Results

To provide an overview of multilayer steganalysis method performance, we measured the processing time for the methods applied in each layer as well as the total processing time required by our method. Each measurement was performed using the methodology described in Section 4.1.

As shown in Table 2, an increased ratio of retransmissions in the network causes an increase in processing time despite the chosen method(s). Processing time increases significantly for lower layers of steganalysis methods, including raw data steganalysis.

Table 2. Steganalysis performance.

Ratio of Retransmissions (%)	Raw Steganalysis Time (s)	1st Layer Detection Time (s)	2nd Layer Detection Time (s)
1	2.79	0.53	0.04
2	4.50	0.89	0.12
3	4.63	1.33	0.16
4	5.91	1.04	0.13
5	6.54	2.01	0.14

In Figure 7, we show the steganalysis time for raw data steganalysis in the retransmission ratio domain. As the chart shows, an increase in the network retransmission ratio causes an increase in the processing time; this increase can be approximated by a linear function. Given that raw data steganalysis for RSTEG means storing, iterating, and comparing retransmitted segments with the original ones, the substantial near-linear increase in processing time is fully legitimate.

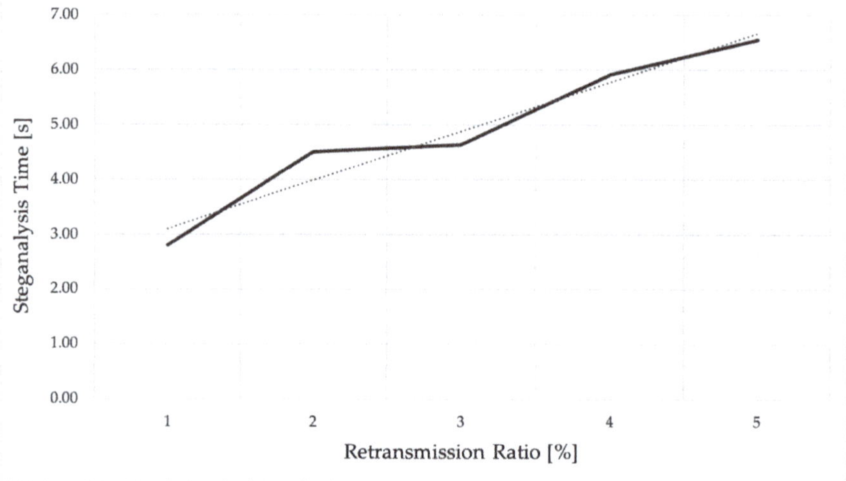

Figure 7. Raw Data Steganalysis time.

In Figure 8, we show the steganalysis time for the first-layer steganalysis in the retransmission ratio domain, which also includes raw data steganalysis for selected traffic. For RSTEG application, the method directs TCP segments belonging to connections that qualified as outliers for further raw data steganalysis, which means payload comparison.

The results also show an increase that can be approximated by a linear function, which makes sense because of the significant overhead required for processing separate connections, anomaly detection, and the potentially higher number of segments directed to lower-layer steganalysis.

In Figure 9, we show the steganalysis time for second-layer steganalysis in the retransmission ratio domain. Second-layer steganalysis involves selectively directing network traffic to first-layer steganalysis as well as raw data steganalysis. In our application, the method analyzes the retransmission ratio in the context of an individual network device, then directs outlier devices to the method that analyzes network connections and directs outlier traffic to payload comparison for retransmitted segments (raw data steganalysis).

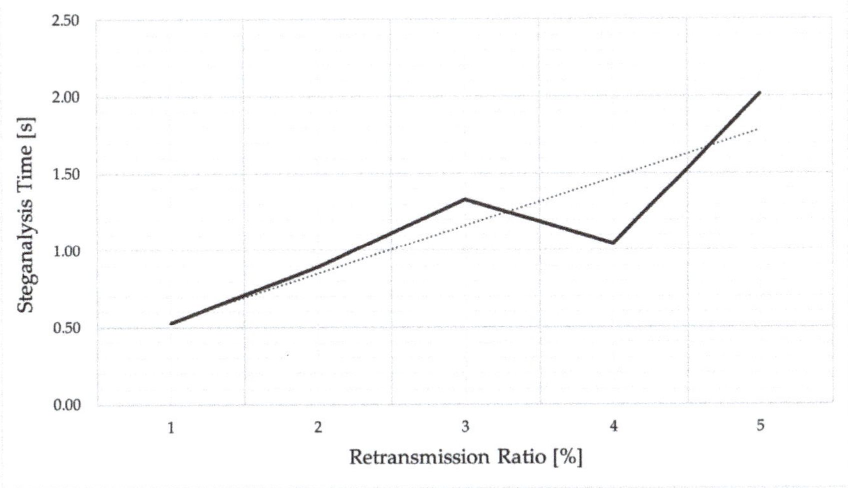

Figure 8. First-layer Steganalysis time.

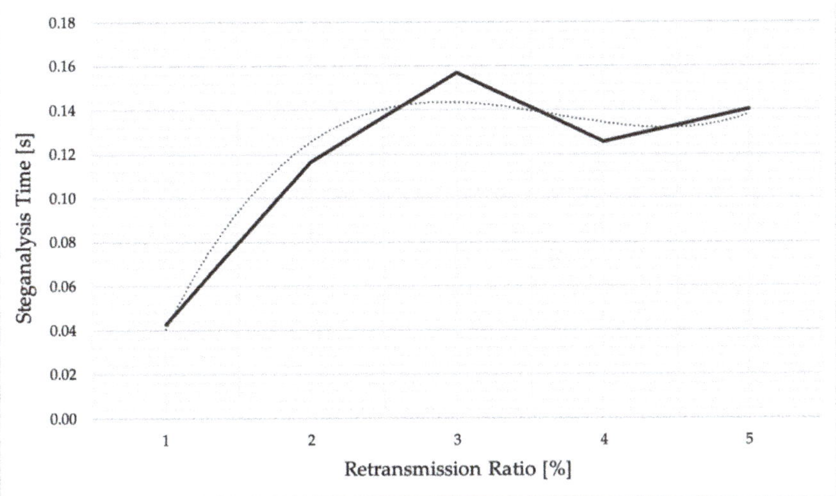

Figure 9. Second-layer steganalysis time.

The results show a non-linear increase in processing time, which can be closely approximated by a third-order polynomial function. Given that the method operates on the highest layer of aggregated metadata, a non-linear increase in processing time is justified. The second-layer method brings the most substantial gain in steganalysis, with an increasing retransmission ratio in our case.

The percentage gain in processing time when multilayer detection is applied is shown in Figure 10 and Table 3. The results show a significant performance gain for higher-layer detection methods (as expected). However, the gain slightly decreases in comparison to the lowest retransmission ratio applied (1%). This is a result of method selection algorithm overhead and aggregation of required metrics.

Table 3. Steganalysis performance gain.

Ratio of Retransmissions (%)	Raw Steganalysis Time (s)	1st Layer Detection Gain (%)	2nd Layer Detection Gain (%)
1	2.79	526%	6552%
2	4.50	506%	3861%
3	4.63	349%	2956%
4	5.91	568%	4716%
5	6.54	325%	4666%

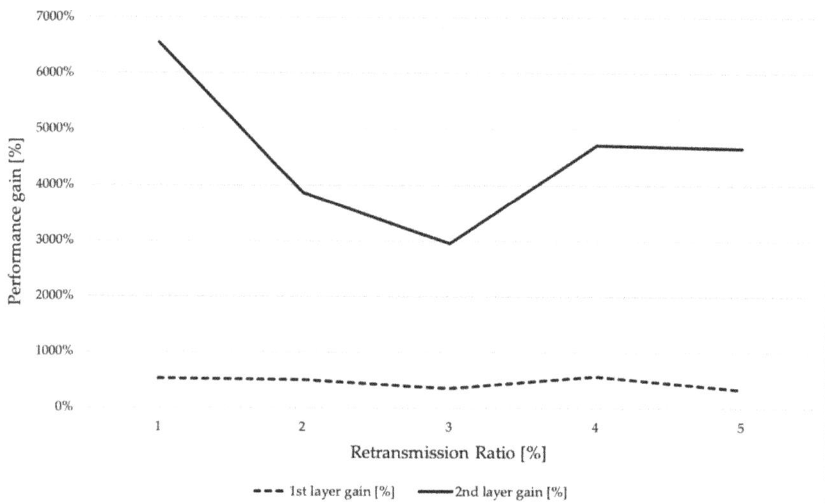

Figure 10. Steganalysis performance gain.

5. Conclusions

Multilayer steganography detection is a method that utilizes a top-down approach for network steganography detection and introduces an intelligent choice of steganographic methods applied to specific network traffic. As a part of the method, we have presented a steganalysis layer selection method that provides an intelligent selection of steganalysis algorithms, preserving the balance between resource consumption and detection performance. To the authors' best knowledge, this is the first generic network steganography detection method that utilizes a top-down approach for a detection method selection algorithm to ensure optimal computation resource allocation.

We have described the method's concept and its key components and discussed the method's applicability for network steganography detection in the context of known data-hiding methods. We also considered steganography detection in real networks in a wider context. The method requires the use of other existing network steganography detection methods for optimum effectiveness. The main novelty of the proposed method is providing a capability for intelligent selection of the best-fit steganalysis method for analyzed network traffic to maintain optimal resource utilization. Other generic detection methods presented so far do not provide orchestration for network steganography detection.

We applied our method for the detection of the RSTEG data-hiding method, presented the proposed detection techniques and assigned them to specific layers. The results demonstrated the method's performance gain over the steganalysis of raw network data. The presented characteristics of performance gain lead us to the conclusion that the method's application for real-time steganalysis is promising as it introduces a non-linear increase in processing time.

We suggest the following areas of future research:

- Performance scaling of required resources;
- Application of the method to other network steganography techniques;
- Application of the method to steganography detection in a broader context not tied to TCP/IP networks.

Author Contributions: M.S. contributed to theoretical formulation, design methodology, dataset development, experiment design and implementation, results interpretation, original draft preparation and revision. The other authors (K.S., J.P.) contributed to project supervision, theoretical formulation, result interpretation, and revision of the initial draft. All authors have read and agreed to the published version of the manuscript.

Funding: This scientific research work was co-financed by the European Union, project name: "The system for identification and monitoring of anomalies and risks in ICT networks". The amount financed by the European Union was EUR 1,044,534.63. The investment outlay value for the entire project was EUR 1,407,526.46. The subsidy was allocated from the European Regional Development Fund, Operational Program "Smart Growth", sub-measure 1.1.1 "Industrial research and development work implemented by enterprises" (grant number: POIR.01.01.01-00-0554/15).

Conflicts of Interest: The authors declare no conflict of interest.

References

1. Tanwar, R.; Malhotra, S.; Singh, K. *Future of Data Hiding: A Walk through Conventional to Network Steganography, in Cyberspace Data and Intelligence, and Cyber-Living, Syndrome, and Health*; Springer Science and Business Media LLC: Berlin, Germany, 2020; Volume 1230, pp. 123–132.
2. Nafea, H.; Kifayat, K.; Shi, Q.; Qureshi, K.N.; Askwith, B. Efficient Non-Linear Covert Channel Detection in TCP Data Streams. *IEEE Access* **2020**, *8*, 1680–1690. [CrossRef]
3. Collins, J.; Agaian, S. Trends Toward Real-Time Network Data Steganography. *Int. J. Netw. Secur. Its Appl.* **2016**, *8*, 1–21. [CrossRef]
4. Seo, J.; Manoharan, S.; Mahanti, A. Network steganography and steganalysis—A concise review. In Proceedings of the 2016 2nd International Conference on Applied and Theoretical Computing and Communication Technology (iCATccT), Bengaluru, India, 21–23 July 2016; pp. 368–371.
5. Mazurczyk, W.; Smolarczyk, M.; Szczypiorski, K. Retransmission steganography and its detection. *Soft Comput.* **2009**, *15*, 505–515. [CrossRef]
6. Lubacz, J.; Mazurczyk, W.; Szczypiorski, K. Principles and overview of network steganography. *IEEE Commun. Mag.* **2014**, *52*, 225–229. [CrossRef]
7. Frączek, W.; Mazurczyk, W.; Szczypiorski, K. Hiding information in a Stream Control Transmission Protocol. *Comput. Commun.* **2012**, *35*, 159–169. [CrossRef]
8. Grabski, S.; Szczypiorski, K. Network steganalysis: Detection of steganography in IEEE 802.11 wireless networks. In Proceedings of the 2013 5th International Congress on Ultra Modern Telecommunications and Control Systems and Workshops (ICUMT), Almaty, Kazakhstan, 10–13 September 2013; pp. 13–19.
9. Goher, S.Z.; Javed, B.; Saqib, N.A. Covert channel detection: A survey based analysis. In Proceedings of the 9 th International Conference on High Capacity Optical Networks and Emerging/Enabling Technologies, Istanbul, Turkey, 12–14 December 2012; pp. 57–65. [CrossRef]
10. Bieniasz, J.; Stepkowska, M.; Janicki, A.; Szczypiorski, K. Mobile agents for detecting network attacks using timing covert channels. *J. Univ. Comput. Sci.* **2019**, *25*, 1109–1130.
11. Lu, S.; Chen, Z.; Fu, G.; Li, Q. A Novel Timing-based Network Covert Channel Detection Method. *J. Phys. Conf. Ser.* **2019**, *1325*, 012050. [CrossRef]
12. Szczypiorski, K.; Tyl, T. MoveSteg: A Method of Network Steganography Detection. *Int. J. Electron. Telecommun.* **2016**, *62*, 335–341. [CrossRef]
13. Mazurczyk, W.; Smolarczyk, M.; Szczypiorski, K. On information hiding in retransmissions. *Telecommun. Syst.* **2011**, *52*, 1113–1121. [CrossRef]
14. Mazurczyk, W.; Lubacz, J. LACK—a VoIP steganographic method. *Telecommun. Syst.* **2010**, *45*, 153–163. [CrossRef]
15. Cabaj, K.; Mazurczyk, W.; Nowakowski, P.; Żórawski, P. Fine-tuning of Distributed Network Covert Channels Parameters and Their Impact on Undetectability. In Proceedings of the 14th International Conference on Availability, Reliability and Security—ARES '19, Canterbury, UK, 26–29 August 2019; pp. 1–8. [CrossRef]

16. Chourib, M. Detecting Selected Network Covert Channels Using Machine Learning. In Proceedings of the 2019 International Conference on High Performance Computing & Simulation (HPCS), Dublin, Ireland, 15–19 July 2019; pp. 582–588.
17. Mazurczyk, W.; Szczypiorski, K.; Jankowski, B. Towards steganography detection through network traffic visualisation. In Proceedings of the 2012 IV International Congress on Ultra Modern Telecommunications and Control Systems, Petersburg, Russia, 3–5 October 2012; pp. 947–954. [CrossRef]
18. Chandramouli, R.; Subbalakshmi, K. Current trends in steganalysis: a critical survey. In Proceedings of the ICARCV 2004 8th Control, Automation, Robotics and Vision Conference, Kunming, China, 6–9 December 2004; pp. 964–967.
19. Krenn, J.R. Steganography and Steganalysis, Internet Publication. Available online: http://www.krenn.nl/univ/cry/steg/article.pdf (accessed on 9 December 2020).
20. Zeng, W.; Ai, H.; Hu, R.; Gao, S. An algorithm of echo steganalysis based on Bayes classifier. In Proceedings of the 2008 International Conference on Information and Automation, Changsha, China, 20–23 June 2008; pp. 1667–1670.
21. Zhao, S.; Chandrashekar, M.; Lee, Y.; Medhi, D. Real-time network anomaly detection system using machine learning. In Proceedings of the 2015 11th International Conference on the Design of Reliable Communication Networks (DRCN), Kansas City, KS, USA, 24–27 March 2014; pp. 267–270.
22. Bieniasz, J.; Sapiecha, P.; Smolarczyk, M.; Szczypiorski, K. Towards model-based anomaly detection in network communication protocols. In Proceedings of the 2016 2nd International Conference on Frontiers of Signal Processing (ICFSP), Warsaw, Poland, 15–17 October 2016; pp. 126–130.
23. Franc, V.; Sonnenburg, S. Optimized Cutting Plane Algorithm for Large-Scale Risk Minimization. *J. Mach. Learn. Res.* **2009**, *10*, 2157–2192.
24. Mazurczyk, W.; Smolarczyk, M.; Szczypiorski, K. Retransmission Steganography Applied. In Proceedings of the 2010 International Conference on Multimedia Information Networking and Security, Nanjing, China, 4–6 November 2010; pp. 846–850.
25. Fisk, G.; Fisk, M.; Papadopoulos, C.; Neil, J. Eliminating Steganography in Internet Traffic with Active Wardens. In *Computer Vision—ECCV 2020*; Springer Science and Business Media LLC: Berlin, Germany, 2002; pp. 18–35.

electronics

Article

A Wireless Covert Channel Based on Dirty Constellation with Phase Drift

Krystian Grzesiak *, Zbigniew Piotrowski and Jan M. Kelner

Institute of Communications Systems, Faculty of Electronics, Military University of Technology, 00-908 Warsaw, Poland; zbigniew.piotrowski@wat.edu.pl (Z.P.); jan.kelner@wat.edu.pl (J.M.K.)
* Correspondence: krystian.grzesiak@wat.edu.pl; Tel.: +48-261-885-509

Abstract: Modern telecommunications systems require the use of various transmission techniques, which are either open or hidden. The open transmission system uses various security techniques against its unauthorized reception, and cryptographic solutions ensure the highest security. In the case of hidden transmissions, steganographic techniques are used, which are based on the so-called covert channels. In this case, the transparency and stealth of the transmission ensure its security against being picked up by an unauthorized user. These covert channels can be implemented in multimedia content, network protocols, or physical layer transmissions. This paper focuses on wireless covert channels. We present a novel method of steganographic transmission which is based on phase drift in phase-shift keying or quadrature amplitude modulation (QAM) and is included in the so-called dirty constellation techniques. The proposed approach is based on the drift correction modulation method, which was previously used in the watermarking of audio-signals. The developed solution is characterized by a variable bit rate, which can be adapted to the used modulation type and transmission conditions occurring in radio channels. In the paper, we present the method of generating and receiving hidden information, simulation research, and practical implementation of the proposed solution using the software-defined radio platform for selected QAM.

Keywords: wireless communications; covert channel; steganography; steganalysis; dirty constellation; wireless postmodulation steganography; phase drift; drift correction modulation; undetectability; security; quadrature amplitude modulation

Citation: Grzesiak, K.; Piotrowski, Z.; Kelner, J.M. A Wireless Covert Channel Based on Dirty Constellation with Phase Drift. *Electronics* **2021**, *10*, 647. https://doi.org/10.3390/electronics10060647

Academic Editors: Juan M. Corchado and Paulo Ferreira

Received: 31 December 2020
Accepted: 8 March 2021
Published: 11 March 2021

1. Introduction

The increase in capacity is one feature of emerging communication systems, including the fifth (5G) and sixth generation (6G) systems. This is due to the use of wider radio channels, their aggregation, or the use of higher frequency bands, i.e., millimeter, terahertz, or optical waves. A wider band of transmitted signals also gives greater possibilities to implement covert data transmission. Hence, the greater interest in searching for new steganographic methods is more evident [1].

Steganography consists in transmitting information to make the act of transmission undetectable. Unlike cryptographic information, whose content is encrypted, the very existence of steganographic information is concealed. Steganographic information, which is also known as covert information, requires a carrier—or, in other words, a cover. The simplest carrier scenario uses photos [2], audio [3] or video signals [4] (multimedia steganography) to hide additional information. One of the important steganography applications includes creating covert channels. The term "covert channel" was coined by Butler W. Lampson as: "... any communication channel that can be exploited by a process to transfer information in a manner that violates the systems security policy ... " [5]. B.W. Lampson focused on the exchange of data between programs. Nowadays, it is assumed that any method of communication used to illegally transmit information, which violates the system security policy, is a covert channel.

Data security is generally ensured by the flow control between the sender and authorized recipient. Though wired networks commonly use firewalls, security can be violated by covert channels. In this case, the information can be embedded by manipulating the packet timing information (i.e., covert timing channel) [6–10] or putting some bits into the packet headers (i.e., covert storage channel) [11,12]. So-called network steganography understood in a broad sense can be applied to both physical layer symbol frames [13], protocols of medium access control (MAC) [14], routing [15], networks [16], or higher layers, e.g., Transmission Control Protocol/Internet Protocol (TCP/IP) [11], Hypertext Transfer Protocol (HTTP) [17], and Domain Name System (DNS) [18]. These methods can be applied in homogeneous wired and wireless networks (e.g., accordant with the IEEE 802.11 standard [12,13]) as well as in heterogeneous ad hoc networks (e.g., [15]). Along with the development of network steganography techniques, we may notice novel solutions of network steganalysis, e.g., [19].

Contrary to the wired network, the wireless physical layer gives further possibilities for implementing the covert communication. In general, the wireless covert channel has its advantages and disadvantages [20]. In wired communications, it must be ensured that the channel is not distorted by network devices on either side of the covert channel. In wireless communication, the range between two points is limited by the transmitter power and the parameters of the receiver, e.g., its sensitivity. In this case, the steganography can relate to both radio and optical (e.g., [21,22]) wired and wireless communications. The further analysis focuses on wireless radio communication steganography.

In wireless communications, we deal with noise, interference, and fading that can seriously degrade transmission capabilities. In this case, it is worth introducing the terms "premodulation" and "postmodulation" steganography [23]. Pemodulation steganography is related to the bit structure change of the transmitted cover information. In contrast, postmodulation steganography refers directly to the physical parameter change of the transmitted waveforms. The proposed method presented in this paper is included in the category of wireless postmodulation steganography.

In the literature, we can find many ways to implement covert channels based on the physical layer. For *N*-ary frequency-shift keying (FSK) signals, E. Szczepaniak et al. proposed hiding information in a frequency drift and offset [24]. For orthogonal frequency-division multiplexing (OFDM) signals, different methods may be used, e.g., virtual carriers [25], the modification of training sequences, the covert-data-dependent shift of the signal carrier frequency, or using changes in the cyclic prefix [26]. The disadvantage of all these solutions is the low bit rates obtained. In contrast, theoretical approaches for the direct sequence spread spectrum (DSSS) and frequency-hopping spread spectrum (FHSS) techniques are shown in [27]. The DSSS and FHSS techniques are used primarily in military communication systems due to the low probability of detection/intercept (LPD/LPI) [28]. However, in [27], B.A. Bash et al. do not show the practical implementation of the proposed solutions. In the literature, we can also find steganographic solutions based on spatial multiplexing. P. Cao et al. propose to use multiple-input multiple-output (MIMO) technologies to create covert channels [29]. In this case, artificial noise modulated from secret messages is distributed as Gaussian channel noise, which increases the undetectability of a hidden transmission. Additionally, to improve its transparency, P. Cao et al. propose to modify the channel state information (CSI) parameters to reduce their correlation [29].

For modulations with constant points in their constellations, such as *N*-ary phase-shift keying (PSK) or quadrature amplitude modulation (QAM), dedicated steganography techniques are applicable. These methods are based on hiding information around the core points of the modulation constellation, and they seem more practical and convenient. For example, [30] describes the embedding of points in a constellation offset from the original points and adopts the term "dirty constellation". A similar solution defined as "constellation shaping modulation" is proposed in [31], although it focuses more on increasing covertness. In comparison to [30], the approach of [31] improves nondetectability at the expense of reliability degradation measured by bit error rate (BER). Creating a covert channel by superimposing pseudonoise asymmetric shift keying (PN-ASK) modulation was proposed in [32]. In this case, covert symbols are mapped by shifting the amplitude of primary symbols to a high order amplitude-phase modulation on a carrier constellation. The main drawback of this solution is limitations to the only phase-modulated cover signals. Moreover, multilevel amplitude modulation of the covert signal causes decreasing security (increasing detectability). Considering the high concentration of radio emissions in the available radio spectrum and the ever-growing number of used transceiver devices, this type of steganographic technique seems to be worth attention, research, and development. Considering the fact that [31] and [32] basically made a comparative analysis of [30], in this paper, we similarly focused on developing a novel idea of dirty modulation.

In this paper, we present a novel dirty modulation that is based on a phase drift and dedicated to the N-ary PSK or QAM signals. A similar solution, but based on the frequency drift in the N-ary FSK signals, is presented in [24]. The idea of hiding information in the drift of radio signal parameters is based on the drift correction modulation (DCM) method [33], which was used to hide information in audio signals. In this case, Z, Piotrowski also used the phase drift in the OFDM signal, which was then psychoacoustic corrected and added to the audio cover [33]. In the developed solution, we hide the information in a determined phase drift around the current constellation point of the transmitted radio signal. The phase difference constituting the drift step and the centroid distance from the constellation points are the parameters of the developed method. These parameters may be selected adaptive to the modulation type or transmission conditions. The modulation choice has a significant impact on the number of points (i.e., transmitted symbols) in its constellation, which translates into the distances of neighboring points on it. The influence of the transmission conditions, i.e., a signal-to-noise ratio (SNR), translates into a detection interference of the symbols and hidden subsymbols in the received signal. During the transmission, the received symbols (i.e., constellation points) change. Hence, phase drift detection, and thus steganalysis of the developed method, is more difficult than other dirty modulations. This is due to the fact that we do not set constant points in the constellation as the place of reading the subsymbol of covert transmission, but we hide the information in the drift step, i.e., the phase difference of the consecutive constellation points. This proves the originality of the proposed solution in comparison with other dirty modulation or wireless postmodulation steganography methods available in the literature.

In the paper, we present a methodology for generating and receiving the covert channels based on the dirty constellation with the phase drift and compare it with [30] as others have done. We want to emphasize that in addition to the simulation analysis typical for this kind of paper, we also present the first lab tests. In this practical implementation, we conducted tests using hardware and covert transmission over a real radio channel. To increase the SNR of the covert signal, multiple repetitions of the hidden subsymbols on the transmitting side and coherent averaging [34] of the successive drift phase differences on the receiving side shall be applied. This reduces the resulting bit rate of the covert transmission in the proposed method. However, this approach allows adaptation to the transmission conditions occurring in the radio channel. The proposed method will be used in future radio communication systems, including 5G networks dedicated to military applications. We plan to use it in the upcoming European Defense Agency (EDA) project, codenamed SOFTANET, for the hidden data layer in the wireless part of a software-defined network (SDN) [35–37]. It is in line with the trend, visible in the literature, of using steganography in 5G systems and networks, e.g., [21,38].

Analyzing the security system trends, including those based on cryptography and steganography, we see numerous threats to the existing techniques. They result from the increasing use of artificial intelligence (AI) algorithms [39,40] and quantum technologies [41,42] in security breach systems. On the other hand, these technologies may also be potential development directions of the security systems. Currently, the literature offers numerous solutions, including steganographic ones, which are based on modern AI (such as machine learning (ML) [43,44]) and quantum technologies [45,46] increasing the robustness, undetectability, and efficiency of emerging security and data transmission systems. In the case of the developed method, in the near future we want to use these ML techniques for a time-varying selection of the DCM parameters, which may increase its robustness, transparency, and the bitrate of the covert transmission.

Based on an approach presented in [47], we want to summarize the contribution of this paper. This research possesses various contributions in the domain of wireless steganography, watermarking, and wide-sense security of future wireless systems.

1. First, a DCM-based novel dirty constellation has been proposed, which can be used in N-ary PSK and QAM signals. The previous DCM solution [33] was dedicated to watermarking audio signals using the OFDM.
2. Second, based on simulation studies, the impact of the parameter variability of the developed method on its detection possibility using statistical analysis techniques has been shown.
3. Third, the efficiency of the developed method and its comparison with another dirty constellation technique [30] have been presented.
4. Lastly, the possibility of the practical implementation of the proposed solution has been shown, which gives a premise for its practical use in hidden data layer creation in SDN for the SOFTANET project and in future wireless systems and networks.

The remainder of the paper is as follows. Section 2 describes the idea of dirty constellation based on [30]. Our solution based on phase drift is presented in Section 3. In Section 4, we introduce the evaluation criteria of the covert channels. Using them, we analyze the developed dirty constellation based on the simulation and measurement approaches in Sections 5 and 6, respectively. Section 7 contains the paper summary.

2. Concept of Dirty Constellation

The main idea for creating covert channels based on the dirty modulation results from the fact that the received signals do not have an ideal constellation. Instead, we see a radio channel and parameter imperfection caused by both the transmitting and receiving devices' influence on blurring (spreading) the received-signal constellations. This effect translates to phase and amplitude distortions and ultimately to an increase in BER.

To increase the bandwidth efficiency, the N-ary PSK or QAM modulations are commonly used in telecommunications together with OFDM access. A time-frequency structure (waveform) of the OFDM signal provides a lot of space and possibilities for creating the covert channel, e.g., [26]. The suggested solutions include the use of the OFDM symbol waveform. When analyzing the QAM signal, the covert channel is provided based on the errors, which are the differences between the theoretical and the real points of the constellation. This results from environmental noise and hardware impairments. The theoretical, finite M number of constellation points corresponding to the N-ary PSK or QAM, in practice, has the form of the finite number of constellation point sets concentrated around the theoretical values without any distortions. Therefore, the channel will remain "hidden" as long as it is perceived as a noisy version of the carrier signal (i.e., the PSK or QAM signal) by the third uninformed party. It is important that after applying the covert information, the carrier remains distorted as little as possible so that the reception of the primary signal should be error-free.

It is the idea of using the dirty constellation proposed in [30]. The bits of covert information are mapped into additional constellation points placed around the base (i.e., carrier) constellation points, which is illustrated in Figure 1. According to the assumption, these additional constellation points are perceived as noise/error by the uninformed user. The value of the covert information symbol is defined in relation to the carrier constellation point.

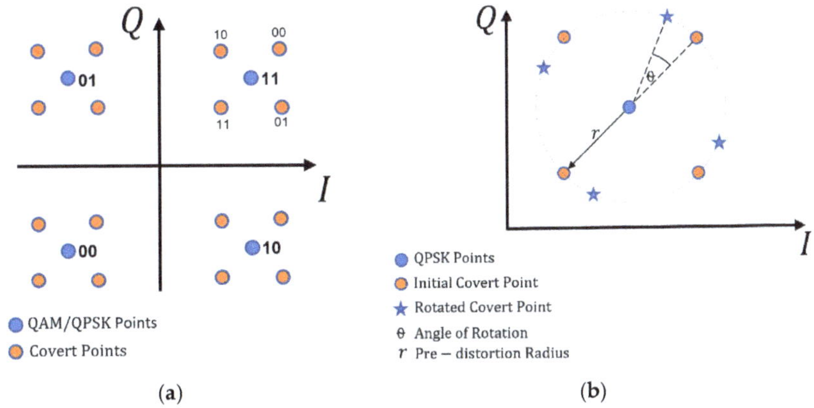

Figure 1. Dirty constellation: (**a**) assigning bits to constellation points; (**b**) chosen constellation point.

The carrier constellation is marked with blue circles, while the constellation related to the covert subsymbols (i.e., covert message bits) is marked with red ones. The probability of detection by adversaries is limited by reducing the predistortion radius, r. In addition, r can be changed (e.g., randomly) within a preset range. A. Dutta et al. for masking purposes used the fact that the QAM signal is transmitted by using the OFDM [30]. Therefore, each unused OFDM subcarrier may be a masking element providing more random (noiselike) in relation to the general character of the covert transmission. Additionally, A. Dutta et al. propose inserting angle rotation, θ, for successive OFDM harmonics, to increase the number of the received QAM constellation states for the subcarrier set. However, it does not change the number of constellation points if only one of the OFDM subcarriers is analyzed.

3. Phase Drift-Based Dirty Constellation

The developed dirty constellation with the phase drift is based on the DCM [33]. The DCM solution was used to create a watermark in audio files. The covert information is represented in monotonic phase changes of the selected signal harmonics. The mth harmonic selected for steganography is expressed by the following formula [33]

$$y_k(t) = A_m \exp \mathrm{j}(2\pi f_m t + \varphi_x + \Delta\chi_m), \tag{1}$$

where A_m and f_m represent amplitude and frequency of the mth signal, respectively, φ_x is an initial phase and $\Delta\chi_m$ is the preset phase drift carrying covert information.

In the DCM detector, the signal is subjected to the phase angle scanning procedure, which results in finding the maximum of the virtual fringe module $\chi_{V\max}$. As shown in [33], the DCM demonstrates good steganographic properties.

In our approach, we adopted the DCM for creating the dirty constellation of the steganographic channel. For further considerations, we assume that the QAM signal (without the use of the OFDM) is the carrier (cover) of the hidden transmission. The proposed method uses a covert symbol represented by the $K \geq 2$ phase drifts of the successive cover constellation points. Therefore, the P symbols of the QAM can carry P/K covert symbols.

For the N-ary QAM, the single constellation point is defined in the complex form:

$$x_n = A_n \exp(\mathrm{j}\varphi_n) \quad \text{for} \quad n = 1, 2, \ldots, N, \tag{2}$$

where A_n and φ_n are the amplitude and phase of the nth constellation point.

A single M-ary covert message adopts the form of a K-component complex vector:

$$y_k = A_m \exp \mathrm{j}(\varphi_x + k\varphi_m) \quad \text{for} \quad k = 1, 2, \ldots, K, \tag{3}$$

where A_m and $\varphi_m = \Delta\chi_m$ are the centroid distance and phase drift step for the proposed dirty constellation, respectively, and φ_x is a random initial phase, which does not change when the covert symbol is in progress.

Steganographic information is created by combining (adding) two successive K cover symbols with successive covert message symbols (i.e., vectors y_k). An example of creating hidden information for a single covert symbol and $K = 3$ is presented in Figure 2.

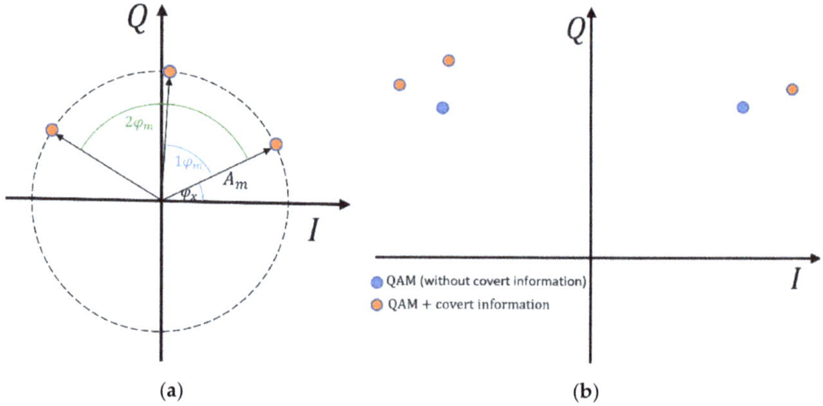

Figure 2. Example of constellation with drift phase: (**a**) vector y_k for $K = 3$; (**b**) quadrature amplitude modulation (QAM) cover (transmitted overt sequence of symbols: 1 + j, 1 − j, 1 − j) and resultant constellation with drift phase applied for preset vector y_k.

Possible phase increments correspond to the M-ary DCM. As a result, we receive new constellation points in the following form:

$$y_l = A_l \exp(\mathrm{j}\theta_l), \tag{4}$$

where A_l and θ_l mean the amplitude and phase of the new point in the constellation relative to the origin of the IQ coordinate system, respectively, which is depicted in Figure 3.

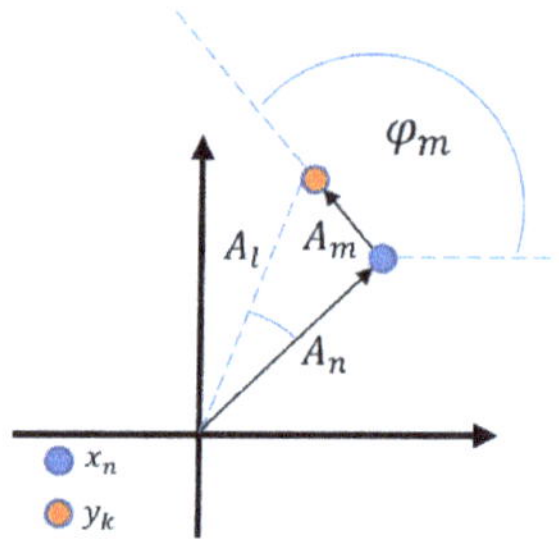

Figure 3. Additional y_k and carrier x_n constellation points.

The parameters A_l and θ_l are limited by the following relationships:

$$A_n - A_m \le A_l \le A_n + A_m, \tag{5}$$

$$-|\arctan(A_m/A_n)| \le \theta_l \le |\arctan(A_m/A_n)|. \tag{6}$$

Equations (5) and (6) mean that the maximum distortions of module A_n and phase φ_n of carrier symbols might be $\pm A_l$ and $\pm\theta_l$, respectively. These values describe the degradation level of the original signal by the DCM. On the one hand, it handicaps the recreation of the cover information. On the other hand, it shows the possibility of detecting stenographic transmission.

The proposed approach allows us to better hide the steganographic channel than the method based on the dirty constellation shown in [30]. This is possible thanks to the random initial phase φ_x, the adaptive selection of the centroid distance A_m and the phase drift step φ_m. In the general case, the parameters A_m and φ_m could also be random within defined ranges. An example of the phase drift-based dirty constellation for long bit strings is presented in Figure 4.

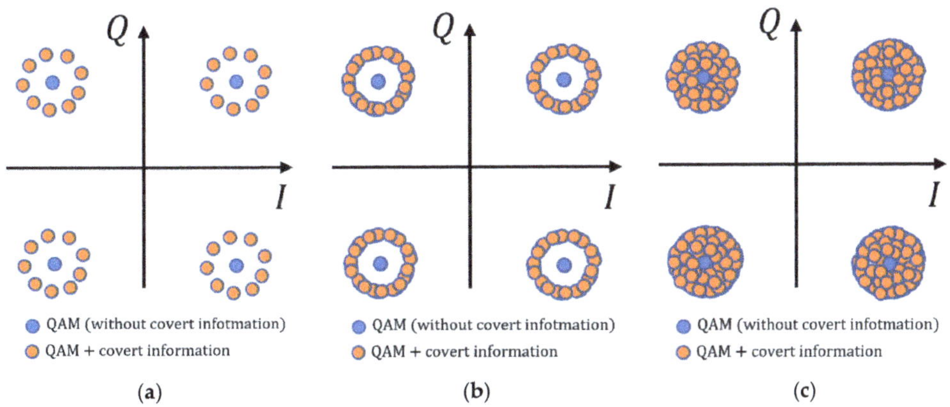

Figure 4. Constellations of carrier and the covert channels applied with: (**a**) fixed initial phase φ_x; (**b**) random φ_x; and (**c**) random φ_x and A_m.

The transceiver (transmitter–receiver) system is shown in Figure 5. This approach is similar to the framework of wireless covert channel proposed in [31] (Figure 4, p. 5).

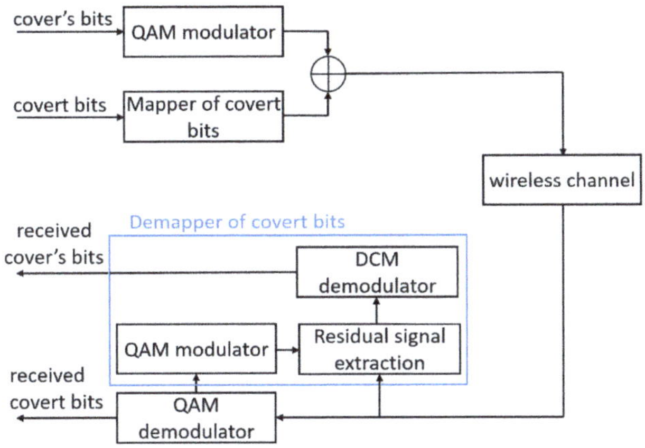

Figure 5. Receiver and transmitter (framework) of phase drift-based covert channel.

In the transmitter system, the fixed phase drift is added to the cover constellation obtained at the output of the QAM modulator. The DCM mapper is used to assign vector y_k. After putting together vector y_k with carrier constellation points x_n, the transmission adopts the steganographic form.

In the receiver system, if necessary, the detection of the overt signal (cover) is performed using a traditional QAM demodulator. By contrast, information from the covert channel is received by using the so-called DCM demapper. In the first instance, this system determines the residual signal, which is the difference between the constellation points of the carrier and the received signal, $\hat{y}_k$. In more detail, it is an assigned complex vector y_k with error e_k caused by the signal transmission through the radio channel

$$\hat{y}_k = y_k + e_k = \hat{A}_m \exp \mathrm{j}(\varphi_x + k\hat{\varphi}_m), \tag{7}$$

where $\hat{A}_m$ and $\hat{\varphi}_m$ are the amplitude and phase of the covert symbols in the received signal.

The DCM demodulator performs the phase angle scanning procedure. In this case, based on the knowledge of the possible angle φ_m values used in the steganographic transmitter, it calculates the U_m values for each φ_m as follows

$$U_m = \sum_{k=1}^{K} \hat{y}_k \exp(-\mathrm{j}k\varphi_m) = \sum_{k=1}^{K} \hat{A}_m \exp \mathrm{j}(\varphi_x + k(\hat{\varphi}_m - \varphi_m)), \tag{8}$$

The maximum value of the module U_m for a given φ_m $(m = 1, 2, \ldots, M)$ corresponds to the assigned symbol of the covert information. It is worth noting that according to Equation (8), the value K can be treated as the number of averages performed for a single covert symbol. By providing coherent averaging [34] of the hidden subsymbols, we obtain an increase in the detection gain of covert transmissions according to [33]. The multiplexing use of the hidden symbols may increase the effectiveness of its correct detection. In a similar way, the oversampling application of various credit-related datasets significantly improves the performance of a credit default prediction model presented in [47].

4. Evaluation Criteria for Covert Channels

To evaluate the covert channel efficiency, we may use the following parameters:

- Covert channel detectability defined by the detection probability,
- Cost understood as the carrier signal distortion,
- Transmission rate and BER in the covert channel.

In steganography, the most important parameter is its transparency, understood as the undetectability of the transmission existence by outside users (i.e., third parties). It should be highlighted that an easily detectable covert channel is completely irrelevant, even if it provides a high transmission rate.

The limit values of the cost function are usually defined for a given cover transmission standard by an error vector magnitude (EVM). The EVM for a signal with a covert channel is the cumulative result of influencing the transmitter systems and distortions introduced by the covert channel. For example, Table 1 contains the EVMs for quadrature PSK (QPSK) and N-ary QAM in the 5G systems [48] (Table 6.5.2.2-1, p. 48).

Table 1. Error vector magnitudes (EVMs) in 5G standard.

Modulation Scheme for Physical Downlink Shared Channel (PDSCH)	Required EVM (%)
QPSK	17.5
16 QAM	12.5
64 QAM	8.0
256 QAM	3.5

There are no clear criteria defining the channel detectability. The presence of a covert channel can be indicated when an exceedingly high EVM level is observed or the high level is more frequent (i.e., higher than assumed). This was detected in the statistical studies, including signal histograms.

Instead of determining the statistical values of the signal at the receiver input by using the classic prisoner problem proposed by Simmons [49] we may use the scheme of a moving observer [50] to assess the detection. In this model, Alice, Wendy, and Bob correspond to the steganographic source (SS), observer (Ob), and steganographic receiver (SR), respectively. Figure 6 illustrates the analyzed scenario.

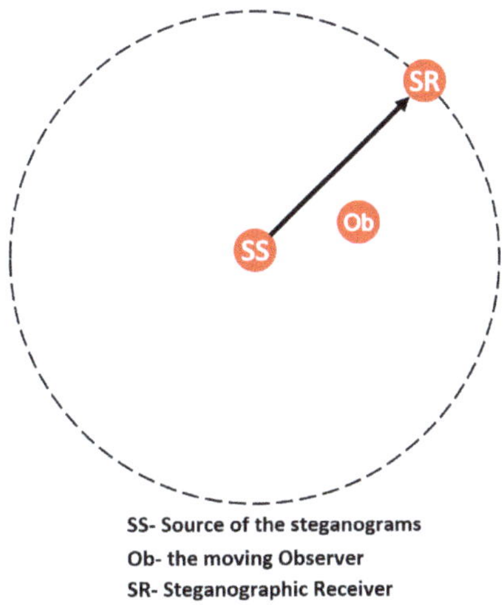

Figure 6. Concept of mobile observer for assessing steganographic method.

In this model, detecting the covert channel depends on the distance from the SS. As the distance increases, the influence of the radio channel intensifies. However, in the case of postmodulation steganography, there is a situation in which the mobile Ob, being near the SR, is able to detect steganographic emission with a high probability. With the increase in SNR, it is easier to identify the regularity of covert information constellation. By moving away from the SS, noise and other propagation phenomena occurring in the real channel make it difficult to detect the emission, as well as its reception, understood as the reproduction of the bitstream [50].

When considering the detectability of the covert channel as a priority, the proposed solution offers several basic properties:

- Random A_m and φ_x, i.e., no fixed constellation points, which provides robustness on statistical steganalysis methods—a major advantage over [30];
- Adaptive choice of A_m allow minimizing the EVM—a major advantage over [32];
- Possible to adaptively adjust the transmission rate by selecting the number K of phase changes—a major advantage over [30–32].

In steganographic signals, the lower EVM provides a lower probability of detecting the covert channel. On the other hand, the lack of fixed constellation points makes the histograms more random. Therefore, it is worth carrying out statistical analysis based on histograms and illustrating the covert channel implementation. In this simple example, we assumed that the cover is the 4 QAM signal, for $M = 4$, i.e., φ_m takes four values according to Equation (3). Figures 7–11 show the impact of random A_m and φ_x on the obtained constellations and histograms.

In Figures 7–9, channel interference impact was not considered. On the other hand, we assumed that the transmitter system did not introduce any impairments, i.e., transmitter EVM is equal to 0%. In real conditions, i.e., EVM > 0, masking the steganographic information is increasing. So, we can see that thanks to the random parameters A_m and φ_x, the distributions of the constellation points in the planes of the IQ quadrature components converge to the Gaussian distributions. Additionally, a real radio channel masks the presence of a covert channel, which is well illustrated by Figures 10 and 11. In these cases, we assume that the energy per bit to noise power spectral density ratio, E_b/N_0, for the signal cover is equal to 20 dB. It is worth emphasizing that the initial phase φ_x has no influence on the resulting bit rate of the covert channel and the obtained BER, but it has a masking function.

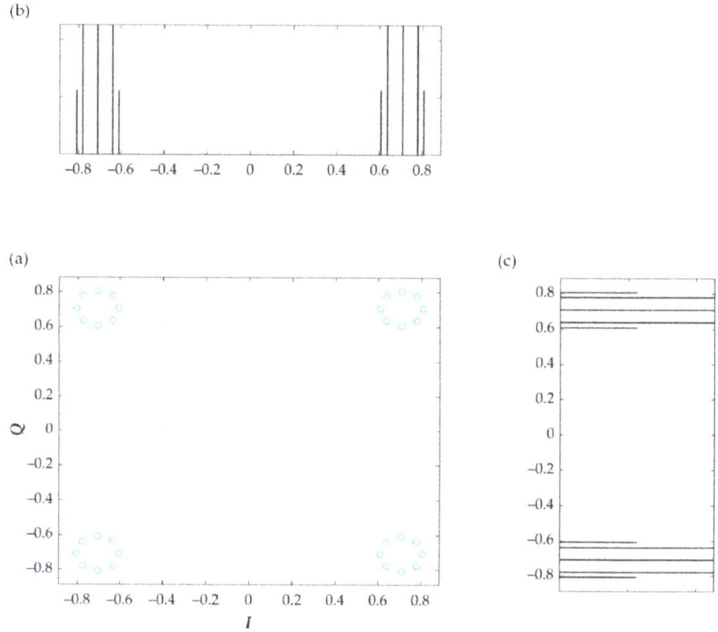

Figure 7. (**a**) Constellation (**b**) in-phase and (**c**) quadrature histograms for 4 quadrature amplitude modulation (QAM) cover and drift correction modulation (DCM) covert channel with $M = 4$ and for fixed parameters.

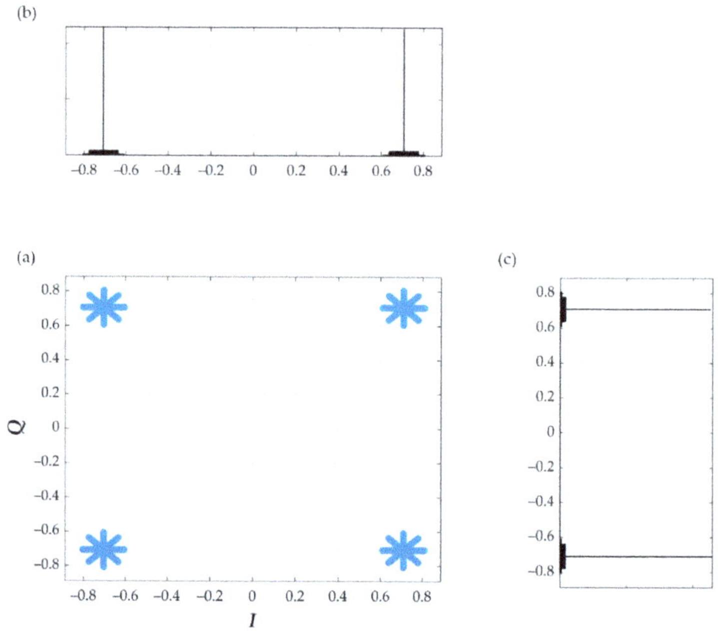

Figure 8. (**a**) Constellation (**b**) in-phase and (**c**) quadrature histograms for 4 QAM cover and DCM covert channel with $M = 4$ and for random A_m.

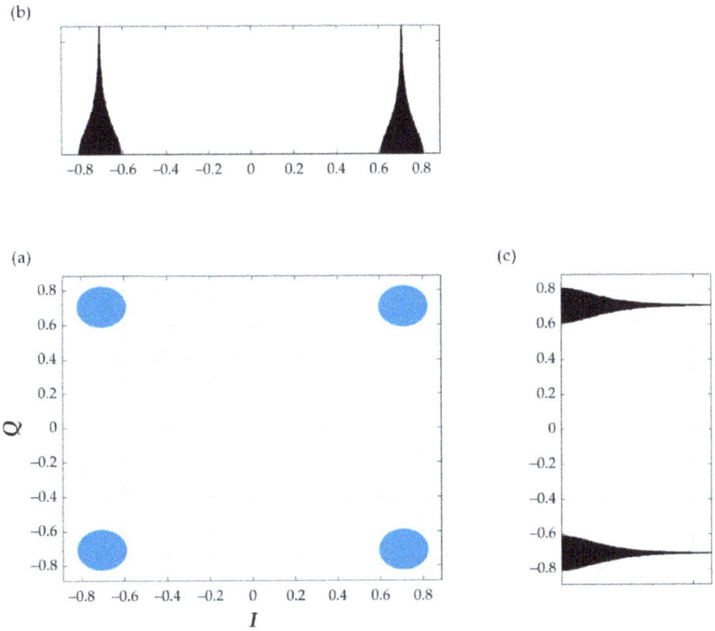

Figure 9. (**a**) Constellation (**b**) in-phase and (**c**) quadrature histograms for 4 QAM cover and DCM covert channel with $M = 4$ and for random A_m and φ_x.

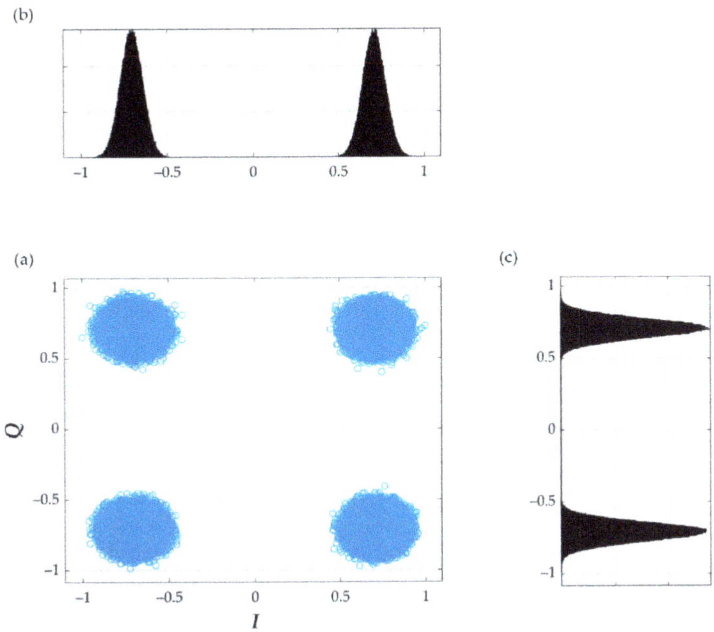

Figure 10. (**a**) Constellation, (**b**) in-phase and (**c**) quadrature histograms for 4 QAM cover and DCM covert channel with $M = 4$ and for random A_m and φ_x, $E_b/N_0 = 20$ dB for cover.

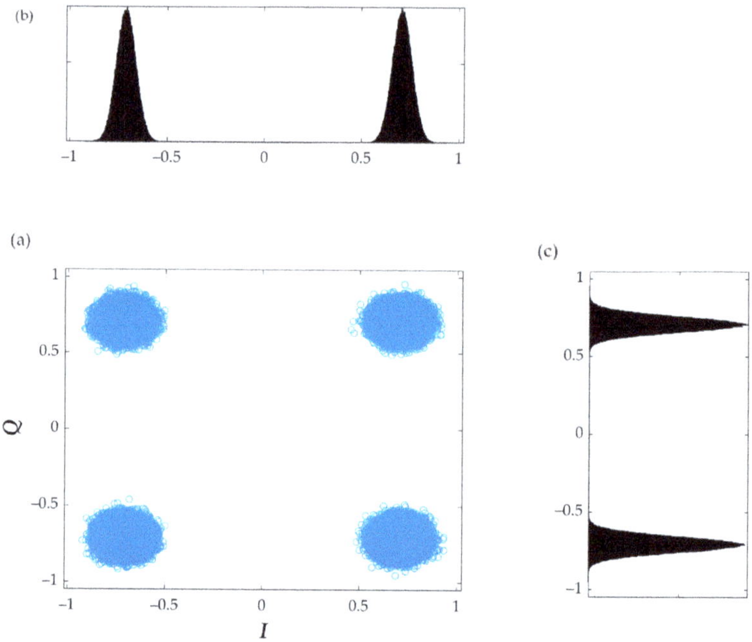

Figure 11. (**a**) Constellation, (**b**) in-phase and (**c**) quadrature histograms for 4 QAM cover and E_b/N_0 = 20 dB, and without DCM covert channel.

5. Simulation Analysis of Dirty Constellations

Based on simulation studies, we compared our solution with the dirty constellation method presented in [30]. In this case, we assumed that the number K of averaging in the DCM modulator corresponds to the number of covert symbols per carrier symbol (SPS) in [30], the cover is the 4 QAM signal, and $M = 4$. For easier comparison, we additionally assumed that the centroid distance A_m and the radius r of covert information dispersion in [30] remain constant. Figure 12 depicts the obtained comparison results of the transmission capabilities of the covert channels and the effect of the covert channels on the carrier QAM signals for EVM = 3%.

The transmission rate obtained for the covert channel is directly related to the number of symbols transmitted by the cover. Assuming the same value of the cover and steganographic information, K-fold averaging or given SPS brings about a K-fold reduction in the number of symbols transmitted over the covert channel. EVM at the level of 3% does not cause any noticeable deterioration of cover properties. In the proposed solution, compared to [30] higher numbers of averages need to be used to obtain similar BER levels for a covert channel.

Introducing random A_m reduces the level of energy per symbol of transmitted covert information, which means that it is necessary to increase the number of averages to provide the assumed BER level. An example for such a case (i.e., random A_m) was presented in Figures 13–16. The cover is provided by 4 QAM, and $M = 4$. In Figure 13, we show BER versus SNR for different K and EVM = 3%. Figure 14 illustrates EVM versus SNR for the carrier signal and covert channel ($K = 30$), whereas Figures 15 and 16 depict the constellations and histograms for $K = 30$, EVM = 3%, SNR > 50 dB and SNR = 38 dB, respectively.

The simulation studies carried out showed that the increase in the number K of averages ensures the BER reduction for a given SNR. On the other hand, introducing the covert information using the proposed approach did not significantly increase the EVM. The difference of a few percent is visible only for SNR > 20 dB. However, in this case,

EVM < 8%. The obtained constellations and histograms show that for good transmission quality (i.e., SNR = 38 dB and SNR > 50 dB), the randomness of the centroid distance ensures the undetectability of the proposed method using statistical steganalysis.

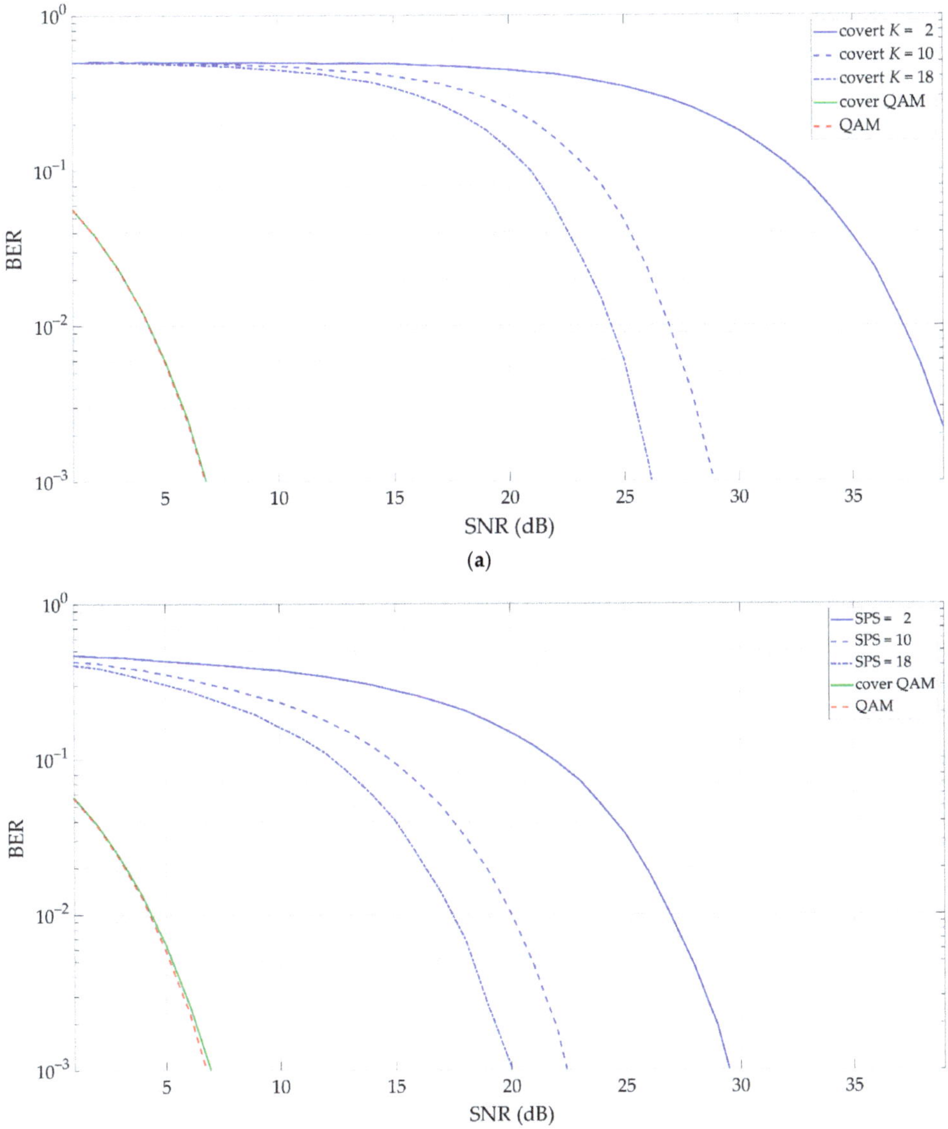

(a)

(b)

Figure 12. Comparison of dirty constellation methods based on bit error rate (BER) versus signal-to-noise ratio (SNR) graphs for EVM = 3%: (**a**) proposed solution; (**b**) solution presented in [30].

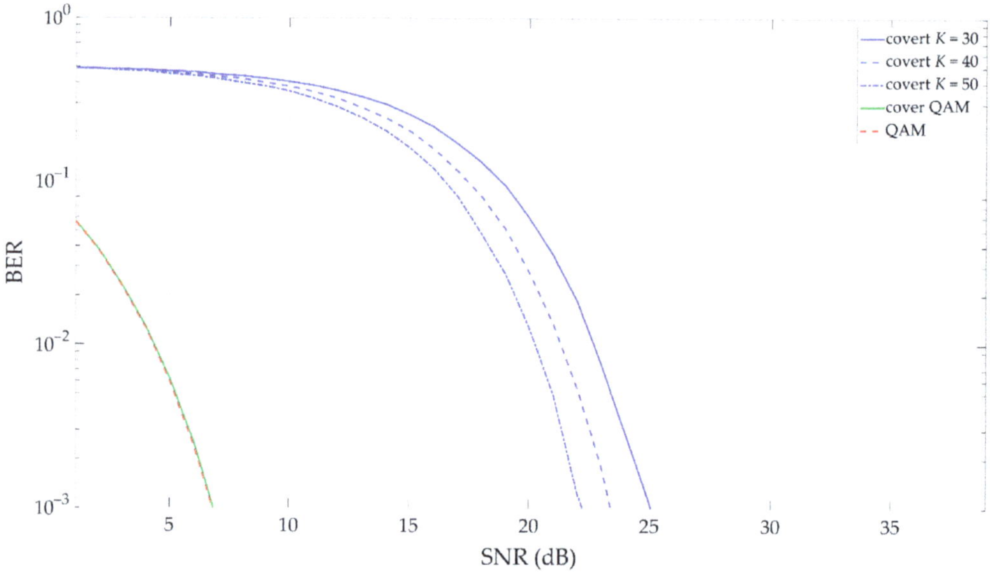

Figure 13. BER versus SNR for different K, random A_m, and error vector magnitude (EVM) = 3%.

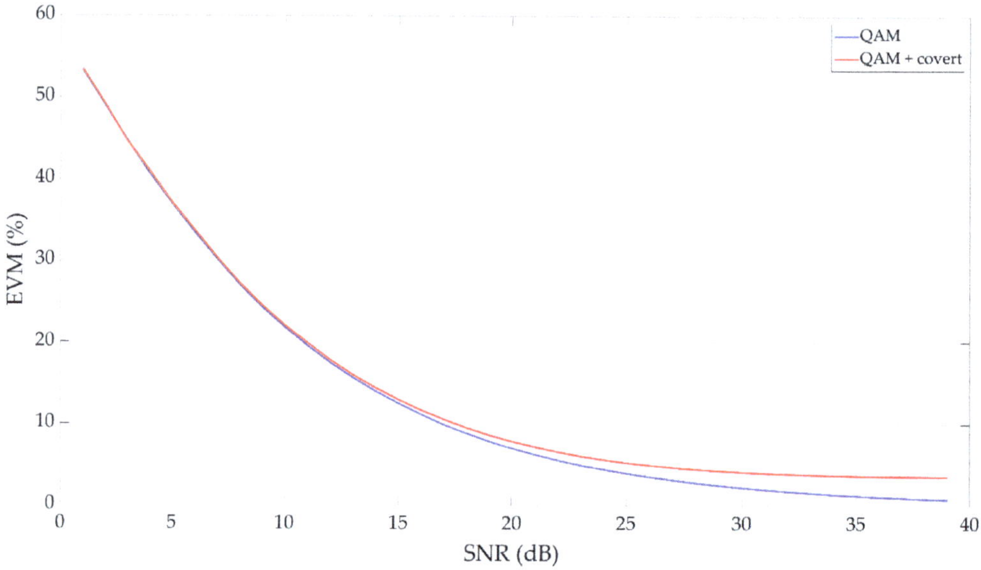

Figure 14. Error vector magnitude (EVM) versus SNR for cover and covert channel (K = 30), and random A_m.

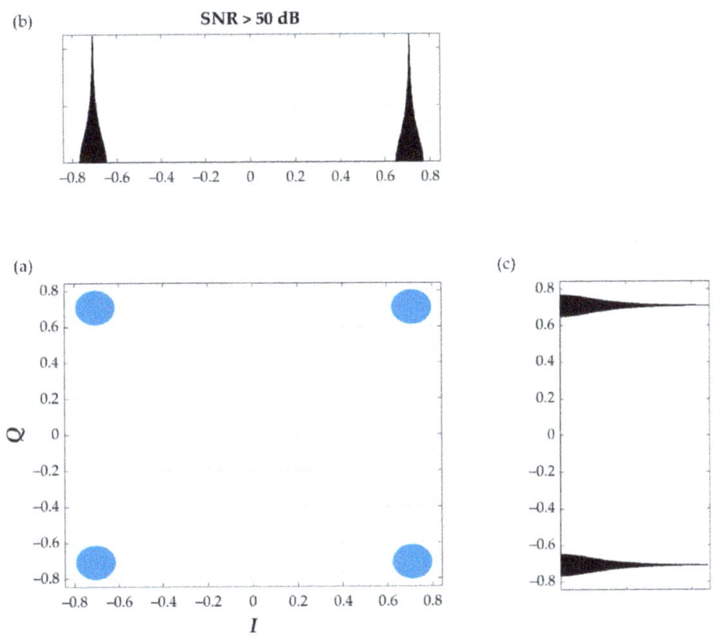

Figure 15. (**a**) Constellation, (**b**) in-phase and (**c**) quadrature histogram for SNR > 50 dB, K = 30, random A_m, and EVM = 3%.

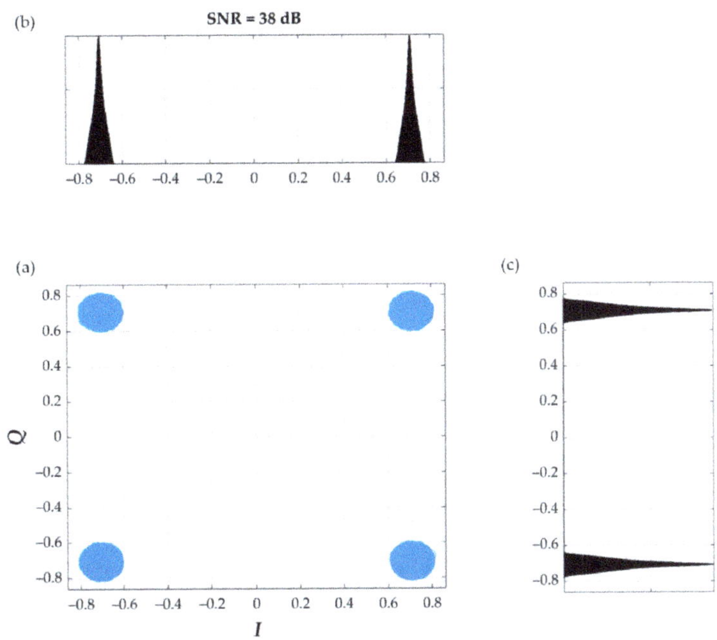

Figure 16. (**a**) Constellation, (**b**) in-phase and (**c**) quadrature histogram for SNR = 38 dB, K = 30, random A_m, and EVM = 3%.

6. Hardware Implementation

The concept of the phase drift-based dirty constellation was implemented by using the Universal Software Radio Peripheral (USRP) hardware platform manufactured by National Instruments (NI, Austin, TX, USA). In this case, we used the USRP-2920 model. USRP was the essential hardware part for generating a radio signal, while the software part was provided by the LabView software with MATLAB scripts installed on a personal computer (PC). An Ethernet network adapter with a bit rate of 1 Gb/s was used to provide communication between the USRP platforms and PC via a switch. Two USRP-2920 were used to implement a test-bed for detectors in the transmitter–receiver system. The prepared test-bed was placed in an office room. The distance between the transmitter and receiver was 5 m. This configuration is illustrated in Figure 17. The parameters of the carrier signal and covert information are presented in Table 2.

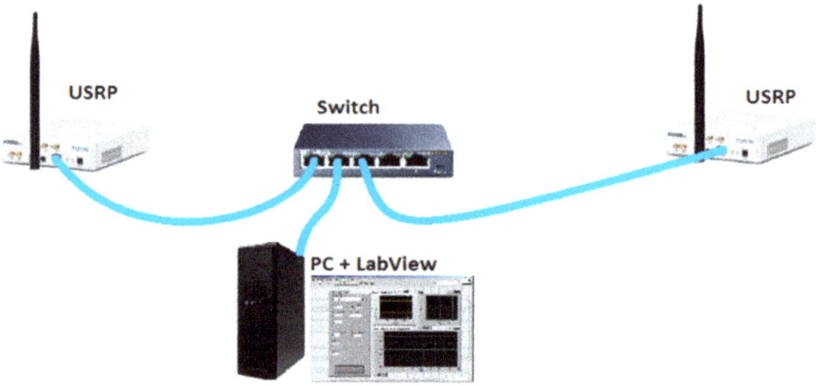

Figure 17. Test-bed based on USRP-2920 used for practical verification.

Table 2. Parameters of radio signal.

Signal	Parameters	Value
Cover (carrier)	Modulation type	4 QAM
	Carrier frequency	850 MHz
	Bandwidth	1 MHz
	Transmission rate	4 MHz
Covert information	M	4
	A_m	$0.05 \cdot A_n$
	φ_m	$\pm\pi/4, \pm3\pi/4$
	K	10 and 25
	Transmission rate	400 kb/s (for K = 10) 160 kb/s (for K = 25)

The results of the experimental research are presented in Figures 18 and 19. In this case, we determined a probability density function (PDF) of EVM and BER versus EVM graphs for two considered K values (see Figure 18). The distribution of the EVM was estimated by using a normalized histogram (see Figure 19).

The carrier constellation is subject to distortions caused by propagation phenomena and impairments introduced by the transceiver systems. We may see that the covert signal detectability reduces if the carrier distortion (i.e., EVM) increases. Tests were performed for two average values: $K = 10$ and $K = 25$. As expected, increasing the number of averages provides the increase in the covert signal detectability. On the other hand, the transmission rate decreases.

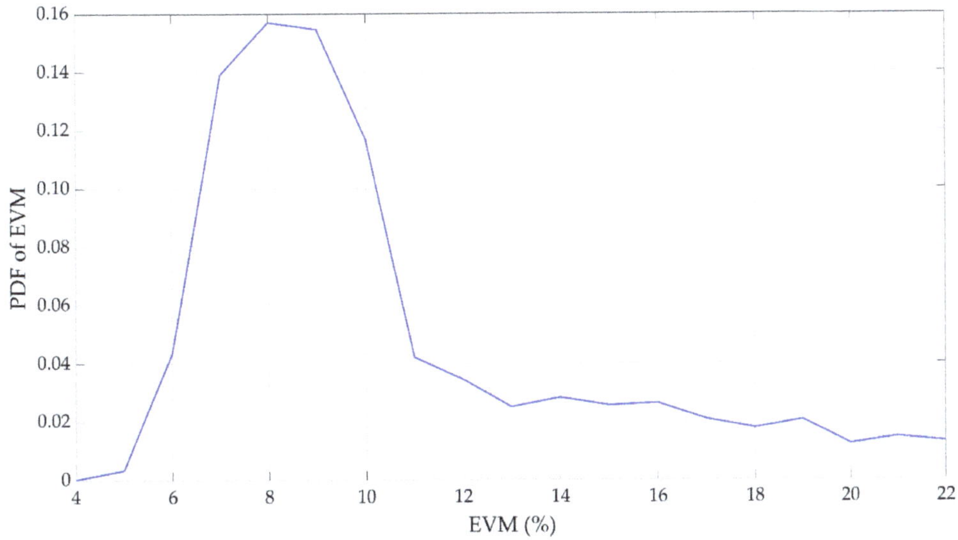

Figure 18. Covert signal detection results: probability density function (PDF) of EVM.

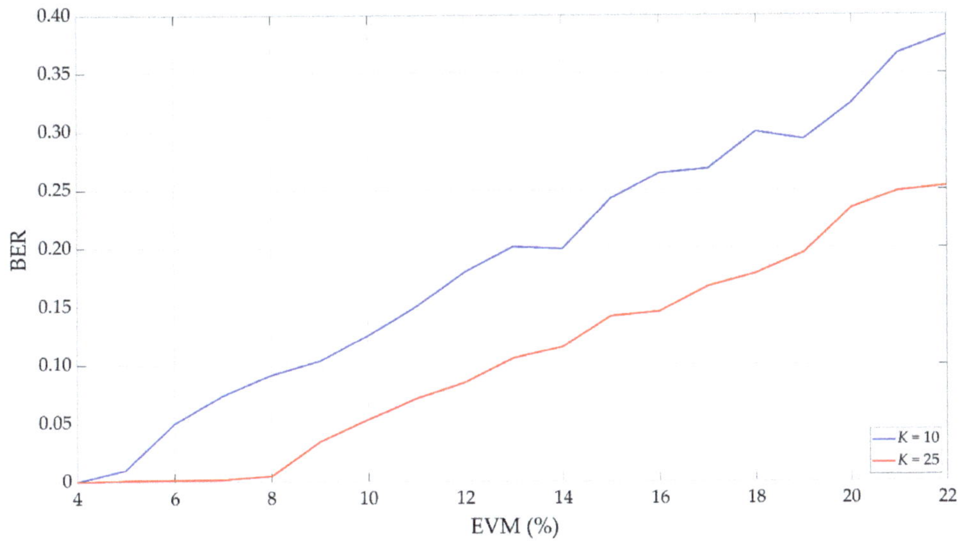

Figure 19. Covert signal detection results: BER versus EVM for selected *K*.

Based on the PDF of EVM, we can conclude that for the analyzed 4 QAM and $A_m = 0.05 \cdot A_n$, the average EVM oscillates between 8–12%. Comparing this value with Table 1, we can expect good transmission quality. On the other hand, increasing the distance between the transmitter and receiver will degrade the transmission quality due to the influence of the propagation conditions occurring in the radio channel. Hence, the adaptive selection of the parameters of the developed method depending on the propagation conditions will determine its future usefulness.

7. Conclusions

In this paper, we presented the novel dirty modulation method based on the phase drift, which is intended to create covert channels in radio transmissions using the N-ary PSK or QAM modulations. The method is based on the DCM approach that was previously used to watermark audio signals. In the proposed solution, a random change of the dirty constellation parameters is possible. It ensures its greater resistance to detection. On the other hand, it is possible to adapt these parameters to the modulation type and propagation conditions in the current radio channel.

In the paper, we described the idea of dirty modulation and developed a solution. Next, we introduced the criteria of the covert channel evaluation. Based on BER and EVM, we analyzed the proposed method using simulation studies and practical implementation, including comparison with other solutions. The obtained results showed that our dirty modulation method could be a valuable supplement to the existing steganographic methods.

In the near future, we want to focus on developing an adaptive method of selecting dirty modulation parameters, including the centroid distance and phase drift step, as well as multiple repetitions, i.e., averaging of secret transmission subsymbols. This will increase the undetectability of the proposed method. We also consider the effectiveness of this method in the case of additional use of the OFDM signal. Additionally, practical implementation of the method in the wireless part of the SDN within the aforementioned SOFTANET project is planned. In the near future, we want to implement ML algorithms (e.g., [39,40,47]) to increase the detection correctness of secret transmissions. These algorithms can be also used for time-variant selection of the developed dirty constellation parameters to improve their undetectability via various steganalysis techniques.

Author Contributions: Conceptualization, Z.P. and J.M.K.; methodology, K.G., Z.P. and J.M.K.; software, K.G.; validation, K.G.; formal analysis, K.G., Z.P. and J.M.K.; investigation, K.G.; resources, Z.P.; data curation, K.G. and Z.P.; writing—original draft preparation, K.G., Z.P. and J.M.K.; writing—review and editing, K.G. and J.M.K.; visualization, K.G.; supervision, Z.P. and J.M.K.; project administration, Z.P.; funding acquisition, Z.P. All authors have read and agreed to the published version of the manuscript.

Funding: This research was funded by the National Centre for Research and Development, grant number CYBERSECIDENT/381319/II/NCBR/2018 on "The federal cyberspace threat detection and response system" (acronym DET-RES) as part of the second competition of the CyberSecIdent Research and Development Program—Cybersecurity and e-Identity.

Data Availability Statement: The data presented in this study are available on request from the corresponding author. The data are not publicly available due to project restrictions.

Acknowledgments: The authors would like to express their great appreciation to the *Electronics* journal editors and anonymous reviewers for their valuable suggestions, which have improved the manuscript quality.

Conflicts of Interest: The authors declare no conflict of interest.

Abbreviations

5G	fifth generation
6G	sixth generation
AI	artificial intelligence
BER	bit error rate
CSI	channel state information
DCM	drift correction modulation
DNS	Domain Name System
DSSS	direct sequence spread spectrum
EDA	European Defense Agency
EVM	error vector magnitude
FHSS	frequency-hopping spread spectrum

FSK	frequency-shift keying
HTTP	Hypertext Transfer Protocol
IEEE	Institute of Electrical and Electronics Engineers
IP	Internet Protocol
LPD	low probability of detection
LPI	low probability of intercept
MAC	medium access control
MIMO	multiple-input multiple-output
ML	machine learning
Ob	observer
OFDM	orthogonal frequency-division multiplexing
PC	personal computer
PDF	probability density function
PDSCH	physical downlink shared channel
PN-ASK	pseudo-noise asymmetric shift keying
PSK	phase-shift keying
QAM	quadrature amplitude modulation
QPSK	quadrature phase-shift keying
SDN	software-defined network
SNR	signal-to-noise ratio
SPS	covert symbols per carrier symbol
SR	steganographic receiver
SS	steganographic source
TCP	Transmission Control Protocol
USRP	Universal Software Radio Peripheral

References

1. Zielińska, E.; Mazurczyk, W.; Szczypiorski, K. Trends in steganography. *Commun. ACM* **2014**, *57*, 86–95. [CrossRef]
2. Duan, X.; Gou, M.; Liu, N.; Wang, W.; Qin, C. High-Capacity Image Steganography Based on Improved Xception. *Sensors* **2020**, *20*, 7253. [CrossRef]
3. Järpe, E.; Weckstén, M. Velody 2—Resilient High-Capacity MIDI Steganography for Organ and Harpsichord Music. *Appl. Sci.* **2021**, *11*, 39. [CrossRef]
4. Cao, M.; Tian, L.; Li, C. A Secure Video Steganography Based on the Intra-Prediction Mode (IPM) for H264. *Sensors* **2020**, *20*, 5242. [CrossRef] [PubMed]
5. Lampson, B.W. A note on the confinement problem. *Commun. ACM* **1973**, *16*, 613–615. [CrossRef]
6. Gianvecchio, S.; Wang, H.; Wijesekera, D.; Jajodia, S. Model-based covert timing channels: Automated modeling and evasion. In *Proceedings of the Recent Advances in Intrusion Detection*; Lippmann, R., Kirda, E., Trachtenberg, A., Eds.; Springer: Cambridge, MA, USA, 2008; Volume RAID 2008, pp. 211–230.
7. Kothari, K.; Wright, M. Mimic: An active covert channel that evades regularity-based detection. *Comput. Netw.* **2013**, *57*, 647–657. [CrossRef]
8. Walls, R.J.; Kothari, K.; Wright, M. Liquid: A detection-resistant covert timing channel based on IPD shaping. *Comput. Netw.* **2011**, *55*, 1217–1228. [CrossRef]
9. Liu, G.; Zhai, J.; Dai, Y. Network covert timing channel with distribution matching. *Telecommun. Syst.* **2012**, *49*, 199–205. [CrossRef]
10. Li, Y.; Zhang, X.; Xu, X.; Tan, Y. A Robust Packet-Dropout Covert Channel over Wireless Networks. *IEEE Wirel. Commun.* **2020**, *27*, 60–65. [CrossRef]
11. Mileva, A.; Panajotov, B. Covert channels in TCP/IP protocol stack—Extended version. *Cent. Eur. J. Comput. Sci.* **2014**, *4*, 45–66. [CrossRef]
12. Frikha, L.; Trabelsi, Z.; El-Hajj, W. Implementation of a covert channel in the 802.11 header. In Proceedings of the 2008 6th International Wireless Communications and Mobile Computing Conference (IWCMC), Crete Island, Greece, 6–8 August 2008; pp. 594–599.
13. Szczypiorski, K.; Mazurczyk, W. Steganography in IEEE 802.11 OFDM symbols. *Secur. Commun. Netw.* **2011**, *9*, 118–129. [CrossRef]
14. Shaukat, K.; Iqbal, F.; Hameed, I.A.; Hassan, M.U.; Luo, S.; Hassan, R.; Younas, A.; Ali, S.; Adeem, G.; Rubab, A.; et al. MAC protocols 802.11: A comparative study of throughput analysis and improved LEACH. In Proceedings of the 2020 17th International Conference on Electrical Engineering/Electronics, Computer, Telecommunications and Information Technology (ECTI-CON), Phuket, Thailand, 24–27 June 2020; pp. 421–426.

15. Hassan, M.U.; Shahzaib, M.; Shaukat, K.; Hussain, S.N.; Mubashir, M.; Karim, S.; Shabir, M.A. DEAR-2: An energy-aware routing protocol with guaranteed delivery in wireless ad-hoc networks. In *Recent Trends and Advances in Wireless and IoT-Enabled Networks*; Jan, M.A., Khan, F., Alam, M., Eds.; EAI/Springer Innovations in Communication and Computing; Springer International Publishing: Cham, Germany, 2019; pp. 215–224. ISBN 978-3-319-99966-1.
16. Frączek, W.; Szczypiorski, K. Perfect undetectability of network steganography. *Secur. Commun. Netw.* **2016**, *9*, 2998–3010. [CrossRef]
17. Graniszewski, W.; Krupski, J.; Szczypiorski, K. The covert channel over HTTP protocol. In Proceedings of the Photonics Applications in Astronomy, Communications, Industry, and High-Energy Physics Experiments 2016, Wilga, Poland, 29 May–6 June 2016; SPIE: Washington, DC, USA, 2016; Volume 10031, p. 100314Z.
18. Szczypiorski, K.; Drzymała, M.; Urbański, M.Ł. Network Steganography in the DNS Protocol. *Int. J. Electron. Telecommun.* **2016**, *62*, 343–346. [CrossRef]
19. Smolarczyk, M.; Szczypiorski, K.; Pawluk, J. Multilayer Detection of Network Steganography. *Electronics* **2020**, *9*, 2128. [CrossRef]
20. Chen, O.; Meadows, C.; Trivedi, G. Stealthy protocols: Metrics and open problems. In *Concurrency, Security, and Puzzles*; Lecture Notes in Computer Science; Springer: Cham, Germany, 2017; pp. 1–17. ISBN 978-3-319-51045-3.
21. Bordel Sánchez, B.; Alcarria, R.; Robles, T.; Jara, A. Protecting Physical Communications in 5G C-RAN Architectures through Resonant Mechanisms in Optical Media. *Sensors* **2020**, *20*, 4104. [CrossRef] [PubMed]
22. Yen, C.-T.; Huang, J.-F.; Zhang, W.-Z. Hiding Stealth Optical CDMA Signals in Public BPSK Channels for Optical Wireless Communication. *Appl. Sci.* **2018**, *8*, 1731. [CrossRef]
23. Moskowitz, I.S.; Safier, P.N.; Cotae, P. Pre-Nodulation Physical Layer Steganography 2013. U.S. Patent Application 13/573,671, 25 April 2013.
24. Szczepaniak, E.; Piotrowski, Z. Radio transmission masking on the example of FSK modulation. In Proceedings of the 2017 21st International Conference on Signal Processing Algorithms, Architectures, Arrangements, and Applications (SPA), Poznan, Poland, 20–22 September 2017; pp. 394–399.
25. Hijaz, Z.; Frost, V.S. Exploiting OFDM systems for covert communication. In Proceedings of the 2010 Military Communications Conference (MILCOM), San Jose, CA, USA, 31 October–3 November 2010; pp. 2149–2155.
26. Classen, J.; Schulz, M.; Hollick, M. Practical covert channels for WiFi systems. In Proceedings of the 2015 IEEE Conference on Communications and Network Security (CNS), Florence, Italy, 28–30 September 2015; pp. 209–217.
27. Bash, B.A.; Goeckel, D.; Towsley, D.; Guha, S. Hiding information in noise: Fundamental limits of covert wireless communication. *IEEE Commun. Mag.* **2015**, *53*, 26–31. [CrossRef]
28. Hero, A.O. Secure space-time communication. *IEEE Trans. Inf. Theory* **2003**, *49*, 3235–3249. [CrossRef]
29. Cao, P.; Liu, W.; Liu, G.; Zhai, J.; Ji, X.; Dai, Y. A novel wireless covert channel for MIMO system. In Proceedings of the 2020 6th International Conference on Artificial Intelligence and Security (ICAIS), Hohhot, China, 17–20 July 2020; Sun, X., Wang, J., Bertino, E., Eds.; Springer: Berlin/Heidelberg, Germany, 2020; Volume 3, pp. 351–362.
30. Dutta, A.; Saha, D.; Grunwald, D.; Sicker, D. Secret agent radio: Covert communication through dirty constellations. In Proceedings of the Information Hiding, Berkeley, CA, USA, 15–18 May 2012; Kirchner, M., Ghosal, D., Eds.; Springer: Berlin/Heidelberg, Germany, 2012; Volume IH 2012, pp. 160–175.
31. Cao, P.; Liu, W.; Liu, G.; Ji, X.; Zhai, J.; Dai, Y. A Wireless Covert Channel Based on Constellation Shaping Modulation. *Secur. Commun. Netw.* **2018**, *2018*, 1–15. [CrossRef]
32. D'Oro, S.; Restuccia, F.; Melodia, T. Hiding data in plain sight: Undetectable wireless communications through pseudo-noise asymmetric shift keying. In Proceedings of the 2019 38th IEEE Conference on Computer Communications (INFOCOM), Paris, France, 29 April–2 May 2019; IEEE: New York, NY, USA; pp. 1585–1593.
33. Piotrowski, Z. Drift Correction Modulation scheme for digital signal processing. *Math. Comput. Model.* **2013**, *57*, 2660–2670. [CrossRef]
34. Lyons, R.G. *Understanding Digital Signal Processing*, 3rd ed.; Prentice Hall: Upper Saddle River, NJ, USA, 2010; ISBN 978-0-13-702741-5.
35. Zaidi, Z.; Friderikos, V.; Yousaf, Z.; Fletcher, S.; Dohler, M.; Aghvami, H. Will SDN Be Part of 5G? *IEEE Commun. Surv. Tutor.* **2018**, *20*, 3220–3258. [CrossRef]
36. Leonardi, L.; Lo Bello, L.; Aglianò, S. Priority-based bandwidth management in virtualized software-defined networks. *Electronics* **2020**, *9*, 1009. [CrossRef]
37. Semong, T.; Maupong, T.; Anokye, S.; Kehulakae, K.; Dimakatso, S.; Boipelo, G.; Sarefo, S. Intelligent Load Balancing Techniques in Software Defined Networks: A Survey. *Electronics* **2020**, *9*, 1091. [CrossRef]
38. Alhaddad, M.J.; Alkinani, M.H.; Atoum, M.S.; Alarood, A.A. Evolutionary Detection Accuracy of Secret Data in Audio Steganography for Securing 5G-Enabled Internet of Things. *Symmetry* **2020**, *12*, 2071. [CrossRef]
39. Shaukat, K.; Luo, S.; Varadharajan, V.; Hameed, I.A.; Xu, M. A Survey on Machine Learning Techniques for Cyber Security in the Last Decade. *IEEE Access* **2020**, *8*, 222310–222354. [CrossRef]
40. Shaukat, K.; Luo, S.; Varadharajan, V.; Hameed, I.A.; Chen, S.; Liu, D.; Li, J. Performance Comparison and Current Challenges of Using Machine Learning Techniques in Cybersecurity. *Energies* **2020**, *13*, 2509. [CrossRef]
41. Roetteler, M.; Svore, K.M. Quantum Computing: Codebreaking and Beyond. *IEEE Secur. Priv.* **2018**, *16*, 22–36. [CrossRef]
42. Zhang, H.; Ji, Z.; Wang, H.; Wu, W. Survey on quantum information security. *China Commun.* **2019**, *16*, 1–36. [CrossRef]

43. Chaumont, M. Deep learning in steganography and steganalysis. In *Digital Media Steganography*; Hassaballah, M., Ed.; Academic Press: Cambridge, MA, USA, 2020; pp. 321–349. ISBN 978-0-12-819438-6.
44. Li, F.; Tang, H.; Zou, Y.; Huang, Y.; Feng, Y.; Peng, L. Research on information security in text emotional steganography based on machine learning. *Enterp. Inf. Syst.* **2020**, 1–18. [CrossRef]
45. Sutherland, C.; Brun, T.A. Quantum steganography over noiseless channels: Achievability and bounds. *Phys. Rev. A* **2020**, *101*, 052319. [CrossRef]
46. Chaharlang, J.; Mosleh, M.; Rasouli-Heikalabad, S. A novel quantum steganography-Steganalysis system for audio signals. *Multimed. Tools Appl.* **2020**, *79*, 17551–17577. [CrossRef]
47. Alam, T.M.; Shaukat, K.; Hameed, I.A.; Luo, S.; Sarwar, M.U.; Shabbir, S.; Li, J.; Khushi, M. An Investigation of Credit Card Default Prediction in the Imbalanced Datasets. *IEEE Access* **2020**, *8*, 201173–201198. [CrossRef]
48. ETSI. *NR; Base Station (BS) Radio Transmission and Reception (3GPP TS 38.104 Version 15.5.0 Release 15)*; European Telecommunications Standards Institute (ETSI), 3rd Generation Partnership Project (3GPP): Sophia-Antipolis, France, 2019.
49. Simmons, G.J. The prisoners' problem and the subliminal channel. In Proceedings of the Advances in Cryptology, Santa Barbara, CA, USA, 22–24 August 1983; Chaum, D., Ed.; Springer: Berlin/Heidelberg, Germany, 1984; pp. 51–67.
50. Szczypiorski, K.; Janicki, A.; Wendzel, S.; Wendzel, S. "The Good, the Bad and The Ugly": Evaluation of Wi-Fi steganography. *J. Commun.* **2015**, *10*, 747–752. [CrossRef]

electronics

MDPI

Article

Multi-Language Spam/Phishing Classification by Email Body Text: Toward Automated Security Incident Investigation

Justinas Rastenis [1,*], Simona Ramanauskaitė [2], Ivan Suzdalev [3], Kornelija Tunaitytė [3], Justinas Janulevičius [1] and Antanas Čenys [1]

[1] Department of Information Systems, Vilnius Gediminas Technical University, Sauletekio al. 11, LT-10223 Vilnius, Lithuania; justinas.janulevicius@vilniustech.lt (J.J.); antanas.cenys@vilniustech.lt (A.Č.)
[2] Department of Information Technology, Vilnius Gediminas Technical University, Sauletekio al. 11, LT-10223 Vilnius, Lithuania; simona.ramanauskaite@vilniustech.lt
[3] Department of Aeronautical Engineering, Vilnius Gediminas Technical University, Sauletekio al. 11, LT-10223 Vilnius, Lithuania; ivan.suzdalev@vilniustech.lt (I.S.); kornelija.tunaityte@stud.vgtu.lt (K.T.)
* Correspondence: justinas.rastenis@vilniustech.lt; Tel.: +370-525-12-333

Abstract: Spamming and phishing are two types of emailing that are annoying and unwanted, differing by the potential threat and impact to the user. Automated classification of these categories can increase the users' awareness as well as to be used for incident investigation prioritization or automated fact gathering. However, currently there are no scientific papers focusing on email classification concerning these two categories of spam and phishing emails. Therefore this paper presents a solution, based on email message body text automated classification into spam and phishing emails. We apply the proposed solution for email classification, written in three languages: English, Russian, and Lithuanian. As most public email datasets almost exclusively collect English emails, we investigate the suitability of automated dataset translation to adapt it to email classification, written in other languages. Experiments on public dataset usage limitations for a specific organization are executed in this paper to evaluate the need of dataset updates for more accurate classification results.

Keywords: spam; phishing; classification; augmented dataset; multi-language emails

Citation: Rastenis, J.; Ramanauskaitė, S.; Suzdalev, I.; Tunaitytė, K.; Janulevičius, J.; Čenys, A. Multi-Language Spam/Phishing Classification by Email Body Text: Toward Automated Security Incident Investigation. *Electronics* **2021**, *10*, 668. https://doi.org/10.3390/electronics10060668

Academic Editor: Krzysztof Szczypiorski

Received: 15 January 2021
Accepted: 7 March 2021
Published: 12 March 2021

1. Introduction

Despite new communication systems and solutions being constantly introduced to the market, email remains in leading positions for both business and personal use. This popularity attracts the attention of persons with malicious intentions—spam and phishing email attacks are one of the most popular cyber-security attacks: in the 3rd quarter of 2020 nearly 50% of email traffic was spam [1]; 98% of cyber-attacks rely on social engineering [2] which is mostly executed by sending phishing emails [3].

Email filtering systems have been improving continuously to follow malicious, unwanted content development to protect the end-users. However, existing solutions are focusing on spam and phishing email filtering out while further analysis and email labeling are not fully developed. Therefore, email-based attacks are either analyzed manually or not investigated at all.

The analysis of cyber-attacks is a must for detecting the attacker and preventing their further malicious activities. The digital information security forensics is a time- and resource-consuming process, therefore automation should be used as much as possible to reduce the investigation time as well as to increase its accuracy [4,5]. One of the first steps in the forensics is classification of obtained data and its prioritization. Taking into account the huge number of unwanted emails, the automated classification of malicious emails would work as initial prioritization of investigating incidents and would work as the initial phase for automated or semi-automated security incident investigation. The prioritization is important as the purpose of spam and phishing attacks are different—spam emails are oriented towards dissemination of advertising, while phishing attacks aim at victims' personal data collecting and its usage for other cyber-attacks. Therefore phishing emails should be investigated as fast as possible, with higher attention to them than spam emails. The automated classification between spam and phishing email would allow appropriate resource allocation.

This paper aims to automate the identification of phishing emails in spam/phishing mixed different language email flow. As a consequence, this would simplify email-based security attack investigation and would lead to a higher degree automation in the forensics process. To achieve this goal several research questions are raised: (i) are existing English language spam/phishing email datasets suitable for spam/phishing email classification in other languages? and (ii) do spam/phishing email text patterns change relating to a specific region and do they have to be updated to achieve a higher classification accuracy?

The further structure of the paper is organized as follows. Related work chapter summarizes existing research in the field of spam or phishing email automated classification as well as datasets, that are usually used to train spam or phishing email detection systems. Based on the existing solutions new research for spam and phishing email classification is presented along with the datasets. The paper does not propose a new classification method; however, it presents research for spam/phishing email following the steps comprising a common classification workflow (data preparation, text augmentation, text classification), applied for solving this specific problem. The performance of the proposed solution is evaluated and experiments on automated email dataset translation as well as the updates needed are investigated. The paper is summarized with conclusions and future work.

2. Related Work

Spam is undesired electronic information spread aiming to cause psychological and monetary harm to the victim [6]. While it can be spread within different channels, a spam email contains an advertisement or irrelevant text, sent by spammers having no relationship with the recipient [7]. While different definitions of spam exist it is mostly related to undesired commercial email, and therefore the end user is unsatisfied by receiving undesired content.

Meanwhile, phishing emails seek to mimic legitimate emails and influence the user to execute some intended actions and reveal their personal information. Phishing attacks are classified as social engineering attacks, where the attacker tries to affect the victim from making rational choices and force the victim to make emotional choices instead [8]. Therefore, phishing attacks are potentially more harmful in comparison to spam mails.

To classify the email automatically, some basic steps are executed: email preprocessing and email classification (with its performance evaluation).

2.1. *Email Preprocessing*

An email has some specific properties which can be used for its classification to spam, phishing, legitimate email (ham), or any other category. An email can be presented in different file formats, therefore the property extraction should be prepared. However, for email classification, some additional processing might be used to obtain some specific features. For example, Ayman El Aassal et al. [9] divide phishing email-related features into two main categories: email features and website features. Email features are related to the data and metadata of the email and can be categorized into header, body, and attachment data. Meanwhile, website features are related to data, which can be gathered from the email body and links in it. Website features are based on the link and the websites the link points to. While most solutions [10–12] rely on the data which can be directly gathered from the email (the link uniform resource locator (URL) presented as internet protocol (IP), not domain name address; the number of different domains in the links; etc.), some solutions [9] go even further and analyze the website itself (the content of the website; script code; etc.) or use some additional tools to validate the URL [13].

To reduce the classification complexity, the number of extracted features is limited and expressed as numerical or binary values [14]. Therefore, different feature selection techniques are used [15,16] to obtain the most important features only and to eliminate non-significant ones. For example, Jose R. Mendez et al. [17] extracts the topic of the email and for spam email identification uses topics rather than the full bag of words of the email text. Sami Smadi et al. [18] uses 22 features, which are calculated, estimated based on a number or existence of some specific patterns; however, term meaning in the email body is not analyzed at all. Meanwhile, Andronicus A. Akinyelu and Aderemi O. Adewumi [19] define 7 features, which are based on the existence or number of some inspected elements in the email and add 2 features based on the existence of specific terms, words in the email body (one to define the direction to click some link; another related to action, which should be done after clicking the link). The proportion of email body content and other features depends on the author. For example, Saeed Abu-Nimeh et al. [20] and Devottam Gaurav et al. [21] use only email body features and by using text-mining their solution gathers the most frequent terms in the email body. To extract the most frequent terms, all hypertext markup language (HTML) code and unwanted terms (stop words), symbols are removed from the email body. Then the terms are processed to get the standard form (stemming). For later analysis, the frequencies or proportion of the specific terms are used as features.

Text analysis is very popular in the latest methods for spam and phishing classification and might include some additional text preprocessing to obtain more accurate classification results. For example, Ayman El Aassal et al. [22] takes into account the data from different datasets that might be associated with the email category, therefore they eliminated as much content as possible (organizations' or universities' names, recipients' names, domain names, signatures, etc.), which could associate it to the dataset. Another solution in email classification is the hierarchical classification [23,24] where, for example, first of all the email body is classified into some semantic categories and based on it the second layer identifies the email category itself.

2.2. *Email Classification Solutions*

Email classification can be implemented as a rule-based [25] system, however, it requires continuous support and updating. Therefore, hybrid [26] or machine learning [27] solutions take over where automated rather than manual rule, decision making logic updates are made. The machine learning solutions allow supervised learning when the model for email classification is designed based on the provided dataset.

In the field of spam, phishing, and ham email classification, the main classification methods are support vector machine (SVM), random forest (RF), decision tree (DT), naïve Bayes (NB), linear regression (LR), k-nearest neighbors (kNN) and other more specific solutions. The summary of classification method usage is presented in Table 1.

As seen, all email classification solutions are focused on the classification of legitimate, ham emails and unwanted, malicious (spam, phishing, or both) emails. The results of presented email classification solutions are high (F-score is 87 or more and even reaches 99.95), however no separation between spam and phishing is analyzed in scientific papers.

The lack of spam and phishing email separation is noticed in email datasets as well. While the Enron dataset is dedicated to legitimate ham emails, the University of California, Irvine (UCI) Machine Learning Repository has a dataset for spam emails, the Nazario dataset stores phishing emails, the SpamAssassin dataset has both spam and ham emails. Those two categories are separated in the SpamAssassin dataset, however, phishing emails are included inside of the spam emails. In most cases, some additional, personal email datasets are used to add variety and an ability to test the proposed solution with real situations, specific to some organization.

Table 1. Summary of recent papers on machine learning email classification solutions.

Paper	Classification Categories	Classification Method	Dataset	MAX F-Score
El Aassal et al. [9]	phishing, ham	SVM, RF, DT, NB, LR, kNN, other	Enron [28], SpamAssassin [29], Nazario [30]	99.95
Li et al. [31]	phishing, ham	DT, NB, kNN	SpamAssassin, Nazario	97.30
Verma et al. [32,33]	phishing, ham	SVM, RF, DT, NB, LR, kNN	SpamAssassin, Nazario	99.00
Sonowal et al. [6]	phishing, ham	RF, other	Nazario	97.78
Gangavarapu et al. [34]	spam + phishing, ham	SVM, RF, NB, other	SpamAssassin, Nazario	99.40
Gaurav et al. [21]	spam, ham	RF, DT, NB	Enron, UCI Machine Learning Repository [35]	87.00
Ablel-Rheem et al. [36]	spam, ham	DT, NB, other	UCI Machine Learning Repository	94.40
Saidani et al. [24]	spam, ham	SVM, RF, DT, NB, kNN, other	Enron	98.90
Jáñez-Martino et al. [37]	spam, ham	SVM, NB, LR	SpamAssassin	95.40
Zamir et al. [23]	spam, ham	SVM, RF, DT, other	SpamAssassin	97.20

Support vector machine (SVM), random forest (RF), decision tree (DT), naïve Bayes (NB), linear regression (LR), k-nearest neighbors (kNN).

3. Research on Text-Based Spam/Phishing Email Classification Solution

While methods for malicious email detection from legitimate emails exist and achieves high accuracy, there are no solutions to classify spam and phishing emails within the malicious email flow. Therefore, in this paper we propose a solution, dedicated to classifying unwanted emails to spam and phishing email categories. The proposed email classification solution incorporates existing classification solutions and is adapted to classify emails of different languages. In Lithuania, the largest portion of emails is written in Lithuanian, English and Russian, therefore the solution will be oriented to these three languages in this paper.

3.1. Email Dataset Preparation

Both spam and phishing emails are undesired for the recipient and sent using very similar techniques. Therefore, the biggest difference between spam and phishing emails is their content. Therefore for spam and phishing email classification, we use email message body only.

We use supervised learning solutions and, therefore, a dataset of labeled spam and phishing emails is needed. The dataset was constructed by integrating three different datasets: (i) the Nazario dataset for list of phishing emails, (ii) the SpamAssassin dataset for a list of spam emails and (iii) an individual spam and phishing email dataset from Vilnius Gediminas Technical University (VilniusTech).

The Nazario dataset was used as it is to represent phishing email examples. Meanwhile, the SpamAssassin dataset includes spam and ham emails. We used the spam emails only; however, after inspecting them some phishing emails were found within the spam emails. Therefore, the dataset was relabeled to indicate spam and phishing emails.

VilniusTech dataset was collected and labeled by VilniusTech information technology specialists and includes emails from the period of 2018–2020.

All datasets were read by getting an email message body only (programming code to extract emails message body were written for each dataset). The emails additionally were preprocessed. Cleanup of email message body text was executed where all HTML, CSS (cascading style sheets), JavaScript code, special symbols were eliminated, leaving unformatted text only. As some emails contained personal information, it was eliminated too. This was done to avoid email message association to a specific dataset—the Nazario dataset has very common reference jose@monkey.org, in the VilniusTech dataset Vilnius Gediminas Technical University is mentioned etc. Therefore, all personal information (recipient's name, email address, organizations name) was replaced with keywords (NAME, EMAIL, ORGANIZATION), and dates (year) were removed from the text. This was done semi-automatically—part of the personal information was removed by using regex expressions and then all emails were revised manually.

Formatting and personal information removal revealed duplication of emails. Multiple instances of the same email templates were noticed and, therefore, unique messages were selected for the dataset while all duplicated versions were removed.

The individual VilniusTech dataset included emails written in different languages. The most popular languages (English, Lithuanian and Russian) were left while very rare cases of different languages (Latvian, German, Spanish, France, etc.) were eliminated from the dataset. Meanwhile, emails from the Nazario and SpamAssassin datasets were in English only. Therefore this dataset was translated (by using automated Google Translate service, integrated via application programming interface (API) into Python code, developed for preparation of the dataset) into Russian and Lithuanian languages. The keywords representing the recipient's personal information were not translated and left as keywords.

During the email filtering of unpopular languages and automated translation, each record in the dataset was assigned a new property—language. This property will not be used for email classification (in this paper), however will be used to form different test cases for the research.

Records from different datasets were combined into one dataset. The number of phishing emails in the combined dataset was much lower in comparison to spam emails (see Table 2). Therefore, random emails were selected from each category to obtain the same number of spam and phishing emails (see Table 2). This reduced the dataset from 3601 record to 1400, where 700 spam and 700 phishing emails are labeled.

Table 2. Summary of prepared spam and phishing dataset.

Initial Dataset	Language	Before Balancing			After Balancing		
		Spam Emails	Phishing Emails	Total	Spam Emails	Phishing Emails	Total
SpamAssassin + Nazario	English	692	182	874	150	150	300
	Lithuanian (translated)	692	182	874	150	150	300
	Russian (translated)	692	182	874	150	150	300
VilniusTech	English	559	205	864	200	200	400
	Lithuanian	40	38	78	35	35	70
	Russian	18	19	37	15	15	30
Total		2693	808	3601	700	700	1400

For text-based classification all message texts were tokenized as separate terms (TF-IDF—term frequency-inverse document frequency) and pruning was applied. We removed very common (over 95% occurrence) and very infrequent terms (below 3% occurrence). The limit of attributes is not applied and reaches about 31,000 attributes (attribute presents relative, rather than the absolute occurrence of the term). The number of attributes was relatively large, however it presented words from three different languages. Taking into account the complexity and variety of word forms in Lithuanian language, the number of attributes was adequate but can be optimized in future.

3.2. Research Methodology and Results

As the dataset includes 700 spam and 700 phishing emails we do not use deep neural networks and concentrate on the usage of the most used classification methods. The research is divided into three main phases (see Figure 1): Figure 1a method selection Figure 1b multi-language-experiment Figure 1c concept-drift-experiment.

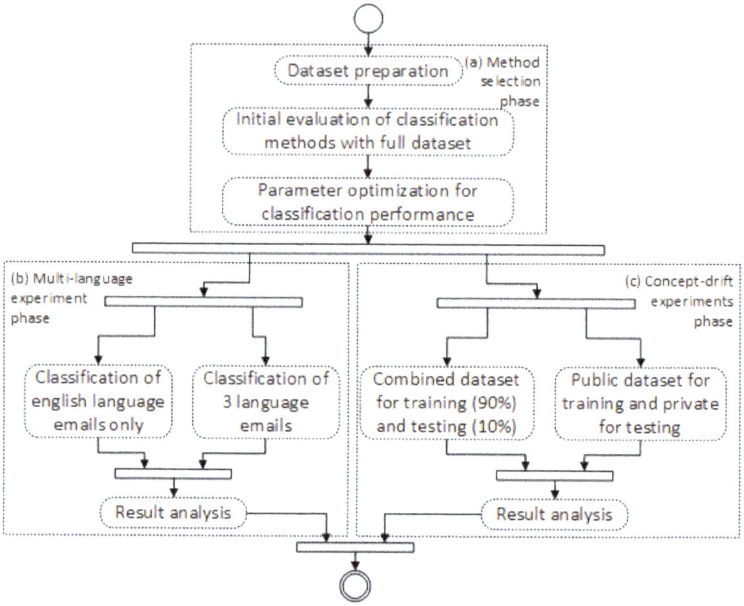

Figure 1. Workflow diagram of the research. (**a**) Method selection phase, (**b**) Multi-language experiment phase, (**c**) Concept-drift experiments phase.

In the first stage naïve Bayes, generalized linear model, fast large margin, decision tree, random forest, gradient boosted trees and support vector Machines methods were selected for the automatic identification of spam/phishing emails. Default settings and the full (balance of 1400 records) dataset was used in this step. The purpose of this step was to obtain the tendencies of classification performance and to select the methods we will be working on further.

For experiment execution, a RapidMiner tool was used to assure equal conditions for all methods (its standard implementation with possible settings). It was running on a 64-bit Windows 10 operating system on HP ProBook × 360 440 G1 Notebook PC with Intel core i3 processor and 8GM of RAM.

The results revealed (see Table 3), that 4 out of 7 analyzed solutions are not suitable to solve this problem as the accuracy does not exceed 60%. While ROC (receiver operating characteristic) curves (see Figure 2) and AUC (area under curve) values show naïve Bayes and decision tree methods are close to random solutions and the results obtained give no value in this situation.

Table 3. Classification methods performance in the initial experiment to classify spam and phishing emails.

Methods	Accuracy, %	Precision, %	Recall, %	F Score, %	AUC, %	Training Time (1000 Rows), s	Scoring Time (1000 Rows), s
Naïve Bayes	59.8	93.0	21.5	34.7	67.7	0.269	14
Generalized Linear Model	82.8	79.6	88.7	83.9	88.9	0.831	10
Fast Large Margin	83.2	79.1	90.7	84.4	92.5	0.157	15
Decision Tree	54.0	100.0	6.1	11.5	52.9	0.419	9
Random Forest	57.2	100.0	12.7	22.4	86.4	5.000	28
Gradient Boost Trees	57.0	93.0	13.7	23.5	98.2	15.000	9
Support Vector Machine	84.0	78.0	95.2	85.6	91.8	2.000	19

Area under curve (AUC).

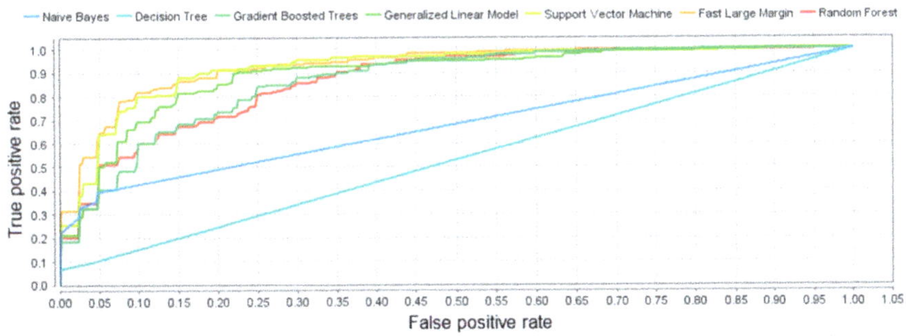

Figure 2. ROC (receiver operating characteristic) curves of different classification methods, used for initial email message classification to spam and phishing.

The support vector machine has the highest accuracy (84.0% ± 1.6%), however is one of the slowest solutions (for 1000 rows it takes 2s for training and 19s for scoring).

The next step of suitable classification method selection phase, a search for the most suitable parameters to increase the spam and phishing email classification performance, was executed with the generalized linear model, fast large margin and support vector machine. Different methods were used to analyze optimal parameters values—grid search, genetic algorithms [38], manual experiments. The best parameters were selected manually from the results obtained.

In this step the best accuracy was achieved with the fast large margin method (which was second in the initial experiment), using L2 SVM Dual solver, cost parameter $C = 1$, tolerance of the termination criteria $\varepsilon = 0.01$, identical class weights, and usage of bias. The cross-validation was executed with automatic sampling type and 10 fold as in the initial experiment. With these parameters, the accuracy increased to 90.07% ± 3.17%, and the confusion matrix of this classificatory is presented in Table 4.

Table 4. Confusion matrix and class prediction as well as class recall values of adjusted parameters for the fast large margin method.

	True Spam	True Phishing	Class Prediction
Predicted Spam	662	101	86.76%
Predicted Phishing	38	599	94.03%
Class recall	94.57%	85.57%	

The obtained configuration is used in parallel (independently) further in multi-language experiments (see Figure 1b,c).

In a multi-language experiment we investigated if the automated dataset translation was suitable for dataset augmentation and application for different language emails. This experiment was oriented to emails of three different languages, where part of the dataset was translated by Google Translate. If we applied the same model to the English language only, the accuracy was 89.2% ± 2.14%. This was the same result as in experiments with three languages and showed that the automated Google translation from English to Lithuanian and Russian languages was a suitable dataset augmentation method to adapt the dataset for spam/phishing email classification for different language emails.

The results similarity can be explained by two facts: (a) in most cases spam and phishing email templates are translated from the English language to other languages and in some cases, it is done with automated translation tools as well, therefore the augmented data in the dataset is similar to the data which would be sent in practice; (b) we use TF-IDF text vectorization where accuracies of separate terms are analyzed, not n-grams and, therefore the influence of translation quality is not as important.

A concept-drift experiment was concentrated on evaluating the need for dataset update. In this experiment, one dataset was used for training and another for testing. We took records from SpamAssassin and Narazio as the training set and VilniusTech email records as the testing set. In this situation, the accuracy decreased by more than 10%—if emails of only the English language were included, the accuracy was 74.94%, while if the augmented/translated SpamAssassin and Nazario datasets were used and tested with all records from VilniusTech dataset, the accuracy was 77.00%.

This shows that there are differences between the datasets which might be influenced by time, region or organization profile (the VilniusTech dataset is constructed from emails, obtained from university email boxes). The accuracy increase by using the augmented dataset can be explained by the increased number of records in the training dataset—there are 300 English language emails in the SpamAssassin and Nazario-based dataset while adding translations of two additional languages increases this to 900 emails.

4. Conclusions and Future Work

Analysis of the existing spam and phishing email classification solutions has revealed that there are multiple papers on this topic; however, all of them are focused on legitimate and malicious (spam and/or phishing) email separation from one email flow. There are no papers on automated spam and phishing email classification solutions. Spam and phishing emails sometimes are difficult to separate and the SpamAssassin dataset includes phishing emails as spam records. However, classification of spam and phishing emails would be beneficial as could be used to inform the user about the danger level of unwanted email as well as to assign priorities to the unwanted emails to investigate the cases.

Existing publicly available spam and phishing email datasets are English language only. This complicates its usage for email classification, which are written in different languages. The proposed solution with automated translation for dataset augmentation, adaptation for other languages prove the classification results do not decrease because of the automated translation—for English-only text, the accuracy was 90.07% ± 3.17% while for multi-language texts (English, Russian and Lithuanian) it was 89.2% ± 2.14%.

By training the spam and phishing classification model with the SpamAssassin and Nazario datasets and testing the model with the VilniusTech collected set of spam/phishing emails, the classification accuracy decreased more than 10% in comparison to a mixed dataset, used both for training and testing. This proves that the dataset should be updated, supplemented with data from the organization to obtain more accurate classification results.

For further directions, a deeper spam/phishing email classification performance analysis could be executed to increase the performance by adapting feature optimization (including header and formatting related features, feature number minimization or application of multi-level classification approaches), and deep-learning solution suitability for this task evaluation.

From the automated security incident investigation perspective, the emails could be classified based not only on spam/phishing classification but on potential thread recognition possibility, prevalence in the organization, and other features as well.

Author Contributions: Conceptualization, J.R. and S.R.; methodology, S.R.; software, J.R. and S.R.; validation, J.R., S.R. and K.T.; formal analysis, S.R.; investigation, J.R. and I.S.; resources, S.R.; data curation, J.R.; writing—original draft preparation, J.R and S.R.; writing—review and editing, J.R and S.R.; visualization, J.R. and S.R.; supervision, J.J.; project administration, A.Č. All authors have read and agreed to the published version of the manuscript.

Funding: This research received no external funding.

Data Availability Statement: Dataset used in this experiment is available. It contains original SpamAssassin and Nazario records (dataset labeled "1"), its translation to Russian and Lithuanian languages (dataset labeled "2") and individual dataset, collected and labeled by VilniusTech information technology specialists during the period 2018–2020 (dataset labeled "3").

Conflicts of Interest: The authors declare no conflict of interest.

References

1. Spam and Phishing in Q3 2020. Available online: https://securelist.com/spam-and-phishing-in-q3-2020/99325/ (accessed on 15 November 2020).
2. 2020 Cyber Security Statistics. Available online: https://purplesec.us/resources/cyber-security-statistics/ (accessed on 15 November 2020).
3. Social Engineering & Email Phishing–The 21st Century's #1 Attack? Available online: https://www.wizlynxgroup.com/news/2020/08/27/social-engineering-email-phishing-21st-century-n1-cyber-attack/ (accessed on 15 November 2020).
4. Carmona-Cejudo, J.M.; Baena-García, M.; del Campo-Avila, J.; Morales-Bueno, R. Feature extraction for multi-label learning in the domain of email classification. In Proceedings of the 2011 IEEE Symposium on Computational Intelligence and Data Mining (CIDM), Paris, France, 11–15 April 2011; pp. 30–36.
5. Goel, S.; Williams, K.; Dincelli, E. Got phished? Internet security and human vulnerability. *J. Assoc. Inf. Syst.* **2017**, *18*, 22–44. [CrossRef]
6. Aassal, A.E.; Moraes, L.; Baki, S.; Das, A.; Verma, R. Anti-phishing pilot at ACM IWSPA 2018: Evaluating performance with new metrics for unbalanced datasets. In Proceedings of the IWSPA-AP Anti Phishing Shared Task Pilot 4th ACM IWSPA, Tempe, Arizona, 21 March 2018; pp. 2–10.
7. El Aassal, A.; Baki, S.; Das, A.; Verma, R.M. An In-Depth Benchmarking and Evaluation of Phishing Detection Research for Security Needs. *IEEE Access* **2020**, *8*, 22170–22192. [CrossRef]
8. Abu-Nimeh, S.; Nappa, D.; Wang, X.; Nair, S. A comparison of machine learning techniques for phishing detection. In Proceedings of the Anti-phishing Working Groups 2nd Annual Ecrime Researchers Summit, Pittsburgh, PA, USA, 4–5 October 2007; pp. 60–69.
9. L'Huillier, G.; Weber, R.; Figueroa, N. Online phishing classification using adversarial data mining and signaling games. In Proceedings of the ACM SIGKDD Workshop on CyberSecurity and Intelligence Informatics, Paris, France, 28 June–1 July 2009; pp. 33–42.
10. Peng, T.; Harris, I.; Sawa, Y. Detecting phishing attacks using natural language processing and machine learning. In Proceedings of the 2018 IEEE 12th international conference on semantic computing (icsc), Laguna Hills, CA, USA, 31 January–2 February 2018; IEEE: New York, NY, USA, 2018; pp. 300–301.
11. Weinberger, K.; Dasgupta, A.; Langford, J.; Smola, A.; Attenberg, J. Feature hashing for large scale multitask learning. In Proceedings of the 26th Annual International Conference on Machine Learning, Montreal, QC, Canada, 14-18 June 2009; pp. 11–1120.
12. Zareapoor, M.; Seeja, K.R. Feature extraction or feature selection for text classification: A case study on phishing email detection. *Int. J. Inf. Eng. Electron. Bus.* **2015**, *7*, 60. [CrossRef]
13. Smadi, S.; Aslam, N.; Zhang, L. Detection of online phishing email using dynamic evolving neural network based on reinforcement learning. *Decis. Support Syst.* **2018**, *107*, 88–102. [CrossRef]
14. Toolan, F.; Carthy, J. Feature selection for spam and phishing detection. In Proceedings of the 2010 eCrime Researchers Summit, Dallas, TX, USA, 18–20 October 2010; IEEE: New York, NY, USA, 2010; pp. 1–12.
15. Verma, R.M.; Zeng, V.; Faridi, H. Data Quality for Security Challenges: Case Studies of Phishing, Malware and Intrusion Detection Datasets. In Proceedings of the 2019 ACM SIGSAC Conference on Computer and Communications Security, London, UK, 11–15 November 2019; pp. 2605–2607.
16. Smadi, S.; Aslam, N.; Zhang, L.; Alasem, R.; Hossain, M.A. Detection of phishing emails using data mining algorithms. In Proceedings of the 2015 9th International Conference on Software, Knowledge, Information Management and Applications (SKIMA), Kathmandu, Nepal, 15–17 December 2015; IEEE: New York, NY, USA, 2015; pp. 1–8.
17. Akinyelu, A.A.; Adewumi, A.O. Classification of phishing email using random forest machine learning technique. *J. Appl. Math.* **2014**, *2014*. [CrossRef]
18. Gangavarapu, T.; Jaidhar, C.D.; Chanduka, B. Applicability of machine learning in spam and phishing email filtering: Review and approaches. *Artif. Intell. Rev.* **2020**, *53*, 5019–5081. [CrossRef]
19. Li, X.; Zhang, D.; Wu, B. Detection method of phishing email based on persuasion principle. In Proceedings of the 2020 IEEE 4th Information Technology, Networking, Electronic and Automation Control Conference (ITNEC), Chongqing, China, 12–14 June 2020; Volume 1, pp. 571–574.
20. Verma, P.; Goyal, A.; Gigras, Y. Email phishing: Text classification using natural language processing. *Comput. Sci. Inf. Technol.* **2020**, *1*, 1–12. [CrossRef]
21. Sonowal, G. Phishing Email Detection Based on Binary Search Feature Selection. *SN Comput. Sci.* **2020**, *1*. [CrossRef] [PubMed]
22. Ablel-Rheem, D.M.; Ibrahim, A.O.; Kasim, S.; Almazroi, A.A.; Ismail, M.A. Hybrid Feature Selection and Ensemble Learning Method for Spam Email Classification. *Int. J.* **2020**, *9*, 217–223. [CrossRef]
23. Zamir, A.; Khan, H.U.; Mehmood, W.; Iqbal, T.; Akram, A.U. A feature-centric spam email detection model using diverse supervised machine learning algorithms. *Electron. Libr.* **2020**, *38*, 633–657. [CrossRef]
24. Gaurav, D.; Tiwari, S.M.; Goyal, A.; Gandhi, N.; Abraham, A. Machine intelligence-based algorithms for spam filtering on document labeling. *Soft Comput.* **2020**, *24*, 9625–9638. [CrossRef]
25. Saidani, N.; Adi, K.; Allili, M.S. A Semantic-Based Classification Approach for an Enhanced Spam Detection. *Comput. Secur.* **2020**, *94*, 101716. [CrossRef]

26. Jáñez-Martino, F.; Fidalgo, E.; González-Martínez, S.; Velasco-Mata, J. Classification of Spam Emails through Hierarchical Clustering and Supervised Learning. *arXiv* **2020**, arXiv:2005.08773.
27. Dada, E.G.; Bassi, J.S.; Chiroma, H.; Adetunmbi, A.O.; Ajibuwa, O.E. Machine learning for email spam filtering: Review, approaches and open research problems. *Heliyon* **2019**, *5*, e01802. [CrossRef] [PubMed]
28. Pérez-Díaz, N.; Ruano-Ordas, D.; Fdez-Riverola, F.; Méndez, J.R. Wirebrush4SPAM: A novel framework for improving efficiency on spam filtering services. *Softw. Pract. Exp.* **2013**, *43*, 1299–1318. [CrossRef]
29. Wu, C.H. Behavior-based spam detection using a hybrid method of rule-based techniques and neural networks. *Expert Syst. Appl.* **2009**, *36*, 4321–4330. [CrossRef]
30. Enron Email Dataset. Available online: https://www.cs.cmu.edu/~{}enron/ (accessed on 22 October 2020).
31. SpamAssassin Dataset. Available online: https://spamassassin.apache.org/ (accessed on 22 October 2020).
32. Nazario Dataset. Available online: https://www.monkey.org/~{}jose/phishing/ (accessed on 23 October 2020).
33. UCI Machine Learning Repository. Available online: https://archive.ics.uci.edu/ml/datasets.php (accessed on 28 October 2020).
34. Asquith, A.; Horsman, G. Let the robots do it!–Taking a look at Robotic Process Automation and its potential application in digital forensics. *Forensic Sci. Int. Rep.* **2019**, *1*, 100007. [CrossRef]
35. Hayes, D.; Kyobe, M. The Adoption of Automation in Cyber Forensics. In Proceedings of the 2020 Conference on Information Communications Technology and Society (ICTAS), Durban, South Africa, 11–12 March 2020; IEEE: New York, NY, USA, 2020; pp. 1–6.
36. Syarif, I.; Prugel-Bennett, A.; Wills, G. SVM parameter optimization using grid search and genetic algorithm to improve classification performance. *Telkomnika* **2016**, *14*, 1502. [CrossRef]
37. Vinitha, V.S.; Renuka, D.K. Feature Selection Techniques for Email Spam Classification: A Survey. In Proceedings of the International Conference on Artificial Intelligence, Smart Grid and Smart City Applications (AISGSC), Coimbatore, India, 3–5 January 2019; Springer: Cham, Switzerland, 2020; pp. 925–935.
38. Mendez, J.R.; Cotos-Yanez, T.R.; Ruano-Ordas, D. A new semantic-based feature selection method for spam filtering. *Appl. Soft Comput.* **2019**, *76*, 89–104. [CrossRef]

electronics

Article

Discussion on IoT Security Recommendations against the State-of-the-Art Solutions

Marta Chmiel †, Mateusz Korona †, Fryderyk Kozioł †, Krzysztof Szczypiorski and Mariusz Rawski *

Institute of Telecommunications, Faculty of Electronics and Information Technology, Warsaw University of Technology, 00-665 Warsaw, Poland; m.chmiel@tele.pw.edu.pl (M.C.); m.korona@tele.pw.edu.pl (M.K.); f.koziol@tele.pw.edu.pl (F.K.); k.szczypiorski@tele.pw.edu.pl (K.S.)
* Correspondence: m.rawski@tele.pw.edu.pl
† These authors contributed equally to this work.

Abstract: The Internet of Things (IoT) is an emerging concept comprising a wide ecosystem of interconnected devices and services. These technologies collect, exchange and process data in order to dynamically adapt to a specific context. IoT is tightly bound to cyber-physical systems and, in this respect, has relevant security implications. A need for IoT security guidelines was identified by the industry in the early 2010s. While numerous institutions across the globe have proposed recommendations with a goal to help developers, distributors and users to ensure a secure IoT infrastructure, a strict set of regulations for IoT security is yet to be established. In this paper, we aim to provide an overview of security guidelines for IoT proposed by various organizations, and evaluate some of the existing technologies applied to ensure IoT security against these guidelines. We gathered recommendations proposed by selected government organizations, international associations and advisory groups, and compiled them into a set of the most common and important considerations, divided into eight categories. Then we chose a number of representative examples from IoT security technologies and evaluated them against these criteria. While none of the examined solutions fulfill all recommendations on their own, the existing technologies introduced by those solutions could be combined to create a design framework which satisfies all the requirements of a secure IoT device. Further research on this matter could be beneficial. To the best of our knowledge, this is the first comprehensive survey to evaluate different security technologies for IoT device security against the compilation of criteria based on existing guidelines.

Keywords: cybersecurity; IoT; data protection; SoC

Citation: Chmiel, M.; Korona, M.; Kozioł, F.; Szczypiorski, K.; Rawski, M. Discussion on IoT Security Recommendations against the State-of-the-Art Solutions. *Electronics* **2021**, *10*, 1814. https://doi.org/10.3390/electronics10151814

Academic Editor: Rashid Mehmood

Received: 6 July 2021
Accepted: 23 July 2021
Published: 28 July 2021

1. Introduction

The present day is a time of unprecedented rapid technology development and the growth of the Internet. The majority of citizens in developed countries are not only smartphone users, but also surround themselves with intelligent devices such as various sensors, smart home appliances or CCTV cameras. These objects, which are capable of collecting, processing and exchanging data via various networks (also without human intervention), make up the Internet of Things (IoT). Researchers estimate that there were 12 billion IoT devices active in 2020 and this number will at least double within five years [1,2]. Unfortunately, people often focus only on the benefits of using IoT devices and tend to underestimate the risks.

Many IoT devices are deployed without sufficient security measures and can be easily exploited by more or less sophisticated attacks [3] with a significant impact on both individuals and society. Remote hijacking of a Jeep on the St. Louis highway is a well-publicized example of a personal IoT security breach [4]. White hat hackers in cooperation with a brave journalist acting as "the victim", were not only able to manipulate car interiors (display, sound system, air conditioning), but also control the engine and brakes—elevating the severity of the incident from a prank to a potentially fatal attack. On the other end of the danger spectrum are hostile actions that affect nationwide systems. Distributed Denial of Service (DDoS) attacks from Mirai malware-based botnets (consisting of thousands of compromised IoT devices) that targeted Internet service providers in France and the USA are a good example of this [5–7].

The original IoT paradigm is changing and system architectures are becoming increasingly edge-focused, moving processing of the data collected by sensors from the cloud into closer edge nodes (fog computing [8–10]) in order to reduce latency and required bandwidth. More and more, applications are expecting IoT nodes to be resilient to network connectivity issues, which means that IoT devices have to retain more intelligence and operation capabilities by themselves. The need for advanced data analysis is driving IoT device implementations in the direction of the entire System-on-Chip (SoC) [11], which consists of multiple interfaces, analog/digital circuits, memories and CPUs running highly functional operating systems, such as Linux. All in all, the attack surface of such highly sophisticated and functional IoT devices has increased greatly.

Many papers have been presented on the subject of IoT security. For instance, Mahmoud et al. [12] in 2015 discussed the security of a robust IoT network, with division into layers (perception, network and application). The authors listed a number of threats and attacks to which such a system is susceptible and, more importantly, raised concerns about existing major gaps in addressing basic security, for example, privacy and confidentiality. In 2019, Mohamad Noor and Hassan published a survey on IoT security research in the years 2016–2018 [13]. The conclusion was far from optimistic, since not much had improved over the years. Similarly, layers were insufficiently secured and not enough effort was put into ensuring comprehensive endpoint security. The rapid growth of IoT technologies was inevitably followed by an equally fast-paced growth of attacks. Insufficient focus on security allowed for the development of new, inventive ways of exploitation. Alladi et al. [14] in 2020 published a case study on vulnerabilities present in consumer devices. Their findings indicate that not only do the manufacturers often neglect proper protection of their devices but also that the users are unaware of the threats posed by, for example, a wireless scale that is in their bathroom. Consequently, many organizations noticed the IoT security problem and took steps to tackle it. Over the last decade, a number of them published documents discussing the importance of secure IoT and proposing guidelines. Some of these recommendations present a very detailed approach, from the design process to the user experience; others focus on just a part of the IoT device's life cycle.

Our work concentrates mainly on hardware and software design and some aspects of later stages of IoT device functioning, such as updating or event logging. The goal of this paper is to analyze how existing solutions for trusted computing, especially dedicated for IoT devices, adhere to these recommendations.

The remainder of this paper is organized as follows. Section 2 discusses our motivation and related work on the subject matter. In Section 3, security guidelines for IoT proposed by various organizations are gathered and compiled into a set of the most common and important. Section 4 contains the analysis of existing technologies addressing IoT security. We chose a number of representative examples and evaluated them against the criteria formulated in Section 3. Finally, Section 5 discusses the results and Section 6 concludes this paper.

2. Motivation and Related Work

IoT security can be analyzed from multiple angles and numerous publications on the subject are available. As a term, IoT is sometimes used to describe particular solutions, often from different ends of the technology spectrum requiring a specific approach to the subject, while sometimes it represents a general concept. The discussion ranges from security problems of specific technologies, such as RFID networks [15], through solutions of growing popularity, such as blockchain, machine learning or artificial intelligence [16], to innovative propositions such as moving target defense [17], aiming at more elaborate structures, such as IoT networks.

National Institute of Standards and Technology (NIST) in 2020 published a document that focuses on defining an IoT device cybersecurity capability core baseline [18]. NIST describes the core baseline as a minimal set of capabilities that an IoT device should be equipped with so that it supports common cybersecurity controls. Advanced security schemes can be built on this basis. However, NIST does not provide advice on how it should be achieved. We decided that this was an interesting perspective and further research on the matter would be beneficial, especially for manufacturers and developers. As a first step, we conducted a literature search and checked whether other organizations provide guidelines regarding IoT device security. Secondly, we examined whether existing, state-of-the-art technologies can be utilized to fulfill the requirements.

We came across multiple survey articles regarding aspects of IoT, from wide-ranging analysis of an entire IoT system [19], to works focusing on protocols [20,21], IoT platforms [22] and frameworks, based on contemporary, commercial examples [23]. In each of these papers, security was considered but the emphasis was rather on existing issues and challenges of discussed solutions, instead of means of protection. To the best of our knowledge, no surveys focusing specifically on IoT *device* security capabilities were available.

Additionally, we researched surveys on IoT security. Our findings showed that the focus of the published work is again more on the challenges than on the solutions. For instance, Macedo et al. in 2019 [24] provided a systematic literature review focused on defining four main aspects of IoT security—authentication, access control, data protection and trust. Their work is addressed to manufacturers, developers, consumer and providers of IoT. Interestingly, the authors recognized a lack of reference architectures to develop secure IoT solutions. Nonetheless, the paper does not present existing guidelines and a link to state-of-the-art technologies. Abdul-Ghani and Konstantas [25] in their work provide an overview of several documents regarding the best practices for securing IoT, published by renowned organizations, such as the Broadband Internet Technical Advisory Group (BITAG) or the IoT Security Foundation (IoTSF). They also recognize the need for standardized security and privacy guidelines for IoT. In contrast to our approach, the analyzed guidelines are not confronted with existing commercial solutions.

Finally, we investigated the availability of articles presenting an overview of contemporary solutions, which could be applicable for securing IoT. In their paper published in 2018, Maene et al. [26] gathered over ten different technologies, dedicated to trusted computing. Similarly to our work, they are compared against a set of criteria. However, these criteria are based on capabilities offered by analyzed solutions, rather than existing guidelines. Even though the trusted computing solutions discussed in this article can, in some cases, be successfully used for IoT applications, there are other technologies catered specifically for this purpose that are in our opinion worth considering.

3. Security Recommendations for Internet of Things

With the number of connected IoT devices growing bigger each year, the question of security has become crucial. The threats and risks related to IoT devices, systems and services are manifold, and evolve rapidly. Hence, it is important to understand what needs to be protected and to develop specific security measures to protect the *things* from cyber threats. While a strict set of regulations on IoT security is yet to be established, a need for guidelines was first identified by the industry in the early 2010s and the discussion has continued since then.

3.1. Existing Guidelines

Numerous institutions across the globe have proposed their recommendations, in order to help developers, distributors and users ensure a secure IoT infrastructure. Government organizations, international associations and advisory groups are aware of the problem and have published many documents on the subject, to name *some* among many more:

- National Institute of Standards and Technology (NIST),
- European Union Agency for Network and Information Security (ENISA),
- GSM Association (GSMA),
- Internet Engineering Task Force (IETF),
- Internet Research Task Force (IRTF),
- IoT Security Foundation (IoTSF),
- ioXt Alliance,
- International Standard Organization (ISO),
- Institute of Electrical and Electronics Engineers (IEEE),
- International Telecommunication Union (ITU),
- Broadband Internet Technical Advisory Group (BITAG),
- Industrial Internet Consortium (IIC),
- Open Web Application Security Project (OWASP),
- Trusted Computing Group (TCG),
- Cloud Security Alliance (CSA),
- GlobalPlatform,
- Internet Society's Online Trust Alliance (OTA).

Our analysis focuses on just a number of them. A time cut-off of 2017 has been adopted for two reasons: rapid IoT industry development might outdate some concepts and, on the other hand, recent publications often reference older ones and align with them in *essential* matters. Furthermore, industry standards issued by renowned organizations or manufacturer associations have been considered over simple brochures or articles. Last but not least, it was important that the given document (or its independent section) primarily concentrates on secure IoT *device* implementation itself, as this is a foundation for deliberations in the next sections.

NIST, part of the U.S. Department of Commerce, in 2020 issued a report, "IoT Device Cybersecurity Capability Core Baseline" (NISTIR 8259A) [18]. The authors define an IoT device cybersecurity capability core baseline, which is a set of device capabilities generally needed to support common cybersecurity features that protect data, systems and ecosystems. The proposed baseline represents a coordinated effort to produce a definition of common capabilities, which is not an exhaustive list. This document highlights activities that aim to improve cybersecurity levels in manufactured products, which in consequence reduces the number of exploited IoT devices.

ENISA created a number of documents on secure IoT development. In 2017, "Baseline Security Recommendations for IoT" [27] was published. The aim of this work was to provide insight into the security requirements of IoT, with a focus on Critical Information Infrastructures. The paper offers a thorough analysis of existing cybersecurity threats, along with a comprehensive set of measures in order to protect IoT systems. The authors developed a series of recommendations based on the results of their research, the views expressed by the experts, and good practices, as well as security measures used in the industry. It is worth noting that this document provides an elaborate list of other security standards regarding IoT, which can be a valuable starting point for further research.

Documents published by the GSMA provide very useful insight and pose questions that IoT designers and network administrators will find useful while discussing system security. The"IoT Security Guidelines" document set [28–30] should especially be considered at the early stages of development, as it asks a series of important questions regarding security which are very helpful during the process. These documents promote a methodology for developing secure IoT services to ensure security best practices are implemented throughout the life cycle of the service. The authors provide recommendations on how to mitigate common security threats and weaknesses within IoT services. The set of documents analyses two ecosystems—service and *endpoint*—but also provides a number of real-life examples.

IETF and IRTF are cooperating, parallel open standards organizations, that focus on short-term and long-term Internet-related research, respectively. They have issued a couple of highly informative drafts regarding IoT security. In 2017, "Best Current Practices for Securing Internet of Things" [31] was published by IETF. This report collects guidelines for IoT designers and developers, written by engineers from Network Heretics, Mozilla and Arm. It offers valuable remarks on low-level IoT development, by discussing the authentication, encryption, and design of a device and firmware. Even though it is now labelled as expired, we find this document provides valuable input into the discussion. In March 2021, another draft was released—"Security Technical Specification for Smart Devices of IoT" [32]—collecting detailed recommendations from hardware to software level and proposing a secure IoT device model. In April 2019, IRTF published "RFC8576—Internet of Things (IoT) Security: State of the Art and Challenges" [33]. In this document, the authors present a list of already existing guidelines regarding IoT security; they report and predict the development of IoT and point out possible challenges, especially with reference to the nature of resource-constrained IoT devices (e.g., in terms of algorithms and protocols that would allow IoT devices to safely operate in a heterogeneous network with powerful, potentially malicious Internet resources). The aforementioned publications are complementary to each other and were issued by cooperating organizations, therefore conclusions drawn from them are presented together.

The IoTSF and ioXt Alliance are composed of industry leaders, manufacturers and government organizations, dedicated to creating a security and privacy standard for IoT—some of these entities belong to both of these organizations. The first consortium published "Secure Design—Best Practice Guides" [34] at the end of 2019. This document highlights the importance of maintaining a chain of trust throughout the hardware and software layers of IoT device. The ioXt Alliance has recently published "ioXt Pledge: The Global Standard for IoT Security" [35], in which eight core principles are defined and described. It considers a wide range of subjects, such as secured interfaces, proven cryptography, software verification/updates and vulnerability reporting mechanisms. The organization offers a certification program and creates a network of authorized laboratories. It also encourages independent researchers to participate in the certification process, by validating that every security requirement is fulfilled.

It is worth highlighting that ISO is currently working on their own guidelines. As of June 2021, ISO/IEC CD 27400 "Cybersecurity—IoT security and privacy—Guidelines" is still under development [36].

3.2. Evaluation Criteria

In this section, we created a set of recommendations with a focus on SoC hardware and software security, deriving from the documents mentioned in Section 3.1. This selection is later used to analyze the state-of-the-art IoT security technologies. The aim of Table 1, presented later in this section, is to collect the most common and important recommendations from the analyzed literature and to provide a solid overview of what is expected from a well-secured IoT device. As already described in Section 2, our starting point was a security core baseline for IoT devices defined by NIST and, as mentioned in Section 3.1, the main focus during criteria analysis was put on secure IoT device implementation. Therefore, high level concepts, such as, for instance, network structure or its security remained out of scope. On the other hand, topics such as the safety of an industrial or automotive IoT node and its capability to operate in various environmental conditions (e.g., temperature, humidity, contact with harsh chemicals) do concern device architecture, but are mostly related to its reliability instead of security. Only the aspects of physical access to the device relevant at the chip level were considered, because those regarding the product level can be very location or application specific.

The analyzed papers have multiple points in common. In Table 1, we collected the most prevailing suggestions and divided them into the following groups on the basis of key functionalities:

- hardware security,
- trust and integrity management,

- data protection and software design,
- device configuration and software update,
- secure interfaces and communication,
- cybersecurity event monitoring and logging,
- cryptography and key management,
- device identification, authentication and strong default security.

The intent was to mimic the process of constructing a secure IoT device by creating a checklist of requirements it has to fulfill. Almost all of the analyzed documents proposed the functional classification of secure IoT device characteristics with some exceptions. A relevant example is GSMA's document [30], where the requirements were distributed by implementation priority (Critical, High, Medium, Low). Categories presented in this paper are similar to the ones recommended by ENISA [27], but the number of groups was reduced, and an appropriate level of granularity was maintained, which allowed for concise comparison of available secure IoT implementations.

Hardware Security

This category collects recommendations for designing IoT devices based on the Root of Trust (RoT) concept and the characteristics such a component should demonstrate.

A Root of Trust is a unit that consists of a computing engine, low level code and data (e.g., cryptographic keys). It provides security services/features necessary to establish trust and security within the platform it is a part of. Its vital characteristics are *immutability* and predictability—the produced results have to be consistent for the same input data. The hardware implementation of an RoT enables the fulfillment of these conditions [37,38].

A Root of Trust can provide the following independent security services—identification, authentication, confidentiality, integrity and measurement (state of the platform), as well as composite services (relying on the independent ones)—authorization, verification, reporting, secure storage and update.

Unsurprisingly, almost every analyzed document provided some guidance on the hardware role in the security of the final product, ENISA [27] and the GSMA [30] being the most elaborate. It proves that protecting an IoT device must start at the hardware level.

Trust and Integrity Management

This section focuses on requirements regarding trust establishment and ensuring integrity, which are fundamental to IoT device security.

An inherently trusted, immutable (hardware) Root of Trust is the trust anchor, from which trust is extended to the whole platform through a secure boot process. A Hardware Root of Trust (HWRoT) utilizes its security services to verify the integrity of subsequently executed software modules (a cryptographic signature of code is checked). Verified software modules then become the next *Chain of Trust* elements. If the integrity verification check fails during one of the secure boot stages, the whole process has to be aborted and the system can only be trusted up to the given *Chain of Trust* level. Depending on which advanced capabilities of the system are available at this point, the device might have to reboot, attempt to return into last known secure state or remain as it is, that is, for reporting purposes.

Apart from NIST and IETF/IRTF publications, every considered paper included recommendations on this matter.

Data Protection and Software Design

This category considers recommendations on data handling and fundamental principles of software architecture. Data confidentiality must be protected through encryption. Additionally, the designer of the system must also ensure that applications processing data operate on minimum privilege and are isolated from each other (e.g., through memory compartmentalization). These subjects were discussed in the majority of analyzed documents.

Device Configuration and Software Update

The question of software updates is highly stressed in the subject matter. All organizations agree on its importance—a lack of identified vulnerability patching and protection against the latest threats is a large security risk in IoT. Thus, keeping the device up-to-date strongly improves its protection. The capability to *securely* update device software by an authorized entity is a must, preferably this process should be automatic and/or remotely available.

Secure Interfaces and Communication

This category analyzes guidance on secure communication, starting from interfaces, through protocols, to data. The usage of device interfaces should be configurable and, by default, only those required should be active. Systems should utilize state-of-the-art, standardized security protocols, with emphasis on using the versions that are intended for IoT, if available. The majority of publications stress the necessity for making only intentional and required connections (preceded by mutual authentication of the peers), which are used to transmit confidential and integral data. Recommendations apply to all layers of IoT devices—hardware, firmware and software.

Cybersecurity Event Monitoring and Logging

Event logging is a key requirement for security management in an IoT device and this category summarizes guidelines on this matter. Adequate level of details aids to solve issues with security incidents, while preventing exposure of sensitive information. The confidentiality and integrity of logs must be protected so that they are reliable—only authorized entities should be able to examine them and they should be unmodifiable.

Furthermore, the GSMA suggests creating a reference model of the IoT device behavior and perform anomaly detection—situations when the device erratically reboots, reconnects to the network or sends multiple poorly-formed messages might indicate a security issue.

Only the ioXt Alliance did not offer guidance on this subject.

Cryptography and Key Management

Each of the evaluated documents provided an insight for this category. It is clearly advised to use standardized, *proven* cryptographic algorithms with secure implementations and sufficient key lengths. Documents highlight the need to support algorithm agility in security protocols—to enable the usage of lightweight cryptography and specialized algorithms for IoT applications, to allow various node types for algorithm and key length negotiation so they can communicate or ensure the ability to change the algorithm or used key length in case it is compromised.

A secure and scalable key management policy must be introduced in the IoT system and every IoT device should be provisioned with a *unique* private key (possibly per application, e.g., device identification, code signature, server communication, etc.). With this approach, every device becomes a separate attack surface and the whole system is secure even if one of them is compromised.

Device Identification, Authentication and Strong Default Security

This section discusses the identity of both the device and its user. The device must be labelled and recognized in a credible way, given that it will operate in a wider network. It should also be protected from undesired access and provide security for user data. As for the software developer and the user, this category includes guidance on password management and general authentication procedures.

Table 1. Synthesized requirements for secure IoT devices.

No.	Requirement	NIST	ENISA	IETF IRTF	GSMA	IoTSF	ioXt
Cat. 1	Hardware Security						
1.1	Utilization of **immutable** Hardware Root of Trust (HWRoT—Trust Anchor)—trusted component that extends the chain of trust to other HW, FW, SW components	-	X	X [32]	X	X	-
1.2	HW provided security features (memory locking, storage for cryptographic keys, secure boot support, device authentication, communication confidentiality and integrity, ...)	X	X	X[32]	X	X	-
1.3	Measures for tamper protection and detection	-	X	-	X	X	-
1.4	Use of proven/cryptographic quality Random Number Generator (hardware-based if feasible)	-	X	X[31,32]	X	-	-
Cat. 2	Trust and integrity management						
2.1	**Secure boot process based on HWRoT.** Boot process initializes the main hardware components, verifies executed code (first-, second-stage bootloader, OS) and results in environment determined to be in an uncompromised state	-	X	-	X	X	X
2.2	Software (code, applications) must be signed cryptographically and verified upon installation or execution	-	X	-	X	X	X
2.3	The ability to restore last known secure state (e.g., firmware rollback, when code is verified as damaged or tampered)	-	X	-	X	X	-
2.4	Software installation control in operating systems (OS), to prevent unauthenticated software and files from being loaded onto it	-	X	-	-	-	-

Table 1. *Cont.*

No.	Requirement	NIST	ENISA	IETF IRTF	GSMA	IoTSF	ioXt
Cat. 3	Data protection and software design						
3.1	Encryption of data storage medium. Possibility of locking or erasing device contents remotely	X	X	X[32]	-	X	-
3.2	Ensuring that data is secure prior to use by the process (input data validation)	-	X	X[32]	-	X	-
3.3	Process privilege minimization - limited permission of allowed actions, applications must operate at the lowest privilege level possible	-	X	X[31]	X	X	-
3.4	Utilization of memory compartmentalization to prevent rogue or compromised applications from accessing memory areas that they are not authorized to use, process isolation	-	X	X[31]	X	X	-
Cat. 4	Device configuration and software update						
4.1	The ability to change the device's FW and SW configuration by authorized entities	X	-	X[32,33]	X	X	-
4.2	The ability to update device's FW and SW through remote or local means by authorized entities	X	X	X[31,32]	X	X	X
4.3	Update file must be encrypted, signed and device verifies its integrity before application	X	X	X[31–33]	X	X	-
4.4	Automatic update capability (checking at frequent, but irregular intervals), enabled by default	X	X	X[31,33]	X	-	X
4.5	Backward compatibility of updates—update should not change network protocol interfaces nor modify user-configured preferences, security and/or privacy settings	-	X	X[31]	-	-	-

Table 1. *Cont.*

No.	Requirement	NIST	ENISA	IETF IRTF	GSMA	IoTSF	ioXt
Cat. 5	Secure interfaces and communication						
	Interface security						
5.1	Ability to disable or logically restrict access (e.g., device/user authentication) to any local and network interfaces	X	X	-	-	X	X
5.2	Deployed device should never contain debugging, diagnostic or testing interfaces that could be abused by adversary (JTAG, CLI, Telnet, etc.)	-	X	X[32]	X	X	X
	Protocol security						
5.3	Ensure that communication security is provided using state-of-the-art, standardized security protocols (e.g., TLS for encryption)	-	X	X[32]	X	X	-
5.4	Mutual authentication of the devices must be performed before trust can be established (prevent man in the middle attack), even when given device leaves and re-joins network	-	X	X[31]	X	X	X
5.5	**Make intentional connections.** Prevent unauthorized connections to product or other devices it is connected to, at all levels of the protocols. IoT devices must provide notice and/or request a user confirmation when initially pairing, onboarding, and/or connecting with other devices, platforms or services.	-	X	X[32]	-	X	X
	Data security						
5.6	Guarantee different security aspects of the transmitted data—confidentiality, integrity, authenticity	-	X	X[31,32]	X	-	-

Table 1. *Cont.*

No.	Requirement	NIST	ENISA	IETF IRTF	GSMA	IoTSF	ioXt
Cat. 6	Cybersecurity event monitoring and logging						
6.1	The ability to log cybersecurity events from device's HW, FW and SW	X	X	X[32]	X	X	-
6.2	The ability to record sufficient details for each logged event to facilitate an authorized entity examining the log and determining issue origin	X	-	X[32]	X	X	-
6.3	The ability to restrict access to the logs so only authorized entities can view them and prevent all entities (authorized or unauthorized) from editing them	X	X	X[32]	-	X	-
6.4	**Avoid security issues when designing error messages.** An error message should give/display only the concise information the user needs—it must not expose sensitive information that can be exploited by an attacker, such as an error ID, the version of the web server, etc.	-	X	X[32]	-	X	-
6.5	Device's behavior should be monitored and compared with model to detect anomalies (e.g., erratic reboots/resets, (dis)connections, different network fingerprint, repeated malformed messages sent, etc.)	-	-	-	X	-	-
Cat. 7	Cryptography and key management						
7.1	Use of Standard Cryptographic Algorithms and Security Protocols, which implementations have been independently reviewed	X	X	X[31–33]	-	X	X
7.2	Security protocols should support algorithm agility (lightweight cryptography for resource constrained devices, ability to change algorithm or use it only with longer key in case it is compromised)	X	X	X[31,33]	-	X	-

Table 1. *Cont.*

No.	Requirement	NIST	ENISA	IETF IRTF	GSMA	IoTSF	ioXt
7.3	Length of the key should provide strong security, make brute-force attack infeasible	-	X	X [31]	-	X	-
7.4	Every device must be instantiated with unique private key(s)	-	X	X [31]	X	X	-
7.5	Secure and scalable management of cryptographic keys (generation, storage, distribution)	-	X	X [31]	X	X	-
Cat. 8	Device identification, authentication, strong default security						
8.1	Device must have a unique physical identifier only authorized entities can access	X	X	-	X	X	-
8.2	Device must be provisioned with unique logical identifier	X	X	X [32]	X	-	-
8.3	Prepare robust authentication and authorization schemes, so device can prove its identity	X	X	-	X	X	-
8.4	Establish strong, device-individual default passwords that has to be changed by the user during initial setup (weak, null or blank passwords are not allowed)	-	X	X [31,32]	X	X	X
8.5	Resistance to keyspace-searching, brute-force or other login abusive attacks by limiting number of invalid login attempts or introducing incrementing delays between them	-	X	X [31,32]	X	-	-
8.6	Availability of two-factor authentication	-	X	X [31]	X	X	-
8.7	Product security shall be appropriately enabled by default. Strong security controls should be something the consumer has to deliberately disable rather than deliberately enable.	-	X	X [31]	-	X	X

4. Review of Existing Solutions

In this section, we analyze some of the current popular trusted computing solutions used for IoT devices against the recommendations presented in Section 3. The solutions chosen are either already mature and currently in use for IoT devices or are emerging concepts that might bring new quality to the topic. The selection of technologies for IoT security includes examples that are more hardware-based (e.g., GEON SoC Platform) and more software-focused, such as Intel Software Guard Extensions (SGX). We chose the most popular solutions available on the market, such as Arm's TrustZone or Intel SGX, the most promising open source ones, such as Keystone and OpenTitan and other representative, usually hardware-based, solutions. In the last group, we targeted solutions that can be integrated into a designed SoC in the form of a standalone IP–Rambus RT, similar but more complex solutions, such as Geon SoC Security Platform, and finally, a solution offered as an integrated circuit (NXP EdgeLock SE050). A short description is provided for each one before the analysis of the IoT security in the context of Table 1. We present our findings for each solution in Table 2. The descriptions are formed from the perspective of a designer who wants to understand the requirements for a secure IoT device. We assumed that the analysis is based only on publicly accessible information, no lab testing nor feasible attacks on devices based on the solutions were performed, and that little to no implementation experience for each of the discussed solutions is required.

4.1. Arm TrustZone

TrustZone is a security extension offered by Arm for application processors (Cortex-A family) and microcontrollers (Cortex-M family), which has become popular recently. There is a growing interest in the subject across academia, and a number of commercial products utilize TrustZone's capabilities, for example, Samsung Knox. Arm's solution is based on a Trusted Execution Environment (TEE) concept and enables the system to fulfill Platform Security Requirements [39]—a set of guidelines defined by Arm. This document provides similar guidelines to those gathered in Section 3, with a strong focus on hardware and firmware security. There are some differences between architecture-specific implementations of TrustZone, but the main idea remains the same—on Cortex-A processors, the secure monitor, a privileged software, implements mechanisms for secure context switching between worlds; on Cortex-M processors, there is no secure monitor and hence the change between the secure and non-secure world is handled by a set of core logic mechanisms. Enabling TrustZone for microcontrollers has allowed for more widespread usage in resource-constrained devices, which has an impact on IoT devices. Pinto and Santos [40] provided a detailed explanation of TrustZone's operation with a distinction between Cortex-A and Cortex-M implementation. The general concept of TrustZone's operation on software and firmware levels is presented in Figure 1.

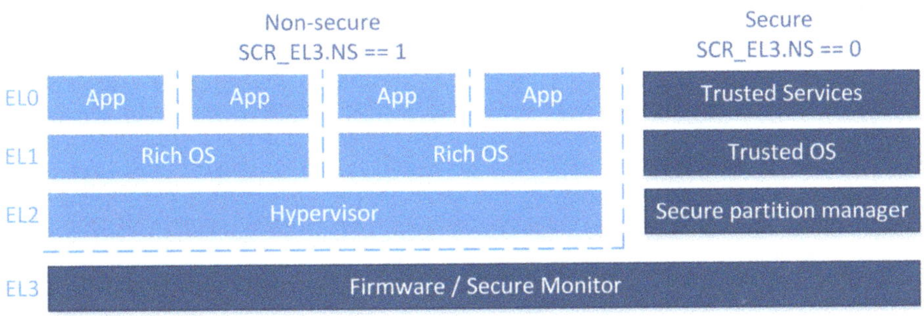

Figure 1. Concept of a TrustZone supported execution environment on software and firmware levels, based on [41]. The division between non-secure and secure worlds, as well as their construction with accentuated levels of operation, are presented. The SCR_EL3.NS is a register bit used to switch between the trusted and not trusted environments. The Secure Monitor, a special processor mode controlling the transition between secure and non-secure states, is presented as a common base for the environments.

TrustZone provides two virtual processors supported by hardware-based access control, resulting in division into secure and non-secure environments. Both execution environments are completely separated in the hardware, thanks to memory isolation and a special processor mode dedicated to monitoring (*secure monitor*). A few additional modules, such as the TrustZone Address Space Controller (TZASC), the TrustZone Memory Adapter (TZMA) and internal interface modifications, are introduced for separation and memory partitioning in TrustZone for Cortex-A. Noticeably, Arm states that the use of TZASC and TZMA is optional in the system. The final decision is left to the designer. Peripherals are flagged as *secure* or *normal*, while the APB-AXI bridge hardware is responsible for access control and rejecting AXI transactions with insufficient permissions. There is no separated trusted path since secure and non-secure transactions are multiplexed on the system bus using the same hardware. Significantly, in TrustZone there are no explicit considerations for Hardware Root of Trust implementation. However, the implementation of Root of Trust has been proposed in the literature by Zhao et al. [42], and a commercial solution, discussed later in this subsection, has also been made available. Some security features are provided by separate hardware modules. For example, remote attestation is achieved by an incorporated hardware component, such as a Trusted Platform Module, responsible for measuring the kernel's integrity and creating unique cryptographic keys. While there is no information on random number generation in TrustZone documents, Arm offers a separate security IP called the True Random Number Generator, advertised as TrustZone compatible. Similarly, secure storage can be implemented by simply denying access to a device from a non-secure world.

From a firmware and software perspective, a privileged instruction called the Secure Monitor Call (SMC) allows for entry to, exit from and general communication between secure and non-secure world applications. This mechanism involves a monitor software with higher privileges than the Rich Execution Environment Operating System (REE OS). Its responsibility is the reliable and protected context switching of the processor [40]. A Non-Secure (NS) bit stored in the Secure Configuration Register (SCR) represents the current context of the processor and the register's state is propagated throughout the entire SoC. Hence, it is possible to use peripheral devices that only allow secure access from the processor. In TrustZone there is no attestation of processes requesting execution in the secure world. In the event of an OS falling into the hands of an adversary, messages that require secure world resources can be crafted and possibly other information may be gleaned from apparently secured processes. A newer TrustZone technology for Cortex-M microcontrollers has some changes to allow for usage in more resource-constrained devices. The division between secure and non-secure worlds is based on memory map partition, and context switching happens due to code exceptions. To support this, a couple of new, dedicated instructions were added.

One could assume that TrustZone alone leaves quite some room for interpretation in terms of security functions implemented in hardware. That being said, Arm proposed the CryptoCell IP-family [43,44], which acts as HWRoT, offering cryptographic acceleration, true random number generator, trusted storage, secure boot support and authenticated debug support, to name a few. CryptoCell-300 is dedicated for Cortex-M implementations, while CryptoCell-700 is Cortex-A oriented [45]. Both families provide broad support for symmetric and asymmetric cryptography, as well as lightweight cryptography [46]. Code integrity and signing, authentication and key management are also mentioned in the documents. Arm claims that this solution is meant for low power, low area designs [43,44]. It may seem that TrustZone, complemented by CryptoCell IP, is a powerful security solution, mostly implemented in hardware.

Analysis for Synthesized Requirements for Secure IoT Devices

Some security flaws inherent to the REE-TEE communication channel have been found in previous years (Black Hat 2014, Black Hat 2015) [40] and have been used in attacks on real devices. Additionally, concerns with cache side-channel attacks arise. The lack of inherent memory encryption also raises questions. TrustZone in itself does not have any recommendations nor requirements for HWRoT implementation other than the need for the existence of a unique device key [40]. This is paramount for the security of an IoT device and cannot be left to wide interpretation. One could assume that Arm's recommendation is to use IP from the CryptoCell-family, though it is not always feasible. On the software execution front, the secure-world approach is not the same as an enclave and allows for a compromised TEE program to interfere with other programs within the TEE. Secure IoT required functionality, as presented in Section 3, is possible in TrustZone, but with much of the implementation left to the designers—secure deployment might be at risk. TrustZone documentation does not provide input on configuration, update or logging mechanisms. We assume it is at the designer's discretion. Finally, TrustZone is entwined with Arm's technology; it is not a universal solution and porting it to different platforms does not seem feasible. On one hand, this can be limiting for some implementations, as not all IoT devices use Arm's solutions. On the other hand, it would be surprising if one of the leaders in the processor market did not have a compatible security solution available.

We can conclude that TrustZone certainly offers an end-to-end security solution, but it requires a good understanding of the framework, some creativity in implementation and support from external IPs. It is worth repeating that this solution is dedicated for Arm infrastructure. The vast number of TrustZone supporting products speaks for itself. TrustZone fulfills most of the security recommendations, but it would not be possible without supplementary hardware, like CryptoCell or applications. TrustZone alone is not an off-the-shelf, ready-to-use solution.

4.2. Intel Software Guard Extensions (SGX) and Security Essentials

Intel SGX [47] is a set of CPU instructions that allow the creation of isolated software containers called enclaves in which code, data and stack of a program are isolated safely from other processes (even with higher privilege levels) through hardware-based access policy control and memory encryption. Unsurprisingly, this solution is meant for Intel architecture. Application code and the hardware it is running on can be attested by a remote entity [48] by verifying measurements of its code and data (named MRENCLAVE), calculated during enclave creation. This process provides increased confidence that the code is running in an enclave and is unmodified by a third party. It provides a mechanism to run a secure code in an unsecured OS, with support from trusted hardware. After successful attestation, an encrypted channel (based on public key schemes) between the remote party and the enclave can be established in order to exchange sensitive data. Data can be sealed using CPU instructions to generate a key (named MRSIGNER) which allows decryption only by the same enclave created in the future.

Additionally, in the Intel security ecosystem, a provisioning functionality is present, which allows trusted entities to update or add software whilst assuring integrity. It is presented in Figure 2. The provisioning process is described below:

- **Creating the enclave**—an untrusted REE application creates an enclave environment in order to protect the software of the trusted provider, during this process the contents and creation parameters are logged as a measurement,
- **Attestation**—the enclave informs the software provider about its readiness to receive new software and the device presents the measurement from before,
- **Provisioning**—a secure channel between the device and provider is created and the data is sent,
- **Sealing and unsealing**—the enclave uses a cryptographic hardware key for encryption before storing the data in memory. Only an identical enclave will be able to decrypt and use the new program data in the future,
- **Software update**—in this step, a new version of a program may ask an older version to unseal the data. After the update process ends, a new seal is created, which renders the old software version (which could be compromised due to flaws patched in the new version) unable to access the new software version data.

From a hardware viewpoint, Processor Reserved Memory (PRM) is isolated by SGX and protected against all memory accesses from outside an enclave, including kernel, hypervisor and system management mode, as well as DMA accesses requested by peripherals. Enclave's code and data are stored in Enclave Page Cache (EPC), providing pages of a 4KB size [49]. The untrusted OS is in charge of assigning EPC pages to enclaves, which creates a potential exposure. The processor is responsible for assuring that each enclave has only one EPC on hand; it is monitored in Enclave Page Cache Metadata (EPCM).

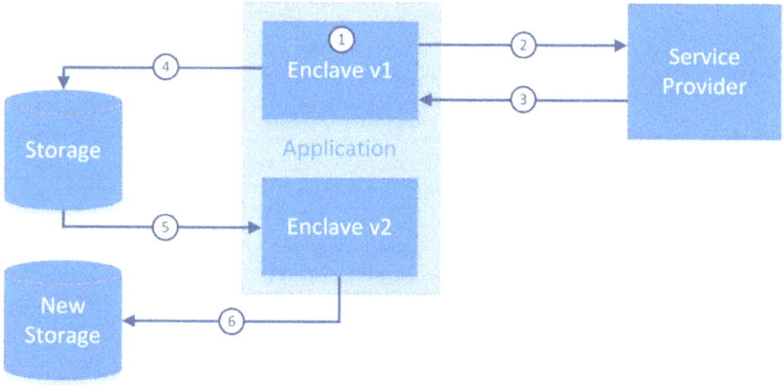

Figure 2. Intel SGX Software Lifecycle; steps are executed in the numerical order, sensitive data are remotely provisioned (3) into the enclave after mutual attestation (2) of the off-platform provider and created enclave (1), after the enclave is destroyed data are sealed in memory (4) in such a way that only an identical enclave can access them again. Software updates (5) are also possible via this scheme and a new seal is created (6) so that an old version of software cannot overwrite the new version. Based on [47].

SGX in itself is generally a TEE implementation, but other Intel modules and technologies allow for wider SoC security such as secure boot capabilities, TPM integration or random number generators [50]. It is advertised that the security extension assists with securing IoT edge device communication [51].

Analysis for Synthesized Requirements for Secure IoT Devices

Intel SGX is a proprietary enterprise solution; therefore all security functionalities are tied in with Intel architecture. This means that other custom application accelerators, interfaces and system bus control are out of the scope for the solution. As such, all SoC security issues have to be solved by custom design decisions, increasing the risk of creating an unsecured device. Costan and Devadas [49] in 2016, one year after the Intel SGX debut, published an in-depth examination of this solution. They exposed a number of potential vulnerabilities to SGX and claimed that "our security analysis reveals that the limitations in SGX's guarantees mean that a security conscious software developer cannot in good conscience rely on SGX for secure remote computation" ([49], [p. 3]). In the five years that have passed since this article was published, a number of new side channel attacks have been discovered [52–55], and although Intel describes these attacks as very difficult to perform in a data center (i.e., physical access to the platform is required), this changes in an IoT context (device present in a public space) and remains alarming. Due to a security by obscurity situation, it is hard to assess the solutions for many requirements in Table 2 and thus no assurance on their completeness can be given, though an optimistic approach is taken. Interestingly, it is possible to change all device cryptographic keys by changing a single register called OwnerEpoch. This process allows for the fast sealing of the device and serves as additional protection of the enclave content. If the OwnerEpoch value is lost, the device is rendered inaccessible.

SGX and the Security Essentials provide multiple good trusted computing basics but is not a solution entirely catered for protection of IoT devices. This can be seen in the Cat. 6 and Cat. 8 sections of Table 2. However, the security level it provides and the critique around it, raises questions about whether it is a good choice at all.

4.3. Keystone

Keystone [56] is an open-source framework designed for creating TEE environments based on unmodified RISC-V architecture. RISC-V Physical Memory Protection (PMP), arbitrarily securing the physical memory locations, and the programmable machine mode (M-Mode) are used to implement the memory protection scheme. The trusted Security Monitor (SM) program is proposed on M-Mode level and its main task is to manage the secure handling of hardware and context switching between enclaves (Figure 3). It should be executed entirely from on-chip memory. This component satisfies typical TEE requirements such as memory isolation and code/configuration attestation. Keystone does not propose direct resource management. This responsibility lays on the secure enclave application developer side. A runtime (RT) component is an enclave-specific platform for secure applications to be executed on, which also communicates with the SM and manages virtual memory chunks assigned to the enclave. The RT should have the functionality of system plugins, interfaces, *libc* implementations, paging and virtual memory management. It can be successfully reused between enclaves or modified as needed. Additionally, because the rich OS is not a part of a typical enclave, the Trusted Computing Base (TCB) is smaller.

From a hardware perspective, implementing Keystone does not require modifications to CPU cores or memory controllers. However, several requirements for the platform are listed in the publication [56]: trusted boot process support, unique authentication key dedicated for this process and a hardware source of randomness. According to the Keystone designers, the Root of Trust can be realized in hardware, but does not have to, which allows for a variety of implementations, for example tamper resistant software (zeroth-order bootloader) [56].

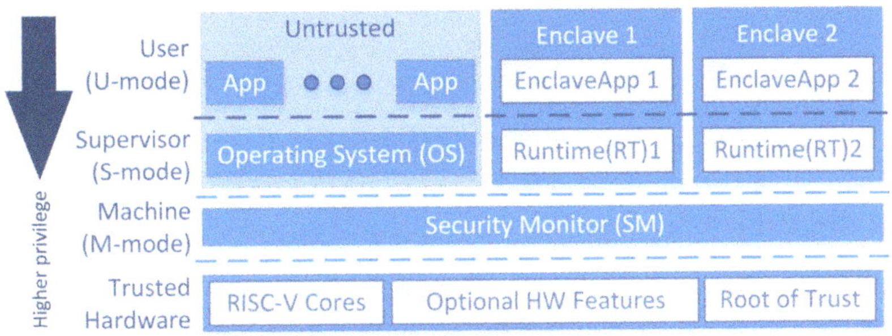

Figure 3. Overview of a Keystone system setup, based on [56]. The distinction between untrusted and trusted execution environments is presented, along with support for multiple secure enclaves and their runtimes. The privilege level for each RISC-V mode is marked. The security Monitor, operating on M-mode, being common for both execution environments, is the manager of context switching. Required trusted hardware is included.

With every CPU reset the Root of Trust executes these steps:

- Measures the SM image loaded,
- Generates a new attestation key based on the randomness source,
- Saves the data in a SM memory location isolated by PMP,
- Sends the cryptographically generated metadata via a public key scheme.

In the RISC-V architecture, only the machine mode has access to hardware resources such as interrupts, system memory, peripherals and devices. The S-mode (supervisor) is used for the OS kernel and the U-mode (user mode) allows execution of typical REE applications. The machine mode is used as a platform for the SM, because:

- It is programmable,
- It allows control over interrupts and exceptions above the OS level,
- PMP mechanism, which allows the enforcement of access policies.

PMP restricts the physical access to memory locations for the S and U modes. Every record in the PMP table has a set of access flags for the programmed memory segment. Each PMP address register encodes a continuous address region and the configuration bits represent write/read/execute (rwx) permissions for the U/S modes. If a mode attempts to access an unassigned memory area, the access is rejected by the hardware. The PMP areas can be dynamically allocated during device operation. During SM image loading, the first PMP area is configured as the highest priority in order to protect its own resources, such as code/stack/data. By default, the OS has access to any memory location that is not part of an assigned PMP area of higher priority. When an application starts an enclave, the OS finds a continuous area of memory, which does not overlap with any other PMP configured area, and then switches the request to the SM that creates a new PMP record. During a context switch from an enclave to the OS or another enclave, the SM takes away all access permissions, but the enclave's address region is preserved. As a result, an enclave is securely isolated from other processes. Upon creation, the enclave's memory is measured and cryptographically signed. The OS uses an interface within the Linux kernel (/dev/Keystone) for enclave creation.

Analysis for Synthesized Requirements for Secure IoT Devices

The paper that introduces Keystone focuses more on the enclave code execution, while treating key management, software updates, device authentication and other problems such as orthogonal. A Root of Trust (either hardware or software) is stated as a necessity along with a trusted boot process, but it is in the hands of the application designer to make sure that secure IoT requirements are in place. Process and application privilege levels are inherent to the RISC-V implementation. Enclave security is supposed to include resistance to side-channel attacks, mapping attacks and syscall tampering attacks. Keystone's authors state that their solution is an improvement on SGX [56], and the measures they took to mitigating the risks that were exposed in Intel's solution seem to confirm that. However, Keystone, being based on a similar concept to SGX, can still demonstrate similar vulnerabilities that may yet to be discovered.

Using concepts presented in Keystone, or even the entire solution, seems like a good starting point, but it is hardly enough in itself to state that a device is secure in the IoT field. It certainly offers some degree of flexibility. Keystone being an open-source solution can also be an advantage. However, as presented in Table 2, there are many important categories that are left completely at the system designer's discretion and it can result in unintentional exposures of the IoT device.

4.4. *OpenTitan*

OpenTitan is undoubtedly one of the solutions to secure IoT devices that should attract designers' attention. This open source Hardware Root of Trust implementation [57] is endorsed by leading non-profit, academic or commercial organizations such as lowRISC, ETH Zürich or Google. The project is fully transparent, its sources are available online and can be inspected by the broader community, which should improve its security.

OpenTitan core is currently under development [58] and multiple features are still missing from the early stage top-level [59]. However, the intentions of its creators are well documented and in the complete form it should be a robust solution for various systems' security as an HWRoT module supporting the secure boot procedure and implementing miscellaneous cryptographic primitives.

OpenTitan acts as an immutable HWRoT with secure boot procedure support. It implements multiple cryptographic primitives (AES, RSA, Elliptic Curve Cryptography—

ECC or keyed-Hash Message Authentication Code based on SHA-256 algorithm) that are used to verify the code of subsequent boot stages and authenticate the device during the ownership transfer process. OpenTitan provides FW to control these primitives, so they can be used for securing communication confidentiality and integrity, but it is a responsibility of higher software layers. OpenTitan implements several tamper protection and detection mechanisms as protection codes or scrambling memory regions that contain secrets. Core implements hardware Random Number Generators that are compliant with international standards (e.g., NIST [60,61]).

OpenTitan provides low-level software that is responsible for first stages of the secure boot process (Silicon Creator level—Figure 4). At the beginning, execution is restricted to the ROM region (which cannot be modified after silicon is manufactured) and then ROM_ext part is loaded from flash (provided that integrity check has passed). At this point, execution is transferred to the entry point of the Silicon Owner code and from now on *it is the Owner who is responsible* for further boot stages security. It is also the end user's software duty to ensure isolation of the applications.

This solution offers the means to securely update its firmware—the integrity of the update block is verified prior to the reboot. The capability to update higher layers of software can be implemented using available cryptographic primitives; however, it is up to the end user.

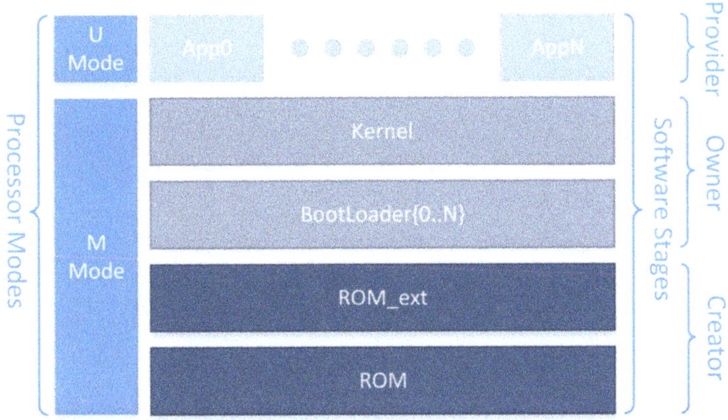

Figure 4. Software stages of the OpenTitan secure boot procedure with respect to particular levels owners, based on [62]. Only low level software layers (ROM, ROM_ext) are provided with the solution (Silicon Creator level), while the end user (Silicon Owner) is responsible for all higher phases of secure boot as well as the isolation of executed applications when the system is already up.

Analysis for Synthesized Requirements for Secure IoT Devices

We attempted to analyze the final set of OpenTitan capabilities against requirements presented in Table 1. The OpenTitan documentation does not mention any means for logging of cybersecurity events, neither in hardware nor in software. Possibly, some mechanisms might be implemented by end user in higher layers of software.

As was already mentioned, OpenTitan offers a rich suite of cryptographic primitives. Algorithm agility may still be improved, especially if it comes to lightweight cryptography, but it might be that OpentTitan does not target resource-constrained devices. Robust key derivation mechanisms and management schemes are available.

In summary, the fully operational OpenTitan core meets the assumptions of its designers—it is a solid foundation for maintaining the trust and integrity of the system, where it is instantiated. However, it still might not be enough to completely secure IoT devices in terms of recommendations from Table 1. It is not an out-of-the box solution for all IoT device security aspects and the potential end user has to implement many missing features.

4.5. NXP EdgeLock SE050

NXP EdgeLock SE050 Plug and Trust Secure Element is presented as a "ready-to-use IoT secure element solution" by the manufacturer [63]. It is an auxiliary security device that connects to host. Optional connections to a sensor mode via another I2C interface and native contactless antenna, granting wireless access are also available (Figure 5). Interestingly, EdgeLock SE050 holds Common Criteria's EAL6+ certificate. It is advertised as suitable for smart cities, smart home, smart industry and smart supply chains. This solution combines a HWRoT with a Java Card OpenPlatform OS (JCOPOS), on which an IoT Applet runs. The Applet supports a wide range of secure functionalities, for example, random number generation, key management, hash operations, Platform Configuration Register (PCR) creation and management [64], and is provided by the manufacturer. In hardware, many configurable cryptographic primitives are included, supporting a wide range of algorithms and operations to choose from: HMAC, CMAC, SHA-1, RSA, ECC, AES. Lightweight cryptography is also available. What distinguishes NXP SE050 from other discussed solutions is the fact that it is a separate integrated circuit in its own packaging. This implicates the need for a dedicated space on the circuit board. It is also the only analyzed technology that supports SmartCard.

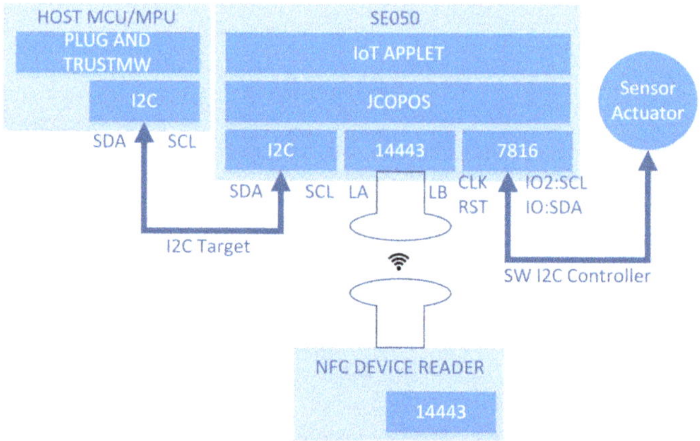

Figure 5. NXP SE050 architecture scheme, based on [63]. The communication between SE050 and Host via I2C bus is presented. The additional interfaces for optional applications, i.e., the antenna and the extra I2C to communicate with Sensor Actuator, are marked. The software/firmware scheme is also present in the form of the IoT Applet, running on Java Card OpenPlatform Operating System (JCOPOS).

Analysis for synthesized requirements for secure IoT devices

Some capabilities required for a secure IoT device hardware are present, but unfortunately as a plug-and-play approach the SE050 can provide a false sense of security, if used inappropriately. No safety is ensured for the host OS, which could in reality be wrongly configured and utterly unsafe with other interface connections that bypass the secure element—each communication channel should be trustworthy to a degree required in a protected IoT device and each communication channel should have the possibility to be disabled if not needed as per Table 1. No approach to software and firmware updating and provisioning is presented. It is hard to assess the safety of the I2C communication channel and possible problems. Additionally, neither software trusted execution nor enclave creation is possible via the applet. It is uncertain whether the host can freely access and use the cryptographic functionalities (including the random number generator) or if they are only available for SE050 internal access. Remarkably, the data sheet states that only the Pseudo Random Number Generator is available [63]. Tamper detection is mentioned, but there is no information on tamper resistance. Moreover, no information is provided about the execution privileges in the Java OS nor about what information could be gathered through the secure logging of modules inside the secure element. If only the secure element interfaces are used for off chip communication, then they should be safe. Even though a wide range of secure functionalities is presented in the data sheet, a variety of use cases is discussed on NXP's website, cryptographic hardware primitives, coherent with standards and certifications, are implemented—this cannot be considered as a plug-and-play secure IoT solution. In fact, software must be developed carefully to ensure a proper level of security whilst using the SE050. By and large, the solution seems more like an extensively functional HWRoT with communication between two operating systems rather than an IC offering complex security for IoT.

4.6. *Beyond Semiconductor GEON SoC Security Platform*

The GEON SoC Security Platform manufactured by Beyond Semiconductor is presented as a processor agnostic solution with essentials for hardware IP security. An IoT application is advertised, including Industrial [65]. The SoC contains a suite of security modules that work together to create a secure IoT device but can also be used independently. This includes a customizable Hardware Root of Trust including secure boot functionality (GEON Secure Boot) and even recovery to a vendor software state in the case of a security breach [66], firmware encryption module (GEON Firmware Encryption) for the integrity and confidentiality of software, a module that stores and generates secret keys (GEON Hardware Security Module (HSM)) and hardware cryptographic operations module, including random number generation. A wide range of cryptographic algorithms is supported, including lightweight. Code authentication and cryptography-supported measurement functionalities are available. Interestingly, GEON SoC Secure Platform offers GEON Secure JTAG that is claimed to offer a complete debug analysis without compromising security [67]. As presented in Figure 6, the communication channel for the SoC Security Platform is realized by AMBA interface (AHB/APB), which may impact the final design of the IoT design.

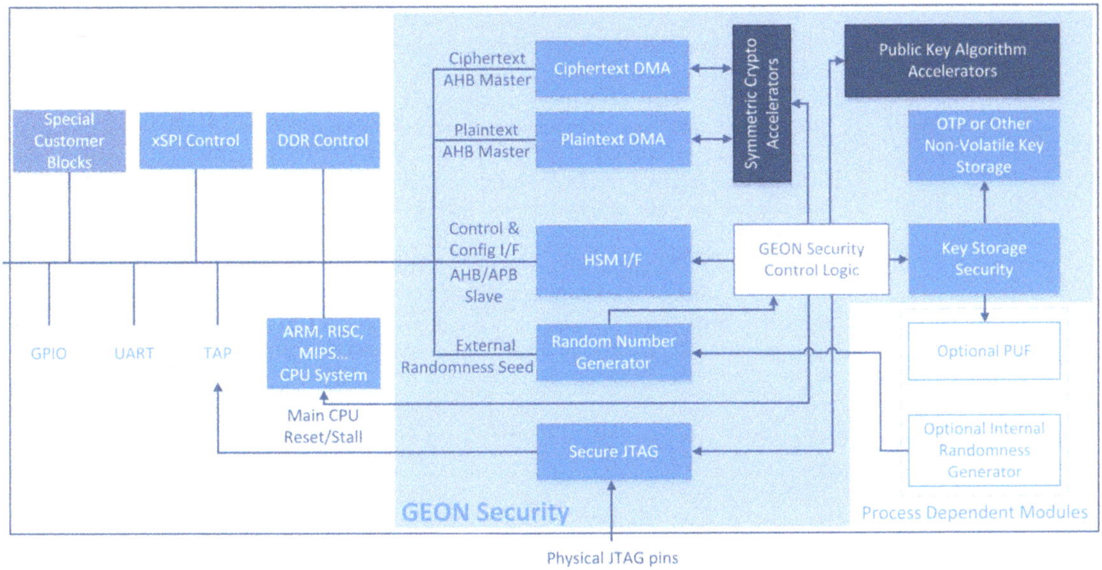

Figure 6. GEON SoC Security Platform hardware architecture, based on [65]. External interfaces and suggested connections to other components of System on Chip are presented.

Analysis for Synthesized Requirements for Secure IoT Devices

Multiple cryptographic ciphers and hash functions are available for use. The debug access is described in some documents mentioned in Section 3 as possibly unsafe and it is advised to disable it after deployment. It should be stressed, though, that these guidelines do not consider a highly secured debug interface and that is the case in the GEON SoC. Agreeably, this functionality can be critical in some applications, and the level of security claimed by Beyond Semiconductors is very promising. Parts of the GEON SoC are described as "flexible" and "disposable" depending on the application, which can obviously be beneficial for system designers, but the lack of awareness can lead to the creation of an unsecured IoT. Additionally, some key secure functionalities could be missing if one of the modules is left behind. No information is provided about cybersecurity event logging and monitoring. In terms of secure updating, firmware downgrade protection is in place; the confidentiality and integrity of software and firmware are also protected. GEON SoC Platform Security does not introduce any interference with other interfaces existing in the system in which it is integrated, mainly no disabling or securing capabilities are present. The security threat of any interface in the IoT device is unaccounted for. It can be considered as a plug in to the main bus of SoC. This solution certainly offers a great variety of hardware-based security features, but the lack of guidance on software and firmware development can be concerning.

4.7. Rambus RT Family

Rambus has a wide range of security IPs in its portfolio. We decided to focus on the Root of Trust IP cores, especially those dedicated for IoT—RT-100 [68], RT-130 [69], RT-140 [70] and RT-260 [71]. It seems that the difference between the aforementioned modules is the supported cryptography and protocols. The rule of operation is the same. These products are advertised as complete Hardware Root of Trust engines, offering cryptographic accelerators (both symmetric and asymmetric, in RT-140 also lightweight), secure boot support, secure asset storage, secure firmware upgrade, device authentication and identity protection, as well as secure debug. RT-140 supports TLS protocols. Remarkably, the Rambus solution is architecture agnostic. It can be integrated in an SoC and is claimed to be compatible with most popular system buses (e.g., AMBA). In order to cooperate with the device, the CPU requires a dedicated API implementation. Figure 7 illustrates the architecture and the location in the system of the Rambus component. Unfortunately, apart from the manufacturer's materials, there is no literature openly available regarding the RT family. Therefore, our research relies solely on product briefs published on the Rambus website. The analyzed documents indicate that Rambus offers the RoT Secure Boot Toolkit, which may be either a firmware or a software extension to communicate with the instantiated hardware module. It interfaces with protected software stored in non-volatile memory and includes a signing tool and boot library.

Analysis for Synthesized Requirements for Secure IoT Devices

Since this is purely a RoT solution, the development of secure software and firmware is left completely to the designer. One of the greatest advantages is the wide range of supported cryptography offered. However, there is very little information about securing memory and data transfers. In conclusion, the Rambus proposition is a foundation for building a secure IoT, not a complete solution.

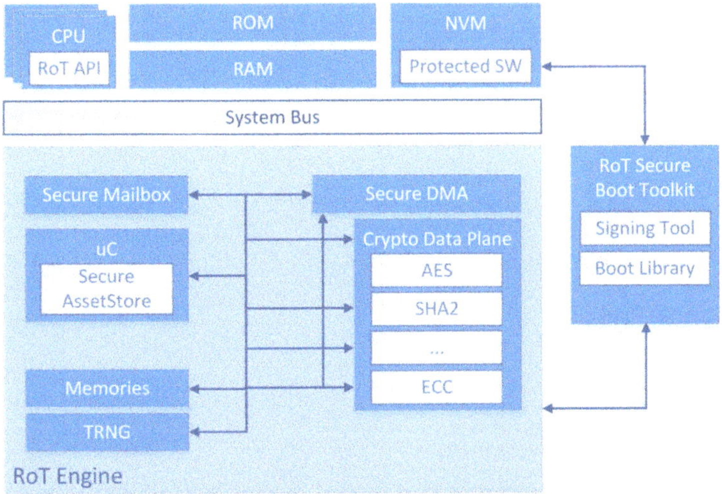

Figure 7. Rambus RT hardware architecture, on RT-100 example, based on [68]. Corresponding software elements (e.g., RoT Secure Boot Toolkit), and the location of this component in a system on chip are presented.

4.8. Summary

Table 2 repeats the requirements from Table 1 and compares the solutions presented in previous sections against them. For clarity, shortened versions of the requirements are presented below; the full description is available in Table 1.

Symbols used in Table 2 have the following meanings:

- ○—difficult to satisfy this secure IoT requirement using a particular solution;
- ◖—can be done, but no explicit recommendation nor solution for use in an IoT device is presented, leaving the design decisions open and potentially unsafe;
- ●—strong recommendation or solution which satisfies the requirement.

For the TrustZone, in some cases, ●* is used. This means that the requirement is fulfilled on the condition that TrustZone is combined with CryptoCell. For NXP the evaluation is based on the use case scenario from the data sheet [63] coined "Plug and Trust"—the integration with a host processing unit.

Table 2. Comparison of the state-of-the-art against the synthesized requirements for secure IoT devices.

No.	Requirement	Arm TrustZone	Intel SGX	Keystone	Open Titan	NXP SE050	GEON SoC	Rambus RT
Cat. 1	Hardware Security							
1.1	Immutable Hardware Root of Trust	●*	●	◖	●	●	●	●
1.2	HW provided security features	●	◖	◖	●	●	●	●
1.3	Tamper protection and detection	●*	●	◖	●	●	◖	◖
1.4	Random Number Generator (hardware-based if feasible)	●*	●	◖	●	◖	●	●
Cat. 2	Trust and integrity management							
2.1	Secure boot process based on HWRoT.	●	●	◖	◖ / ●	◖	●	●
2.2	Software (code, applications) signed cryptographically and verified upon installation or execution	●*	●	◖	◖	◖	●	◖
2.3	Restore last known secure state (e.g., firmware rollback, when code is verified as damaged or tampered)	●*	◖	◖	●	○	●	○
2.4	Software installation control in OS, to prevent unauthenticated software and files from being loaded onto it	●*	◖	◖	◖	○	●	◖
Cat. 3	Data protection and software design							
3.1	Encryption of data storage medium.	◖	●	◖	◖	◖	◖	◖
3.2	Input data validation	◖	●	◖	◖	◖	◖	○
3.3	Process privilege minimization	◖	◖	●	◖	○	○	○
3.4	Memory compartmentalization, process isolation	●	●	●	◖	○	○	○

Table 2. *Cont.*

No.	Requirement	Arm TrustZone	Intel SGX	Keystone	Open Titan	NXP SE050	GEON SoC	Rambus RT
Cat. 4	Device configuration and software update							
4.1	Changing the device's FW and SW configuration by authorized entities	●*	●	◖	●	◖	●	◖
4.2	Updating device's FW and SW through remote or local means by authorized entities	◖	●	◖	●	◖	◖	◖
4.3	Update file must be encrypted, signed and device verifies its integrity before application	●*	●	◖	●	◖	●	●
4.4	Automatic update capability	◖	◖	◖	◖	◖	◖	◖
4.5	Backward compatibility of updates	◖	◖	◖	◖	◖	◖	◖
Cat. 5	Secure interfaces and communication							
	Interface security							
5.1	Ability to disable or logically restrict access to any local and network interfaces	◖	◖	◖	◖	○	○	○
5.2	Deployment without debugging, diagnostic or testing interfaces that could be abused by adversary	◖	◖	◖	◖	◖	●	◖
	Protocol security							
5.3	Communication security using state-of-the-art, standardized security protocols (e.g., TLS for encryption)	●*	●	◖	◖	●	◖	●
5.4	Mutual authentication of the devices before trust can be established (prevent man in the middle attack)	◖	◖	◖	◖	◖	◖	◖
5.5	Prevent unauthorized connections to product or other devices it is connected to, at all levels of the protocols.	◖	◖	◖	◖	◖	◖	○
	Data security							
5.6	Confidentiality, integrity and authenticity of the transmitted data	●*	◖	◖	◖	◖	◖	◖

Table 2. *Cont.*

No.	Requirement	Arm TrustZone	Intel SGX	Keystone	Open Titan	NXP SE050	GEON SoC	Rambus RT
Cat. 6	Cybersecurity event monitoring and logging							
6.1	Logging cybersecurity events from device's HW, FW and SW	◖	◖	◖	○/◖	○	◖	◖
6.2	Recording sufficient details for each logged event	◖	◖	◖	○/◖	○	◖	◖
6.3	Restricting access to the logs to authorized entities only and prevent all entities from editing them	◖	◖	◖	○/◖	◖	◖	◖
6.4	Avoid security issues when designing error messages.	◖	◖	◖	◖	◖	◖	◖
6.5	Device's behavior should be monitored and compared with model to detect anomalies	◖	◖	◖	◖	◖	◖	◖
Cat. 7	Cryptography and key management							
7.1	Use of Standard Cryptographic Algorithms and Security Protocols	●*	◖	◖	●	●	●	●
7.2	Security protocols should support algorithm agility	●*	◖	◖	◖	●	●	●
7.3	Length of the key should provide strong security, make brute-force attack infeasible	◖	◖	◖	●	◖	◖	◖
7.4	Every device must be instantiated with unique private key(s)	●*	◖	◖	●	◖	◖	●
7.5	Secure and scalable management of cryptographic keys (generation, storage, distribution)	●*	◖	◖	●	◖	●	●
Cat. 8	Device identification, authentication, strong default security							
8.1	Unique physical identifier only authorized entities can access	●*	◖	●	●	◖	◖	◖
8.2	Device must be provisioned with unique logical identifier	◖	◖	◖	●	◖	◖	◖

Table 2. *Cont.*

No.	Requirement	Arm TrustZone	Intel SGX	Keystone	Open Titan	NXP SE050	GEON SoC	Rambus RT
8.3	Robust authentication and authorization schemes	◖	◖	◖	●	◖	◖	◖
8.4	Strong, device-individual default passwords that has to be changed by the user during initial setup	◖	◖	◖	◖	◖	◖	◖
8.5	Limiting number of invalid login attempts or introducing incrementing delays between them	◖	◖	◖	◖	◖	◖	◖
8.6	Availability of two-factor authentication	◖	◖	◖	●	◖	◖	◖
8.7	Product security shall be appropriately enabled by default.	◖	◖	◖	◖	◖	◖	◖

5. Discussion

In the current IoT security landscape, many institutions and entities are defining security requirements, but no industry-wide standard has been agreed upon. Device designers and vendors have their own proprietary solutions, which address some issues but miss the target in others. To the best of our knowledge, no solution addresses all the requirements presented in Table 1 out-of-the-box which generally leaves the solving of security problems mainly in the hands of the designing entity with the potential to create unsecured IoT devices.

Additionally, there is no industry wide standardization of the *level* of security needed, which would aid the designers of IoT devices. A different security level is required for traffic light control in a smart city than for an "intelligent" and connected toothbrush. This should also be part of the discussion and possibly even have a certification entity. Table 1 could be divided into different security levels and additionally be configurable within some of the requirements, that is, the key strength or chosen encryption algorithm should be implicitly standardized for different security needs.

Some solutions presented in Section 4 are, in theory, processor/architecture agnostic, but others like Arm TrustZone or Intel SGX are tied in with a specific architecture, which makes the security improvements dependent on the designing entity and the agility of their bug-fixing process. Additionally, the solutions in the current environment are generally divided into more hardware or software heavy, even though a combined solution of strong TEE and software process isolation with hardware cryptographic implementations and Hardware Root of Trust functionalities would seem to be the most secure. In our opinion, the discussed open-source solutions, Keystone and OpenTitan, could work together to create a highly secure prototype of a framework for IoT device protection, whilst being publicly attestable and with all the benefits of independence.

Many solutions in the present state-of-the-art fulfill a subset of secure IoT device requirements, but none adheres to all of them. No common approach is present for many application-level IoT requirements (Table 1), such as the return to a secure state, interface disabling in the SoC, automatic update by default and the logging of IoT device security events and others. This is an issue during the design of such devices and could have a real-world impact in the future. All current solutions focus on the wider trusted computing paradigm or on providing a hardware cryptographic acceleration, rather than being dedicated explicitly to IoT devices.

A comprehensive design framework is needed to assist in the design of secure IoT devices. Such a solution should take into consideration the most important security requirements shown in Section 3. We believe that it would, in a way, simplify the technology development and mitigate the problem of time-to-market/functionality trade-off in opposition to security. It should be possible to add, remove or configure IoT security on a block system level in accordance with the target application of the device.

6. Conclusions

The IoT ecosystem poses new security challenges that extend beyond traditional data security. There is a need for IoT security guidelines, and numerous institutions across the globe have proposed their recommendations, aiming to help ensure a secure IoT infrastructure. Device designers and vendors have their own proprietary solutions, which address some issues, but miss the target in others. In this paper, selected recommendations have been analyzed and compiled into a set of the most common and important considerations, divided into eight categories. The evaluation of representative examples from IoT security technologies against these criteria shows that there are solutions with the potential to meet all these recommendations, but at the moment no solution addresses all requirements in an out-of-the-box capacity, which allows for further research in this field.

Author Contributions: Conceptualization, M.C., M.K., F.K. and M.R.; methodology, M.C., M.K. and F.K.; investigation, M.C., M.K. and F.K.; writing—original draft preparation, M.C., M.K. and

F.K.; writing—review and editing, K.S. and M.R.; visualization, M.K.; supervision, M.R.; project administration, M.R.; funding acquisition, M.R. All authors have read and agreed to the published version of the manuscript.

Funding: This research was funded by The Polish National Centre for Research and Development under project No. CYBERSECIDENT/456446/III/NCBR/2020.

Conflicts of Interest: The authors declare no conflict of interest.The funders had no role in the design of the study; in the collection, analyses, or interpretation of data; in the writing of the manuscript, or in the decision to publish the results.

Abbreviations

Abbreviations

The following abbreviations are used in this manuscript:

AC	Air Conditioning
AES	Advanced Encryption Standard
AHB	AMBA High-performance Bus
AMBA	ARM Advanced Microcontroller Bus Architecture
APB	AMBA Advanced Peripheral Bus
API	Application Programming Interface
AXI	AMBA Advanced eXtensible Interface
CCTV	Closed-circuit Television
CLI	Command Line Interface
CMAC	Cipher-based Message Authentication Code
CPU	Central Processing Unit
DDoS	Distributed Denial of Service
DMA	Direct Memory Access
ECC	Elliptic-curve cryptography
FW	Firmware
GSM	Global System for Mobile Communications
HMAC	keyed-Hash Message Authentication Code
HW	Hardware
HWRoT, RoT	(Hardware) Root of Trust
I2C	Inter-Integrated Circuit communication bus
IC	Integrated Circuit
IoT	Internet of Things
IP	Intellectual Property
JCOPOS	Java Card OpenPlatform Operating System
JTAG	Joint Test Action Group
M-mode	RISC-V Machine Mode
NIST	National Institute of Standards and Technology
OS	Operating System
REE	Rich Execution Environment
RFID	Radio-frequency Identification
ROM	Read-only Memory
RSA	Rivest–Shamir–Adleman (public key cryptosystem)
S-mode	RISC-V Supervisor Mode
SGX	(Intel) Software Guard Extensions
SHA-1	Secure Hash Algorithm 1
SoC	System-on-Chip
SW	Software
TCB	Trusted Computing Base
TEE	Trusted Execution Environment
TLS	Transport Layer Security
TPM	Trusted Platform Module
U-mode	RISC-V User Mode

References

1. Ericsson. IoT Connections Outlook. Available online: https://www.ericsson.com/en/mobility-report/dataforecasts/iot-connections-outlook (accessed on 24 June 2021).
2. Lueth, K.L. State of the IoT 2020: 12 billion IoT Connections, Surpassing Non-IoT for the First Time. Available online: https://iot-analytics.com/state-of-the-iot-2020-12-billion-iot-connections-surpassing-non-iot-for-the-first-time/ (accessed on 24 June 2021).
3. Neshenko, N.; Bou-Harb, E.; Crichigno, J.; Kaddoum, G.; Ghani, N. Demystifying IoT Security: An Exhaustive Survey on IoT Vulnerabilities and a First Empirical Look on Internet-Scale IoT Exploitations. *IEEE Commun. Surv. Tutor.* **2019**, *21*. [CrossRef]
4. Greenberg, A. Hackers Remotely Kill a Jeep on the Highway—With Me in It. Available online: https://www.wired.com/2015/07/hackers-remotely-kill-jeep-highway/ (accessed on 24 June 2021).
5. Graff, G.M. How a Dorm Room Minecraft Scam Brought down the Internet. Available online: https://www.wired.com/story/mirai-botnet-minecraft-scam-brought-down-the-internet/ (accessed on 24 June 2021).
6. Ben Herzberg, I.Z.; Bekerman, D. Breaking Down Mirai: An IoT DDoS Botnet Analysis. Available online: https://www.imperva.com/blog/malware-analysis-mirai-ddos-botnet/ (accessed on 24 June 2021).
7. Green, A. The Mirai Botnet Attack and Revenge of the Internet of Things. Available online: https://www.varonis.com/blog/the-mirai-botnet-attack-and-revenge-of-the-internet-of-things/ (accessed on 24 June 2021).
8. Bonomi, F.; Milito, R. Fog Computing and its Role in the Internet of Things. *Proc. MCC Workshop Mob. Cloud Comput.* **2012**. [CrossRef]
9. Gupta, N.; Rodrigues, J.J.P.C.; Dauwels, J. (Eds.) *Augmented Intelligence toward Smart Vehicular Application*; CRC Press: Boca Raton, FL, USA, 2020; pp. 7–13. [CrossRef]
10. Park, H.; Zhai, S.; Lu, L.; Lin, F.X. StreamBox-TZ: Secure Stream Analytics at the Edge with TrustZone. In Proceedings of the 2019 USENIX Annual Technical Conference (USENIX ATC 19), Renton, WA, USA, 10–12 July 2019; pp. 537–554.
11. Fournaris, A.P.; Alexakos, C.; Anagnostopoulos, C.; Koulamas, C.; Kalogeras, A. Introducing Hardware-Based Intelligence and Reconfigurability on Industrial IoT Edge Nodes. *IEEE Des. Test* **2019**, *36*, 15–23. [CrossRef]
12. Mahmoud, R.; Yousuf, T.; Aloul, F.; Zualkernan, I. Internet of things (IoT) security: Current status, challenges and prospective measures. In Proceedings of the 2015 10th International Conference for Internet Technology and Secured Transactions, London, UK, 8–10 December 2016; pp. 336–341. [CrossRef]
13. binti Mohamad Noor, M.; Hassan, W.H. Current research on Internet of Things (IoT) security: A survey. *Comput. Netw.* **2019**, *148*, 283–294. [CrossRef]
14. Alladi, T.; Chamola, V.; Sikdar, B.; Choo, K.K.R. Consumer IoT: Security Vulnerability Case Studies and Solutions. *IEEE Consum. Electron. Mag.* **2020**, *9*, 17–25. [CrossRef]
15. Tsiropoulou, E.E.; Baras, J.S.; Papavassiliou, S.; Qu, G. On the Mitigation of Interference Imposed by Intruders in Passive RFID Networks. In *Decision and Game Theory for Security*; Zhu, Q., Alpcan, T., Panaousis, E., Tambe, M., Casey, W., Eds.; Springer International Publishing: Cham, Switzerland, 2016; pp. 62–80.
16. Mohanta, B.K.; Jena, D.; Satapathy, U.; Patnaik, S. Survey on IoT security: Challenges and solution using machine learning, artificial intelligence and blockchain technology. *Internet Things* **2020**, *11*, 100227. [CrossRef]
17. Zhu, Q.; Başar, T. Game-Theoretic Approach to Feedback-Driven Multi-stage Moving Target Defense. In *Decision and Game Theory for Security*; Das, S.K., Nita-Rotaru, C., Kantarcioglu, M., Eds.; Springer International Publishing: Cham, Switzerland, 2013; pp. 246–263.
18. Fagan, M.; Megas, K.N.; Scarfone, K.; Smith, M. IoT device cybersecurity capability core baseline. In *Technical Report*; National Institute of Standards and Technology: Gaithersburg, MD, USA, 2020. [CrossRef]
19. Fortino, G.; Savaglio, C.; Spezzano, G.; Zhou, M. Internet of Things as System of Systems: A Review of Methodologies, Frameworks, Platforms, and Tools. *IEEE Trans. Syst. Man Cybern. Syst.* **2021**, *51*, 223–236. [CrossRef]
20. Tournier, J.; Lesueur, F.; Mouël, F.L.; Guyon, L.; Ben-Hassine, H. A survey of IoT protocols and their security issues through the lens of a generic IoT stack. *Internet Things* **2020**, 100264. [CrossRef]
21. Sinche, S.; Raposo, D.; Armando, N.; Rodrigues, A.; Boavida, F.; Pereira, V.; Silva, J.S. A Survey of IoT Management Protocols and Frameworks. *IEEE Commun. Surv. Tutor.* **2020**, *22*, 1168–1190. [CrossRef]
22. Babun, L.; Denney, K.; Celik, Z.B.; McDaniel, P.; Uluagac, A.S. A survey on IoT platforms: Communication, security, and privacy perspectives. *Comput. Netw.* **2021**, *192*, 108040. [CrossRef]
23. Ammar, M.; Russello, G.; Crispo, B. Internet of Things: A survey on the security of IoT frameworks. *J. Inf. Secur. Appl.* **2018**, *38*, 8–27. [CrossRef]
24. Macedo, E.L.; De Oliveira, E.A.; Silva, F.H.; Mello, R.R.; Franca, F.M.; Delicato, F.C.; De Rezende, J.F.; De Moraes, L.F. On the security aspects of Internet of Things: A systematic literature review. *J. Commun. Netw.* **2019**, *21*, 444–457. [CrossRef]
25. Abdul-Ghani, H.A.; Konstantas, D. A comprehensive study of security and privacy guidelines, threats, and countermeasures: An IoT perspective. *J. Sens. Actuator Netw.* **2019**, *8*, 22. [CrossRef]
26. Maene, P.; Götzfried, J.; De Clercq, R.; Müller, T.; Freiling, F.; Verbauwhede, I. Hardware-Based Trusted Computing Architectures for Isolation and Attestation. *IEEE Trans. Comput.* **2018**, *67*, 361–374. [CrossRef]

27. European Union Agency for Network and Information Security. Baseline Security Recommendations for IoT in the Context of Critical Information Infrastructures. Available online: https://op.europa.eu/en/publication-detail/-/publication/c37f8196-d96f-11e7-a506-01aa75ed71a1/language-en (accessed on 30 June 2021).
28. GSM Association. IoT Security Guidelines Overview Document—Version 2.2. Available online: https://www.gsma.com/iot/wp-content/uploads/2020/05/CLP.11-v2.2-GSMA-IoT-Security-Guidelines-Overview-Document.pdf (accessed on 30 June 2021).
29. GSM Association. IoT Security Guidelines for IoT Service Ecosystem—Version 2.2. Available online: https://www.gsma.com/iot/wp-content/uploads/2020/05/CLP.12-v2.2-GSMA-IoT-Security-Guidelines-for-Service-Ecosystems.pdf (accessed on 30 June 2021).
30. GSM Association. IoT Security Guidelines for Endpoint Ecosystem—Version 2.2. Available online: https://www.gsma.com/iot/wp-content/uploads/2020/05/CLP.13-v2.2-GSMA-IoT-Security-Guidelines-for-Endpoint-Ecosystems.pdf (accessed on 30 June 2021).
31. Moore, K.; Barnes, R.; Tschofenig, H. Best Current Practices for Securing Internet of Things (IoT) Devices. Internet-Draft draft-moore-iot-security-bcp-01. *Internet Eng. Task Force* **2017**, Work in Progress.
32. Wang, B.; Liu, S.; Wan, L.; Li, J.; Wang, X.T. Technical Requirements for Secure Access and Management of IoT Smart Terminals. Internet-Draft draft-wang-secure-access-of-iot-terminals-01. *Internet Eng. Task Force* **2021**, Work in Progress.
33. Garcia-Morchon, O.; Kumar, S.; Sethi, M. Internet of Things (IoT) Security: State of the Art and Challenges. Available online: https://www.rfc-editor.org/info/rfc8576 (accessed on 30 June 2021).
34. Foundation, I.S. Secure Design—Best Practice Guides—Release 2. Internet of Things (IoT) Security: State of the Art and Challenges. Available online: https://www.iotsecurityfoundation.org/wp-content/uploads/2019/12/Best-Practice-Guides-Release-2_Digitalv3.pdf (accessed on 30 June 2021).
35. ioXt Alliance. ioXt Pledge—The Global Standard for IoT Security. Available online: https://static1.squarespace.com/static/5c6dbac1f8135a29c7fbb621/t/5fb43a05bda7c3689ea5bd32/1605646853608/ioXt+Pledge+Book_Secure.pdf (accessed on 30 June 2021).
36. International Organization for Standardization. Cybersecurity—IoT Security and Privacy—Guidelines. Available online: https://www.iso.org/standard/44373.html (accessed on 30 June 2021).
37. Trusted Computing Group. Trusted Computing Group Glossary. Version 1.1. Available online: https://trustedcomputinggroup.org/wp-content/uploads/TCG-Glossary-V1.1-Rev-1.0.pdf (accessed on 30 June 2021).
38. GlobalPlatform, Inc. GlobalPlatform Security Task Force Root of Trust Definitions and Requirements. Available online: https://globalplatform.org/wp-content/uploads/2018/06/GP_RoT_Definitions_and_Requirements_v1.0.1_PublicRelease_CC.pdf (accessed on 30 June 2021).
39. Arm. Arm® Platform Security Requirements 1.0. Available online: https://developer.arm.com/documentation/den0106/latest/ (accessed on 30 June 2021).
40. Pinto, S.; Santos, N. Demystifying Arm TrustZone: A comprehensive survey. *ACM Comput. Surv.* **2019**, *51*, 36. [CrossRef]
41. Arm. Learn the Architecture: TrustZone for AArch64. Available online: https://developer.arm.com/documentation/102418/0100 (accessed on 24 June 2021).
42. Zhao, S.; Zhang, Q.; Hu, G.; Qin, Y.; Feng, D. Providing Root of Trust for ARM TrustZone using On-Chip SRAM. In Proceedings of the 4th International Workshop on Trustworthy Embedded Devices—TrustED '14, Scottsdale, AZ, USA, 3 November 2014; pp. 25–36.
43. Arm. CryptoCell-300 Family. Available online: https://developer.arm.com/ip-products/security-ip/cryptocell-300-family (accessed on 24 June 2021).
44. Arm. CryptoCell-700 Family. Available online: https://developer.arm.com/ip-products/security-ip/cryptocell-700-family. (accessed on 24 June 2021).
45. Wallace, J. Arm CryptoCell-312: Simplifying the Design of Secure IoT Systems. 2016. Available online: https://community.arm.com/developer/ip-products/system/b/embedded-blog/posts/arm-trustzone-cryptocell-312-simplifying-the-design-of-secure-iot-systems (accessed on 24 June 2021).
46. Nordic Semiconductor. CRYPTOCELL—ARM® TrustZone® CryptoCell 310. 2021. Available online: https://infocenter.nordicsemi.com/index.jsp?topic=%2Fps_nrf52840%2Fcryptocell.html (accessed on 24 June 2021).
47. Anati, I.; Gueron, S.; Johnson, S.; Scarlata, V. Innovative technology for CPU based attestation and sealing. In *Proceedings of the 2nd International Workshop on Hardware and Architectural Support for Security and Privacy*; ACM: New York, NY, USA, 2013; Volume 13, p. 7.
48. Intel SGX-Remote Attestation. Available online: https://software.intel.com/content/www/us/en/develop/topics/software-guard-extensions/attestation-services.html (accessed on 30 June 2021).
49. Costan, V.; Devadas, S. Intel SGX Explained. Available online: https://eprint.iacr.org/2016/086.pdf (accessed on 30 June 2021).
50. Intel SGX—Security Essentials Solution Brief. Available online: https://www.intel.com/content/dam/www/public/us/en/documents/solution-briefs/intel-security-essentials-solution-brief.pdf (accessed on 30 June 2021).
51. Intel SGX—IoT Edge Devices. Available online: https://software.intel.com/content/www/us/en/develop/topics/software-guard-extensions.html (accessed on 30 June 2021).

52. Brasser, F.; Müller, U.; Dmitrienko, A.; Kostiainen, K.; Capkun, S.; Sadeghi, A.R. Software Grand Exposure:{SGX} Cache Attacks are Practical. Available online: https://www.usenix.org/conference/woot17/workshop-program/presentation/brasser (accessed on 30 June 2021).
53. Van Bulck, J.; Minkin, M.; Weisse, O.; Genkin, D.; Kasikci, B.; Piessens, F.; Silberstein, M.; Wenisch, T.F.; Yarom, Y.; Strackx, R. Foreshadow: Extracting the keys to the intel {SGX} kingdom with transient out-of-order execution. In Proceedings of the 27th {USENIX} Security Symposium ({USENIX} Security 18), Baltimore, MD, USA, 15–17 Auguest 2018, pp. 991–1008.
54. Chen, G.; Chen, S.; Xiao, Y.; Zhang, Y.; Lin, Z.; Lai, T.H. Sgxpectre: Stealing intel secrets from sgx enclaves via speculative execution. In Proceedings of the 2019 IEEE European Symposium on Security and Privacy (EuroS&P), Stockholm, Sweden, 17–19 June 2019; pp. 142–157.
55. Murdock, K.; Oswald, D.; Garcia, F.D.; Van Bulck, J.; Gruss, D.; Piessens, F. Plundervolt: Software-based fault injection attacks against Intel SGX. In Proceedings of the 2020 IEEE Symposium on Security and Privacy (SP), San Francisco, CA, USA, 17–21 May 2020; pp. 1466–1482.
56. Lee, D.; Kohlbrenner, D.; Shinde, S.; Asanović, K.; Song, D. Keystone: An open framework for architecting trusted execution environments. In Proceedings of the Fifteenth European Conference on Computer Systems, Heraklion, Greece, 27–30 April 2020; pp. 1–16.
57. OpenTitan—Open Source Silicon Root of Trust. Available online: https://opentitan.org/ (accessed on 30 June 2021).
58. OpenTitan—Hardware Dashboard. Available online: https://docs.opentitan.org/hw/ (accessed on 30 June 2021).
59. OpenTitan—Earl Gray Top Level Specification. Available online: https://docs.opentitan.org/hw/top_earlgrey/doc/ (accessed on 30 June 2021).
60. Barker, E.; Kelsey, J.; National Institute of Standards and Technology. NIST Special Publication 800-90A, Revision 1: Recommendation for Random Number Generation Using Deterministic Random Bit Generators. Available online: https://csrc.nist.gov/publications/detail/sp/800-90a/rev-1/final (accessed on 30 June 2021).
61. Barker, E.; Kelsey, J.; National Institute of Standards and Technology. NIST Special Publication 800-90C (Second Draft): Recommendation for Random Number Generator (RBG) Constructions. Available online: https://csrc.nist.gov/publications/detail/sp/800-90c/draft (accessed on 30 June 2021).
62. OpenTitan—Logical Security Model. Available online: https://docs.opentitan.org/doc/security/logical_security_model/ (accessed on 30 June 2021).
63. NXP. SE050 Plug and Trust Secure Element. Available online: https://www.nxp.com/docs/en/data-sheet/SE050-DATASHEET.pdf (accessed on 30 June 2021).
64. NXP. AN12413-SE050 APDU Specification—Application Note—Rev. 2.12. Available online: https://www.nxp.com.cn/docs/en/application-note/AN12413.pdf (accessed on 30 June 2021).
65. Beyond Semiconductor. GEON SoC Security Platform. 2021. Available online: https://www.beyondsemi.com/116/geon-security-platform/ (accessed on 30 June 2021).
66. Beyond Semiconductor. GEON Secure Boot. 2021. Available online: https://www.beyondsemi.com/117/geon-secure-boot/ (accessed on 30 June 2021).
67. Beyond Semiconductor. GEON Secure JTAG. 2021. Available online: https://www.beyondsemi.com/119/geon-secure-jtag/ (accessed on 30 June 2021).
68. Root of Trust RT-100. Foundational Security for SoCs and FPGAs. Available online: https://go.rambus.com/root-of-trust-rt-100-product-brief (accessed on 30 June 2021).
69. Root of Trust RT-130. Foundational Security for SoCs and FPGAs for IoT Servers, Gateways and Edge Devices. Available online: https://go.rambus.com/root-of-trust-rt-130-product-brief (accessed on 30 June 2021).
70. Root of Trust RT-140. Foundational Security for SoCs and FPGAs for Cloud-Connected Devices. Available online: https://go.rambus.com/root-of-trust-rt-140-product-brief (accessed on 30 June 2021).
71. Root of Trust RT-260. Foundational Security for SoCs and FPGAs for Secure MCU Devices. Available online: https://go.rambus.com/root-of-trust-rt-260-product-brief (accessed on 30 June 2021).

electronics

MDPI

Article

Analysis and Implementation of Threat Agents Profiles in Semi-Automated Manner for a Network Traffic in Real-Time Information Environment

Gaurav Sharma [1,*], Stilianos Vidalis [1], Catherine Menon [1], Niharika Anand [2] and Somesh Kumar [3]

1 School of Computer Science & Engineering, University of Hertfordshire, Hatfield Al10 9AB, UK; s.vidalis@herts.ac.uk (S.V.); c.menon@herts.ac.uk (C.M.)
2 Indian Institute of Information Technology Lucknow (IIITL), Lucknow 226002, India; niharika@iiitl.ac.in
3 ABV-Indian Institute of Information Technology & Management, Gwalior 474015, India; somesh@iiitm.ac.in
* Correspondence: g.gaurav@herts.ac.uk

Citation: Sharma, G.; Vidalis, S.; Menon, C.; Anand, N.; Kumar, S. Analysis and Implementation of Threat Agents Profiles in Semi-Automated Manner for a Network Traffic in Real-Time Information Environment. *Electronics* **2021**, *10*, 1849. https://doi.org/10.3390/electronics10151849

Academic Editor: Krzysztof Szczypiorski

Received: 2 July 2021
Accepted: 28 July 2021
Published: 31 July 2021

Abstract: Threat assessment is the continuous process of monitoring the threats identified in the network of the real-time informational environment of an organisation and the business of the companies. The sagacity and security assurance for the system of an organisation and company's business seem to need that information security exercise to unambiguously and effectively handle the threat agent's attacks. How is this unambiguous and effective way in the present-day state of information security practice working? Given the prevalence of threats in the modern information environment, it is essential to guarantee the security of national information infrastructure. However, the existing models and methodology are not addressing the attributes of threats like motivation, opportunity, and capability (C, M, O), and the critical threat intelligence (CTI) feed to the threat agents during the penetration process is ineffective, due to which security assurance arises for an organisation and the business of companies. This paper proposes a semi-automatic information security model, which can deal with situational awareness data, strategies prevailing information security activities, and protocols monitoring specific types of the network next to the real-time information environment. This paper looks over analyses and implements the threat assessment of network traffic in one particular real-time informational environment. To achieve this, we determined various unique attributes of threat agents from the Packet Capture Application Programming Interface (PCAP files/DataStream) collected from the network between the years 2012 and 2019. We used hypothetical and real-world examples of a threat agent to evaluate the three different factors of threat agents, i.e., Motivation, Opportunity, and Capability (M, O, C). Based on this, we also designed and determined the threat profiles, critical threat intelligence (CTI), and complexity of threat agents that are not addressed or covered in the existing threat agent taxonomies models and methodologies.

Keywords: threat agents; motivation; opportunity; capability; user profiling; implicit; modeling; real-time user monitoring; complexity threat agent; threat assessment

1. Introduction

Identifying the potential cybersecurity threat capability in real-time is a crucial activity. It helps provide practical information about the threat in a network that allows cybersecurity practitioners to take suitable action to mitigate the risk in a network [1]. Elaborating all the information about the potential cybersecurity threats of an organisation is typically achieved manually by the existing models and methodology. Threat assessment is implemented in an automated manner with the help of machine learning techniques and various real-time models [2]. The behaviours of threat agents are erratic, and the goals of threat agents change with time. Threat agent groups change their behaviour to penetrate a network based on motivation, opportunity, and capability [3,4]. The motivation of the threat agent constantly changes with time depends on the financial gain, revenge from an organisation, etc., and the type of environment targeted. Profiling is a process that generates a profile for the threat agents based on the historical information extracted from the Packet Capture Application Programming Interface (PCAP) files captured in a network with the help of penetration testing phases. The profile can be populated by having suitable, ample, and precise information about the threat agent like behaviour, source I.P. address, destination I.P. address, number of open ports, number of packets generated, location of the threat agent, and time spent on the network with minimal user intervention [5]. The user has minimal intervention because of the footprints captured by the capturing data tool like LibPcap, WinPcap, PCAPng, NPcap, etc., during threat assessment in the form of PCAP files that cannot be altered by the potential threat agent while traversing an organisation's network. The threat agent cannot alter because once they generate the packets in the network, they cannot erase the footprint of generating the packets because of the accessing property of the network. This research attempts to recognise the aspects of profiling and deliver solutions by implementing the profiling of threat agents. Threat profiling is an essential aspect of performing threat assessment for an organisation. Suppose we have the threat profile for the historically identified threat agents from the network of an organisation. In that case, we can use these profiles as references while executing the threat assessment for the situational awareness data captured from the network. The model can address the recent threat agent effectively identified from a network with optimised complexity.

It has been accepted that continuous threat assessments practice mitigate the risks for any organisation and business [6]. However, in the modern, socially driven, virtual computing era, threat assessments are hindered by a lack of resources, complexity, and data size [7]. Information Environments are large heterogeneous infrastructures, hosting a large amount of data collected from different types of sensors and platforms [8]. To cope with a large amount of data, decision aid tools should understand the situational awareness property of data and threat assessments required for an organisation. University computer emergency response team (CMU-CERT) groups determined three critical groups of threat agents, i.e., the technology of organisation sabotage, compromising with intellectual property, and data stream fraud [9]. The number of growing cases highlighted by internet media in recent years revealed that both business organisations and government organisations suffered a similar experience. In contrast, the priority information has been filtrated by the organisation's internal users and shared with the threat agents [10]. The threat agents require serious attention from both users and organisations.

Referencing to the COVID-19 pandemic nowadays, organisations and businesses share their file and documents frequently with the help of the internet to run their business. It is now standard practice for users of the organisation to have admittance to large repository documents which are electronically warehoused on distributed file servers. Many organisations offer company laptops and desktops to the users for work while using e-mail to organise and schedule/rescheduling meetings. Amenities such as video conferencing are repeatedly used for holding meetings throughout the world, and users of an organisation are continuously connected to the internet. The electronic nature of the files and records of an organisation on the internet makes it easier for the threat agents to attack the organisation. On the advantageous side of continuous threat assessment, an organisation can easily capture the activity logs of the internal threat agent while analysing their captured packets [11]. However, practically analysing such activity logs is infeasible due to the high volume of activities performed by the user every day.

In this work, we present an efficient model for threat detection and analysis based on the conception of anomaly detection. The proposed model implements the threat agent profiles from the PCAP files and determines the cyber threat intelligence based on evaluating motivation, opportunity, and capability of threats. With the help of these profiles, comparisons can be populated that show the current observations fluctuate from the previous observations. To assess the performance of the tactic, we extracted the valuable information from the PCAP files in a semi-automated manner, and output has been generated in the form of an Excel sheet which consists of various attributes of threat agents identified in the next to the real-world information environment. The system executed expressively soundly for detecting the attacks, and the visualisation of reports enabled us to remember which attributes help determine M, O, C factors for the threat agents. This paper illustrates all the threats identified in a network captured during the penetration testing against the ESXi server of the University of Hertfordshire, UK.

The rest of this paper is as follows. Section 2 discusses the related work. Section 3 labels the necessities of analysis, the experimental set-up of the proposed system, and describes how to evaluate motivation, capability, and opportunity of threat agents. Section 4 presents the actual results from practical experimentation of the system, and Section 5 concludes this paper.

2. Related Work

The field of threat agents profiling and analysis of cyber threat intelligence has recently received ample attention. Researchers have proposed an assortment of different models and methodologies designed to detect or prevent attacks [12,13]. Likewise, Vidalis et al. [8] briefly addresses the TAME (Threat Assessments Model For EPS) methodology for threat assessments in real-time informational environments and provides a high-level overview of its phases and process while performing threat assessments. They compare the TAME (Threat Assessments Model For EPS) methodology with other existing methods based on the number of parameters as sting, effectiveness, and understanding of information security from the threat. TAME is the upgraded version of METEORE 2000 for the micropayment system (MPS). In the initial phases, the authors analyse the number of methodologies like Alberts 1999, 2001, Baker 1998, Bayne 2002, Blyth 2003, Dimitrakos 2001, Forte 2000, Hancock 1998, Jones 2002, Nichols 2001, etc., and they found that all are working on the waterfall model principle, but such approach is not suitable for the Micro Payment System (MPS). So, they developed a new methodology i.e., TAME (Threat Assessments Model For EPS) which has ability to resolve the issues related to Micro Payment System (MPS). TAME (Threat Assessments Model For EPS) is working simultaneously in four phases named as:

(a) Scope of Assessments.
(b) Threat Agent and Vulnerability Analysis.
(c) Scenario Construction and System Modelling.
(d) Evaluation.

According to these phases, TAME determined how much security is required for a particular organisation and business of the system. All four stages are working simultaneously, and one input from a phase becomes the output of another degree. Similarly, the vice-versa of inputs and outputs are generated from the TAME, and it depends on the requirements of threat assessments. The authors conclude the TAME by using the assessor as an asset for better understanding and analysing an organisation's systems.

Morakis et al. [14] measure vulnerabilities and their exploitation cycle by various tools such as COPS, NESSUS, SYSTEM SCANNER, RETINA, NET RECON, WHISKER, and CYBER COB. In this work, the authors address a problem faced by a large amount of data in the informative environment is cyber-attacks. The authors propose a vulnerability tree analysis to address such issues faced by several organisations for a long time. They believe in constructing knowledge information concerning a specific domain in an object-oriented hierarchy tree and building a formal model to analyse them concerning possible scenarios of attacks faced by the computer systems. The primary purpose of this is to provide a depth classification of vulnerabilities, find why such attacks happened on a particular data/asset, and analyse footprints and scenarios of threat agents to exploit vulnerabilities. The main aim of the vulnerability tree analysis is to identify the attacks in the early stages and address them before severe damage to real-world informational systems. Here, the authors illustrate the various tools capable of analysing the vulnerability of complex organisational environments; such tools are COPS, NESSUS, SYSTEM SCANNER, RETINA, NET RECON, WHISKER, and CYBER COB, etc. However, these are not adequate in today's modern electronic era of cyber-crime because they cannot address hazards like fault-tree analysis, checklists, event-tree analysis, cause-consequences analysis, etc. To cope with such hazards, the authors combine these tools of vulnerabilities tree analysis with object-oriented trees (O.O.) and adequately address such hazards concerning Boolean Mathematics.

Gerald L. et al. [15] briefly explain about threat agents regarding how they can have unauthorised access to the computer systems of real-world informational environments and from where they got the motivation, capability, and opportunity to perform such damage in the networks systems. Here, they also illustrate the threat agents and their attributes, function, and impact on a network of informational systems. The authors also analyse the digital attacks that occured in 2002 in several countries. They identify that the threat agents of real-world informational environments consist of:

(a) Threat agent catalogue.
(b) Historical data.
(c) Technical report enterprises.
(d) Reports of business environments.
(e) Reports of physical environments.
(f) Recent knowledge/information.
(g) Current knowledge of stakeholders.
(h) Current knowledge of the staff.
(i) List of stakeholders.

The authors evaluate the capabilities, motivation, opportunities, and impact with the help of 3-dimension matrix mathematics. They assess each factor with the help of metrics and ESA (Empowered Small Agents) threat agents. They identify that because of threat agents in 2002, the European union's worldwide economic damage is USD 35 million. So, as the damage cost is relatively more, the system security officer needs to require all knowledge and information about the threat agents or risk management to secure the system from damage done by cyber-attacks in informational environments.

Adetorera Sogbesan et al. [16] developed a model to identify the MERIT (Management & Education of Risk of Insider Threat) based on the study of insider threat concerning the institute of CERT/USSS. This MERIT provides the facility to mitigate the insider threat of an organisation, and the key finding is to make the case study of individual threat agents, i.e., collision threat. MERIT models the case studies on the insider threat for an organisation, and based on that, threat assessments have been conducted to determine the impact of danger on the business. They also show some figures for losses based on studies done by USSS/CERT. They categorise the insider attack based on the ex-employee, or the financial gain of any vital position held by an employee in an organisation. Based on the number of organisations, 69% of companies measured stated data theft events (not external attacks). These threats were originated from inside the organisation. At the same time, a massive 91% of companies testified not having operative detection systems for recognising an insider threat. The MERIT model has a limitation/shortcoming in analysing compressive pattern analysis based on motivation and behavioural characteristics. The motivation factor of collusion attack is not able to be addressed by the MERIT model. This model is not able to explain the capability of an insider threat.

Casillo, M. et al.'s study [17] "Embedded Intrusion Detection System for Detecting Attacks over CAN-BUS" designs a model based on AIC (availability, integrity, and confidentiality). The authors address the issues related to cyber-attacks on the automotive vehicle system. They introduce the automotive IDS embedded method for the CAN (controlled area network) BUS. Referencing the Bayesian network approaches, identifying malicious messages to the connected devices to the vehicles is accomplished. In this paper, the authors identify the snag for the IoT devices connected to automotive vehicles and their attacks while using automation. They suggest machine learning approaches, particularly the Bayesian network approach to cope with the cyber-attacks on the CAB bus. The authors used the CARLA simulator to provides the solution. The PYTHON library and several APIs were cast off for clustering the data and FPGA techniques for developing the model's architecture.

Lombardi, M. et al.'s research [18] "EIDS: Embedded intrusion detection system using machine learning to detect attack over the can-bus" introduced an IDS approach to identify the threats in the automated vehicles, particularly CAN (controlled area network) bus. The authors cast off the development of an IDS approach with the help of machine learning techniques through the Bayesian network approach to detect possible attacks on the CAN bus. The main benefit of developing an IDS approach was using the embedded framework for designing and determining the non-linear messages flow. The castigate faced by the connected IoT devices and the intelligent device for self-driving vehicles was identified with the help of an introduce IDS approach in the research.

These related works draw an intense observation that access to a real-world data stream is enormously challenging. Thus, researchers synthesise data into several groups based on the threat agents identified in a network. The existing model and methodology did threat assessment manually, due to which their complexity is exorbitant. This research predominantly wants to epitomise the volume and variety of data analysed in a modern real-world information environment and display how this could be pooled to form an overall threat assessment for each PCAP file. We also want to exhibit a wide range of threat scenarios as epitomised by our data collected from a real-world in a specific environment and show how our profiling and CTI system of threat agents would detect the different attacks based on the patterns identified.

3. Experiment Set-Up and Evaluation of MCO Attributes

The work described in this research has been carried out as part of a more comprehensive interdisciplinary project that includes computer security researchers and cyberpsychology experts. CTI data–driven threat agent profiling can be used for determining the motivation, opportunity, and capabilities attributes of threat agents under the context of a continuous threat assessment [19]. The threat remains of budding apprehension to governments and businesses organisation, and it becomes an acute necessity for practical tools to help mitigate the threat posed. The modern risk assessment models recognise a need to perform several threat assessments to identify and analyse various threats in the contemporary information environment. If we conduct iterative threat assessment for the network, then with the help of designing the profiling prepared by practitioners, a new type of threat agents identified in situational awareness data will be addressed quickly. The continuous threat assessments help generate the paradox of warning to the cyber operations performed in the information environment. This paper identifies the research gap in semi-automated information environments, which consists of large heterogeneous infrastructures, hosting a large amount of data collected from different types of platforms or environments [20]. The different types of platforms mean different kinds of environment and the conditions used by the threat agent to attack the particular network. To identify the solution for such a large amount of data, decision aid tools should understand situational awareness and critical intelligence feeds of the threats in real-time information environments.

In the modern knowledge-based, socially driven, virtual computing era, threat assessments are hindered by lack of resources, complexity, and data size. Information environments are large heterogeneous infrastructures, hosting a large amount of data collected from different platforms with the help of many tools. The purpose of the research paper is to introduce a novel approach that will enable us to take advantage of the vast amount of data collected by the large number of platforms designed to identify suspicious traffic, malicious intentions, and network attacks in an automated manner. State of the art on threat assessment models and methodologies will be considered in this project, while procedural and technology issues will be resolved by applying cyber analytics principles [21].

3.1. Experimental Environment of the System

Figure 1 shows the testing set-up through which we execute the penetration testing against the specific condition of the platform or environment. The number of VPNs used to connect with the REDNET network and connect through the firewall saves the data from unauthorised access. Further, REDNET connects to DMZ (Demilitarized Zone), the number of V.M.s, and public I.P. of staff to control the activities. BLUENET connects to the user's V.M.'s I.P.s, ESXi server, UH CSC WIFI (University of Hertfordshire Wi-Fi), and public I.P. of staff. In this environment, the PCAP files are collected from the server with the help of the Wireshark tool [22]. Other tools like SolarWinds Deep Packet Inspection and Analysis, Paessler Packet Capture, ManageEngine NetFlow Analyzer, Omnipeek Network Protocol Analyzer, TCPdump, and WinDum, etc. are also available. Still, Wireshark is more efficient in extracting useful information from PCAP files and provides the advantage of saving the information in CSV formats. Figure 1 shows the source of the attack I.P. address and the destination of the attack I.P. address through which penetration is executing on the network. The role of DMZ is to stop the hacker at the threshold point, and henceforth, no one is allowed to do access excluded the administrator of the server [23]. The BLUENET refers to the internal security team that defends against real-world attackers. Red Teams of REDNET are internal/external entities dedicated to testing the effectiveness of a security program by emulating the tools and techniques of likely attackers in the most realistic way possible.

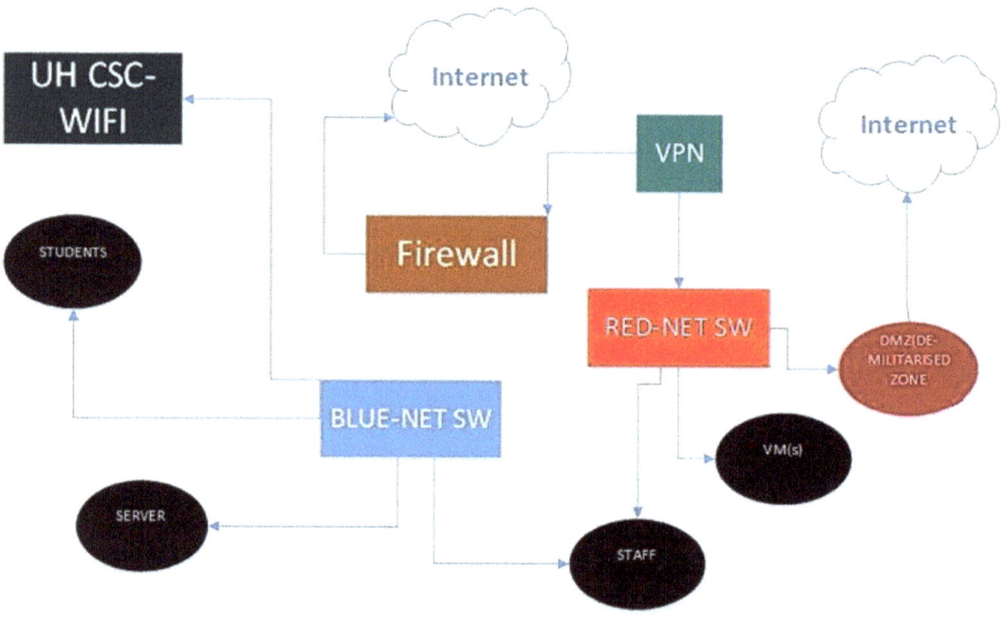

Figure 1. Penetrating Testing Setup at Cybersecurity Laboratory.

3.2. The Architecture of System

The primary purpose of Figure 2 is to understand how the attacker groups generate traffic in the network, increase a delay time to upload the web page and extract useful information from the server such as user credentials, webpages I.P. addresses, and accessing the files from the databases. The architecture in Figure 2 shows that the ESXi server consists of RED, BLUE, and BLACK NET HP-DL380 ESXi VM WARE CD, DNS, DHCP, which is further connected to the Blue ESXi security zone, and DMZ (Demilitarised Security Zone). In this server, all the data and information of the University of Hertfordshire are available, and a dedicated environment installed on V.M.s is available for the attackers. Black ESXi connected to 27 x juniper srx240 and srx340 firewalls via 27 x lab system multiple images of the environment and dedicated interface in red, blue, and black networks. DMZ's role is to stop the hacker at the threshold point to control further damage by the attacker groups.

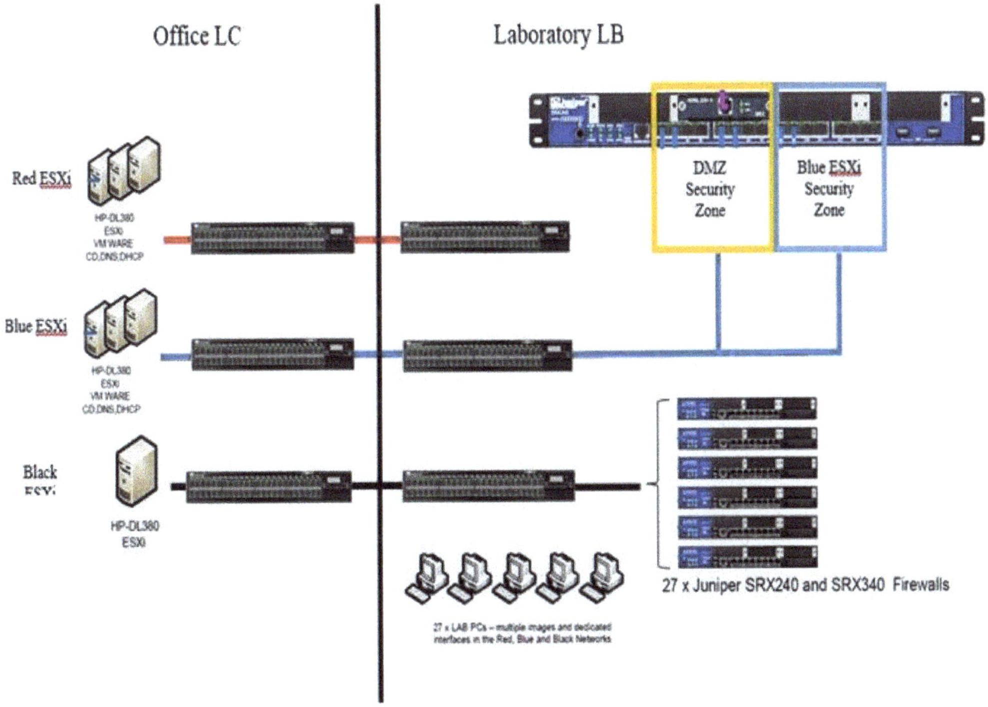

Figure 2. Architecture of System.

3.3. Evaluation of Motivation, Capability, and Opportunity

The threat assessment is a continuous process to collect the PCAP files from the network in an informative environment. The evaluation of the impact of threat agent groups on the organisation or the business, determining the value of assets, vulnerability identification, and threat agent's footprint attributes play a prominent role in the calculation [24]. In Figure 3, the representation of main characteristics in a 3-dimensional matrix is shown, which needs to be addressed by the model while performing threat assessments of the real-time network.

A threat assessment is a statement of threats related to vulnerabilities of company assets and threat agents and a message of believed capabilities that those threat agents possess. In Equation (1), the function threat can be calculated with the help of the threat agent's motivation, capability, opportunity, and the impact of the successful attacks on an organisation of the nation.

$$Threat = f\ (Motivation,\ Capability,\ Opportunity,\ and\ Impact) \quad (1)$$

The threat can be evaluated in the above Equation (1) when the extracted attribute from the PCAP files is analysed. Then, based on the analysis of characteristics, motivation evaluation can be achieved. Similarly, when the model identifies the open port and the vulnerable ports from the extracted attributes, opportunity can be evaluated. In the same way, the model amalgamating all the information of motivation and opportunity leads to assess the capability and impact on the assets by the threats. So, the function (f) can be evaluated using motivation, opportunity, capability, and impact of acquisitions.

$$F(X) = f(Cap,\ Opp,\ Mto,\ V(VIA))\ Y + f(Vulnerability)Asset + Impact + T \quad (2)$$

From Equation (2), the function $F(X)$ represents the threat assessment of the model for all the captured files, *Cap* stands for capabilities, *Opp* is an opportunity of the threat agent, *Mto* is motivation, $V(VIA)$ stands for the value of intangible assets, Y is for threat assessments, and T stands for time complexity.

The threat assessment can be evaluated by amalgamating all results determined by the function for the motivation, opportunity, capability of threat agents, and value of intangible assets of environments. Similarly, vulnerability exploitation of assets concerning the CVE list available on the Nation Institute of standard and technology (NIST) database, the impact of threat agents on an organisation's assets, and the time complexity to evaluate all the parameters of the threat agents can be assessed.

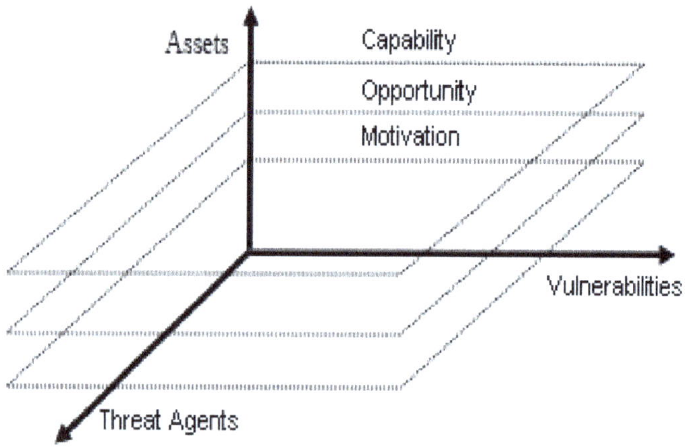

Figure 3. Three-Dimensional Matrix and 3D Representation of Threat Assessment.

3.3.1. Motivation

The evaluation of motivation for threats is the problematic part. It could be determined with the help of analysis of hacktivism branded attacks by groups of assessment models and the network's vulnerability in next to real-time semi-automated information environments. The motivations of attackers are constantly changing, and it could be noticed by the growing rate of hacktivism attacks by different groups of people. It can also see differences in unique motivations based on each group. Motivation is the degree to which a threat agent is prepared to implement a threat. The motivational factors are the elements that drive a threat agent to consider attacking a computer system. Some common motivations for threats include [25]:

a. Profit (direct or indirect).
b. Direct grudge.
c. Fun / Reputation.
d. Further access to partner/connected systems.
e. Political.
f. Secular.
g. Personal gain.
h. Religious.
i. Revenge.
j. Power, terrorism.
k. Curiosity.

3.3.2. Capability

The capability of threats is determined by analysing risk assessment models and the network vulnerability in a next to real-time semi-automated information environment [15].

$$\text{Risk} = (\text{Threat}) + (\text{Vulnerability}) + (\text{Consequences}) \quad (3)$$

In Equation (3), the risk of the threat agent can be evaluated by the combination of threats, the vulnerability identified for the threat concerning the CVE list of the NIST database and identified consequences of the threat agents.

$$\text{Threat} = \text{Intent} \times \text{Capability} \quad (4)$$

Similarly, in Equation (4), the capability of the threat will be evaluated by the multiplication of intention of the threat agents determined by the model and the overall capability of the threat agent. Further, vulnerability exploitation is achieved with the help of several kali Linux tools such as NESSUS, SAINTS, WHISKER, SARA, etc. The initial phase of the automatic version of the threat assessment model is collecting the DataStream/ PCAP files from the server, which has been achieved by the administration of the server between 2012 to 2019. This data mainly consists of PCAP files, which will be extracted in a semi-automatic manner with the help of a machine learning PYTHON tool library available on Tensorflow. The information extracted from these PCAP files having some unique attributes such as Time (in min), Highest Protocol, TCP protocol, Source I.P. Address, Destination I.P. Address, Source port, Destination port, Total Packet Length, City, Region, Country, Latitude, Longitude, and Internet Service Provider. The large number of PCAP files collected from the server will be converted into a large number of Excel sheets based on the unique attributes. These Excel sheets consist of all the valuable information available about the threat in the PCAP files, such as time spent on the network, location of their I.P.s, and environment used by them while penetrating the server.

A large amount of information about the threats can be profiled based on their activities performed on the network or specific environment or Protocol used to achieve their goal. We use all this information to extract all critical threat intelligence (CTI) from these threats to determine the threats' capability, opportunity, and motivation. This CTI can also be used to identify the new threat in-network and extracted all information by taking previously identified CTI as a reference. As shown in Figure 3, the motivation of these threat agent groups can be calculated based on the environment used by them, the type of activities executing during the process, factors responsible for digging information, and data from the server.

In the first phase of the model, an algorithm was executed against the PCAP files captured from the ESXi server and extracted the unique attributes from the PCAP files I.P. addresses, such as time (in min), Highest Protocol, TCP protocol, Source I.P. Address, Destination I.P. Address, Source port, Destination port, Total Packet Length, City, Region, Country, Latitude, Longitude, and Internet Service Provider. When the model has all this information about the attacker, the next phase model extracts the location of the threat agents from where they generate the traffic in the network. The model considers only those threat agents for location identification who have generated more than 1000 packets in the network. The model considers the threshold point based on the level of skill or knowledge the threat agent showing while traversing the network. Likewise, if considered less than 1000 packets generated I.P. address of threat agent, then the exploitation of vulnerable port is significantly less or can be ignorable. It is the primary reason for a semi-automatic model to provide the optimised time complexity for threat assessment of an organisation.

3.3.3. Opportunity

Similarly, opportunity can be calculated by identifying the number of open ports, the number of protocols that have unrestricted access and would be vulnerable, and what other factors help a hacker do unauthorised access to the server. The model will evaluate all this information by initialising the PCAP file captured by the model. The model will determine the open ports with the help of various tools like NMAP, NS-LOOKUP, DIPScan, etc. The combination of all the information about such attributes led to the evaluation of the opportunity of the threat agent groups.

4. Results and Discussion

4.1. State-of-the-Art Algorithms

Many different models are used to perform threat assessment for a network in an informational environment on specialised datasets, where some of the datasets are discussed in the previous section. Here, we illustrate all the threats identified in a network captured during the penetration testing against the ESXi server of the University of Hertfordshire. To provide an overview of the current state-of-the-art ML approaches used to perform the threat assessment, we group all the identified threats from a network based on their profile maintenance concerning the PYTHON program run against the DataStream/PCAP files captured in the experiment. Similarly, the critical threat intelligence [26] feed is evaluated from the group of threat agents based on their footprints extracted during the analysis phase of the experiment. This overview is further divided into two main categories, i.e., traditional extraction of information from the PCAP files and machine learning techniques applied on the information extracted from the PCAP files to generate the footprints used by the threat agents during traversing network of the server.

The PYTHON script provides the accuracy and the unique attributes of the threat agents for precision, false-positive rate (FPR), anomaly detection rate (ADR), and fault-measure as initially reported [27]. Secondly, we calculated the performance of the threat agent followed by our proposed three-dimensional metrics, i.e., motivation, opportunity, and capability. Figure 4 shows that the input is an enormous number of heterogeneous PCAP files captured during the experiment. The potential output generated with analysis of PCAP files is the unique number of Excel sheets which consist of information about the threat agents such as time (in min), Highest Protocol, TCP protocol, Source I.P. Address, Destination I.P. Address, Source port, Destination port, Total Packet Length, City, Region, Country, Latitude, Longitude, and Internet Service Provider. The specific attributes for each experiment run against the PCAP files can be retrieved from https://github.com/Gauravsbin/Excell-sheets-of-pcap-files-and-results-of-Threat-Assessment-analysis (accessed on 8 May 2021) [28]. Furthermore, with the help of these unique attributes, we can determine the capability and opportunity of the threat agents [29]. Based on the footprints followed by the threat agents during the analysis, we can determine the motivation factor for attackers.

Some of the captured PCAP files were corrupted during the experiment, and the PYTHON program list of crashed files generated during the investigation can be fetched as shown in Figure 5. We also checked all these crashed files manually and with other analysis tools. We found the same result that no information can be extracted from these files. There may be some capture issue or the connection lost on the hacker's end during the network establishment. The time complexity to generate the unique I.P.s with information attributes can also be evaluated from this experiment. This is the unique feature of this model as compared to the existing model and methodologies. This could happen because of the use of semi-automatic approaches for threat assessment of networks next to the real-time informational environment.

4.2. Workflow and Comparative Experiments

As per the previous discussion, the output is generated in the form of Excel sheets with the unique attribute of threat agents in a semi-automatic manner. So, to determine the motivation, opportunity, and capability of threat agent groups, we applied machine learning techniques on the previous phase's output to provide a semi-automatic feature to the model [30]. This novel approach helps us optimise the threat assessment's complexity against the network of influential organisations. This paper also shows the process of using ML libraries of PYTHON on TensorFlow and automatic techniques of the JUPYTER notebook to identify the unique tuples of DataStream/PCAP files. This approach mainly depends on the chronological order of packets in PCAP files. Here, we first make groups of all the unique I.P.s extracted from raw PCAP files captured from the network with the help of Wireshark. The grouping of all unique I.P.s based on their attributes and characteristic features was identified during the analysis and implementation of DataStream.

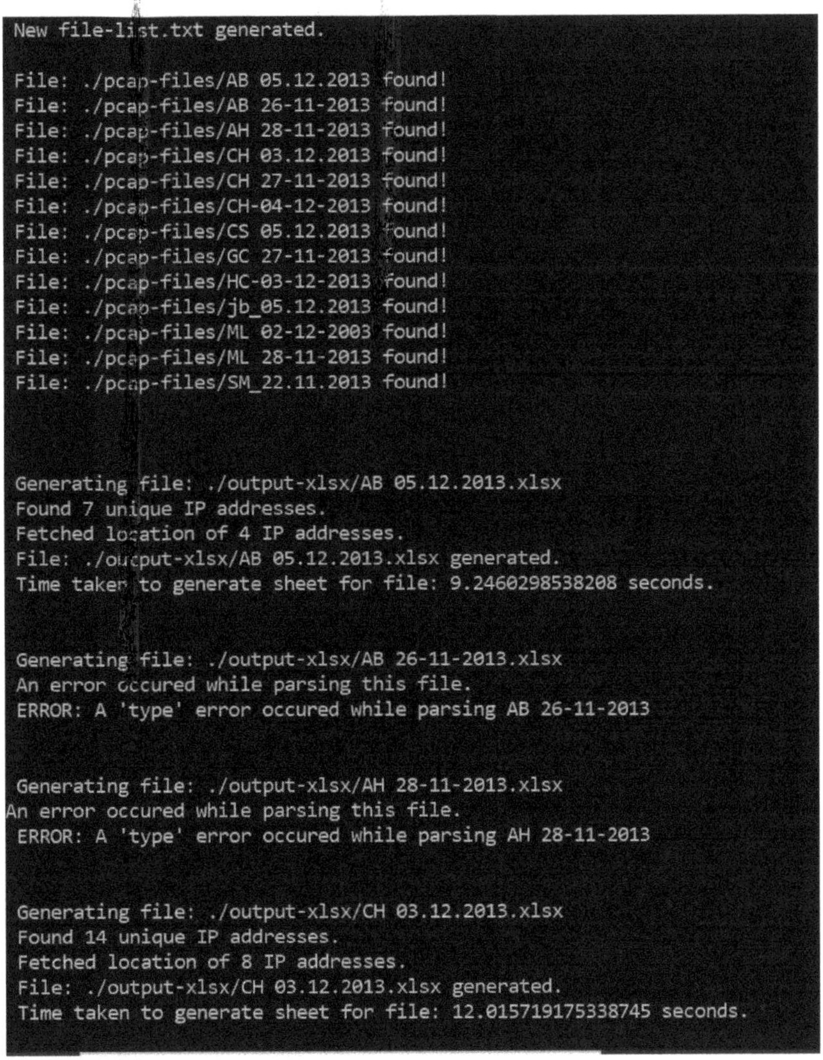

```
New file-list.txt generated.

File: ./pcap-files/AB 05.12.2013 found!
File: ./pcap-files/AB 26-11-2013 found!
File: ./pcap-files/AH 28-11-2013 found!
File: ./pcap-files/CH 03.12.2013 found!
File: ./pcap-files/CH 27-11-2013 found!
File: ./pcap-files/CH-04-12-2013 found!
File: ./pcap-files/CS 05.12.2013 found!
File: ./pcap-files/GC 27-11-2013 found!
File: ./pcap-files/HC-03-12-2013 found!
File: ./pcap-files/jb_05.12.2013 found!
File: ./pcap-files/ML 02-12-2003 found!
File: ./pcap-files/ML 28-11-2013 found!
File: ./pcap-files/SM_22.11.2013 found!



Generating file: ./output-xlsx/AB 05.12.2013.xlsx
Found 7 unique IP addresses.
Fetched location of 4 IP addresses.
File: ./output-xlsx/AB 05.12.2013.xlsx generated.
Time taken to generate sheet for file: 9.2460298538208 seconds.


Generating file: ./output-xlsx/AB 26-11-2013.xlsx
An error occured while parsing this file.
ERROR: A 'type' error occured while parsing AB 26-11-2013


Generating file: ./output-xlsx/AH 28-11-2013.xlsx
An error occured while parsing this file.
ERROR: A 'type' error occured while parsing AH 28-11-2013


Generating file: ./output-xlsx/CH 03.12.2013.xlsx
Found 14 unique IP addresses.
Fetched location of 8 IP addresses.
File: ./output-xlsx/CH 03.12.2013.xlsx generated.
Time taken to generate sheet for file: 12.015719175338745 seconds.
```

Figure 4. Workflow for raw PCAP file traffic-based feature extraction and experimental results for Unique I.P. addresses with Time complexity.

Similarly, the potential output generated in the previous phase is used as potential input for the second phase of analysis and implementation. Such a process is known as the profiling of threat agents. As in the previous stage, we generated the Excel sheet for each captured PCAP file consist of helpful information like ports open. They are operating on that layer: time spent on the network, location of the threat agent, etc.

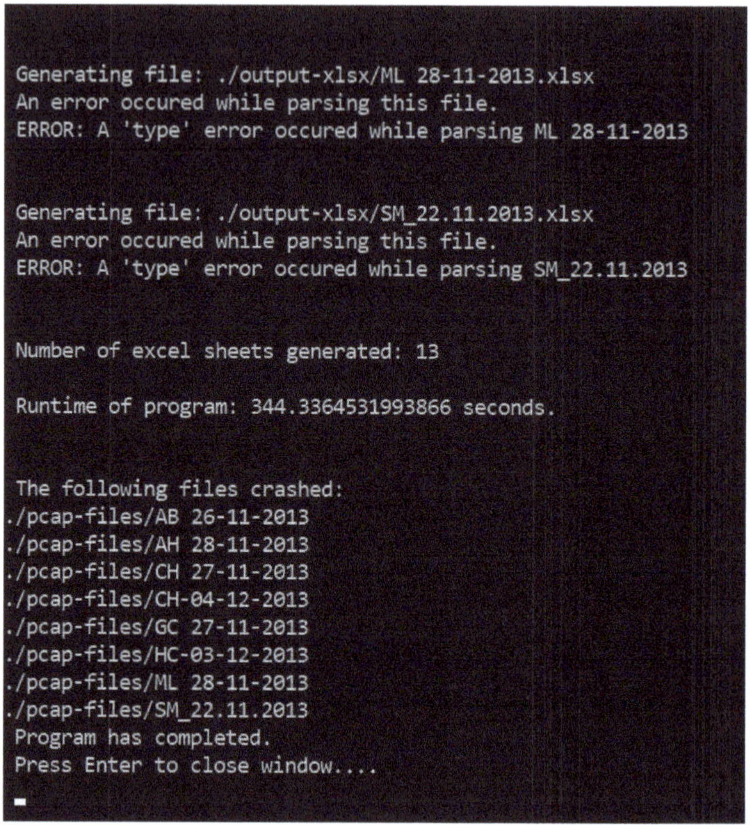

Figure 5. Workflow for raw PCAP file and experimental results for Unique I.P. addresses with Time complexity.

Based on this analysis, we make one more IPYNB file (Interactive Python Notebook) known as the Jupyter notebook. Jupyter is a free, open-source, interactive web tool known as a computational notebook. Researchers can combine software code, computational output, explanatory text, and multimedia resources in a single document. A Jupyter Notebook document is a JSON document, following a versioned scheme, containing an ordered list of input/output cells which can have code, text (using MARKDOWN), mathematics, plots, and rich media, usually ending with the IPYNB extension [31–33]. This file consists of an algorithm performing data clustering of Unique I.P.s found in the Excel sheet of the previous phase. The data clusters of I.P.s form based on the number of I.P.s facing a particular type of attack. This specific type of attack is determined based on the number of factors identified during the analysis. The IPYNB file is collecting all the unique I.P.s as input and extracting the information like on which layer they are operating, what type of ports and protocols are compromised when they are attacking the source I.P.s of end-users, and what information they extracted from the particular environment of the V.M.s, etc. Based on the analysis, the model designed the group of all the threat agents into particular categories concerning their attacking behaviours identified during the analysis.

Figures 6–8 show the histogram of the bar chart with the help of the IPYNB algorithm for each Excel sheet generated during the first phase. Note that we have demonstrated the experimental results of only three PCAP files, and similarly, we can show this for the other PCAP file. There are two parts to the outputs generated by the. IPYNB file. In the first part, three histograms are generated for every file in the output Excel sheet, and the second part develops the histograms on the cumulative data of all the files in the folder. For every file in the output Excel sheet, three histograms have been generated, and all these three histograms consist of common data at the y-axis, i.e., the number of unique I.P.s. Figures 6a, 7a and 8a show the protocols being used by the attackers and the number of unique I.P.s using these protocols. Figures 6b, 7b and 8b show the ports on the host targeted and the number of unique I.P.s that targeted them. This histogram highlights the vulnerable ports. Figures 6c, 7c and 8c show the time spent as a function of the number of unique I.P.s. This histogram highlights how much time an attacker will usually spend to attack a host. These histograms for the protocols, ports, and time spent on the network will help evaluate the three main attributes for the threat agents, i.e., motivation, opportunity, and capability. Once we identify the port open during the network access, we can determine the opportunity for the groups of threat agents used during the penetration of the network. In the same way, the above histograms will help us identify the protocols accessed by the threat agents, evaluate the hacker's potential capability, and level of skills acquired by threat actors.

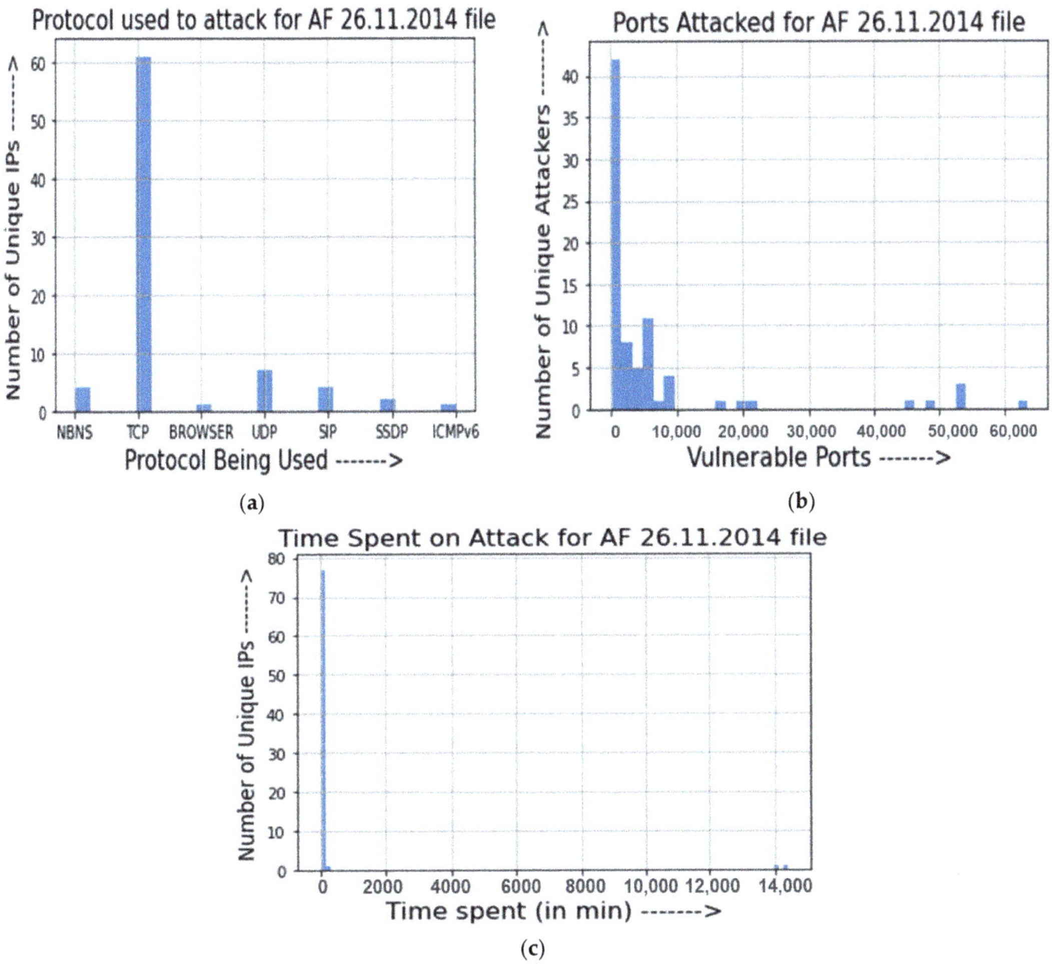

(a) (b) (c)

Figure 6. Experimental Results for PCAP file (AF 26.11.2014). (**a**) Number of Unique I.P.s vs. Protocol being used; (**b**) Number of Unique Attackers vs. Vulnerable Ports; (**c**) Number of Unique I.P.s vs. Time Spent.

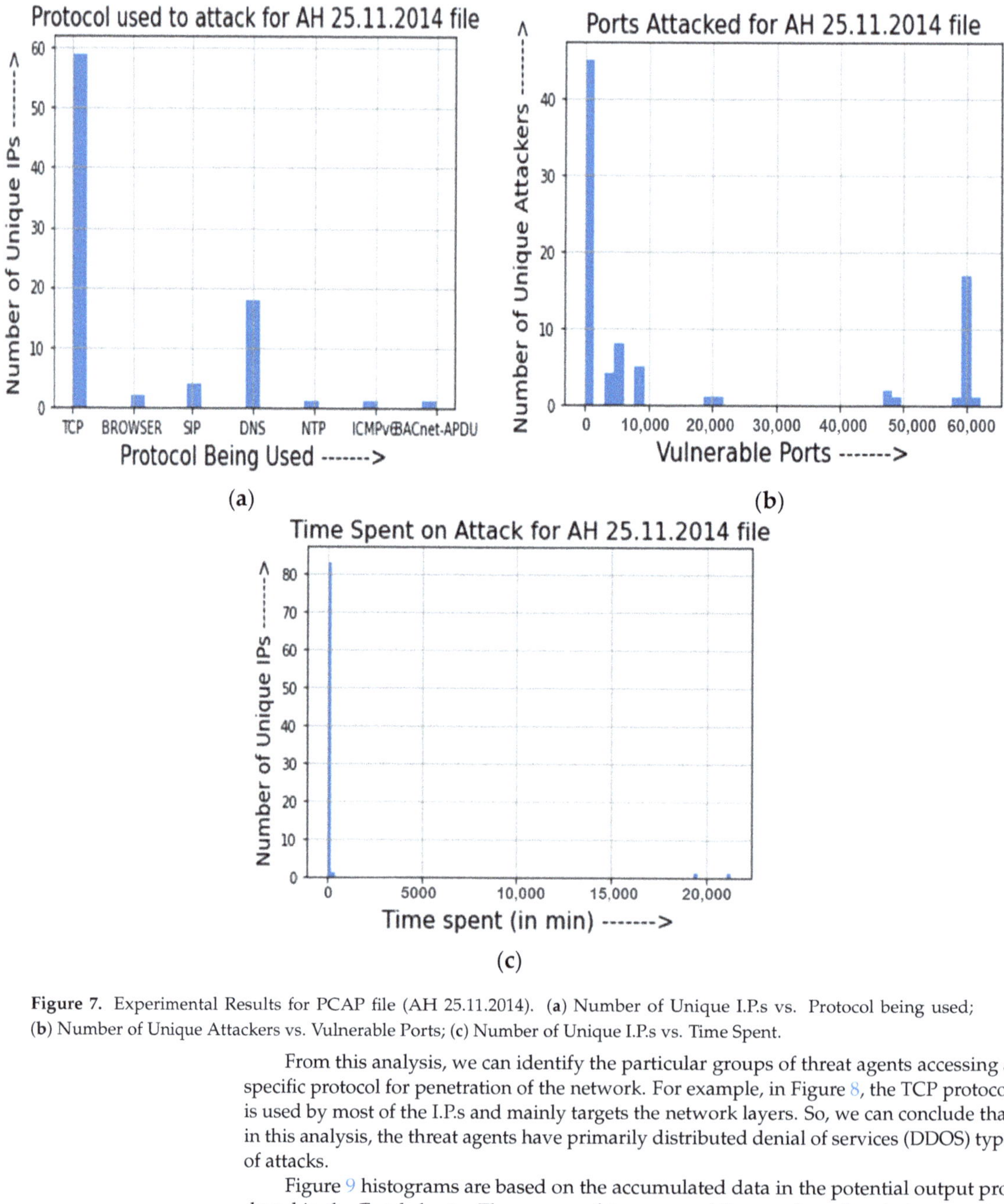

Figure 7. Experimental Results for PCAP file (AH 25.11.2014). (**a**) Number of Unique I.P.s vs. Protocol being used; (**b**) Number of Unique Attackers vs. Vulnerable Ports; (**c**) Number of Unique I.P.s vs. Time Spent.

From this analysis, we can identify the particular groups of threat agents accessing a specific protocol for penetration of the network. For example, in Figure 8, the TCP protocol is used by most of the I.P.s and mainly targets the network layers. So, we can conclude that in this analysis, the threat agents have primarily distributed denial of services (DDOS) type of attacks.

Figure 9 histograms are based on the accumulated data in the potential output produced in the Excel sheets. They are used to represent the number of packets generated for traffic during penetration testing, protocols, or layers being used by threat agents and targeting vulnerable ports for achieving the goal. Figure 9a shows how many packets are sent to which port on the host machine, and Figure 9b shows the volume of packets for every Protocol used to attack the host.

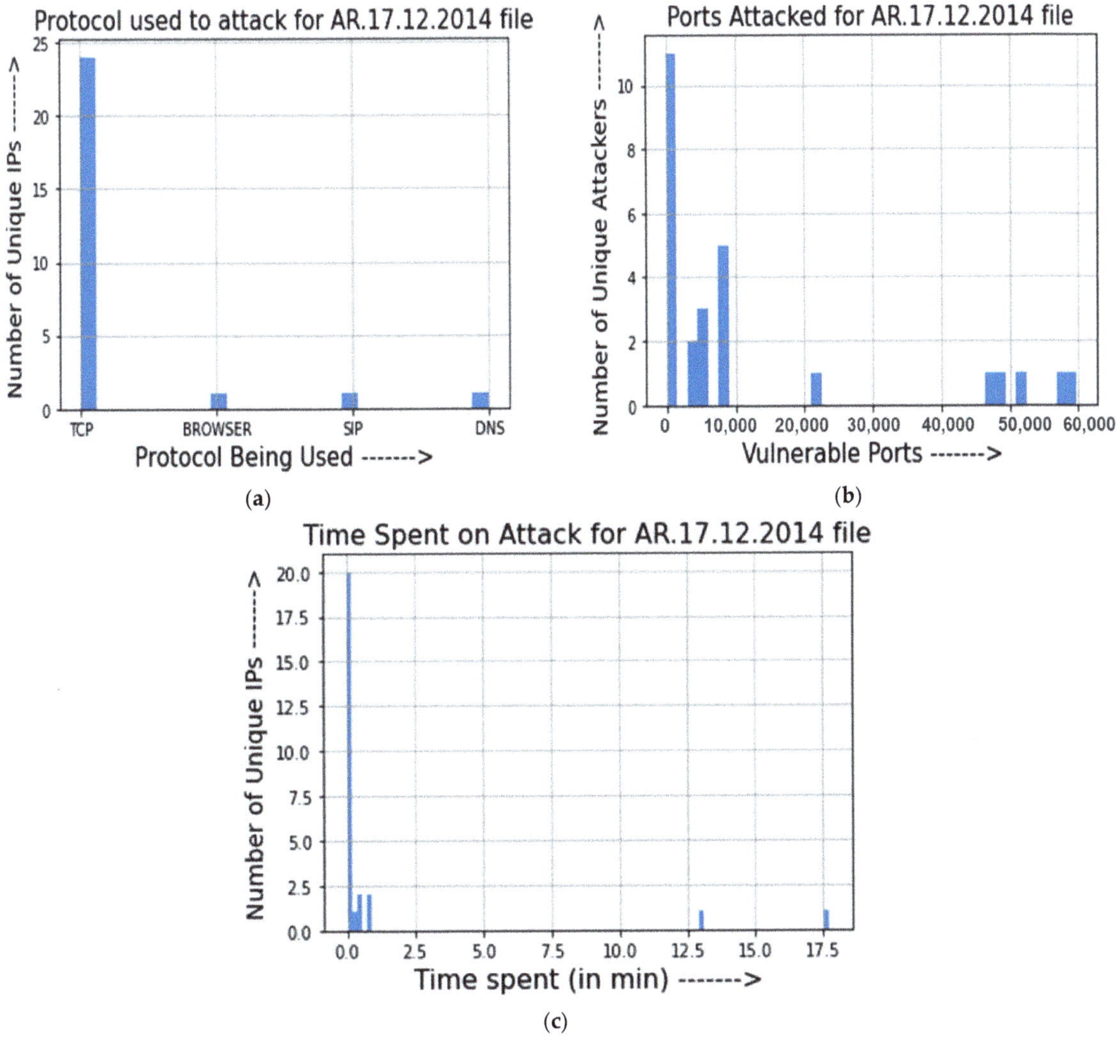

Figure 8. Experimental Results for PCAP file (AR 17.12.2014). (**a**) Number of Unique I.P.s vs. Protocol being used; (**b**) Number of Unique Attackers vs. Vulnerable Ports; (**c**) Number of Unique I.P.s vs. Time Spent.

Figure 10 represents the histogram between the total data collected from each unique I.P., whole time spent on the network, and protocols used to attack the network. Figure 10a highlights the amount spent by the attacker for every Protocol used to attack the host. In Figure 10b, the data points for time spent are highlighted in blue, whereas the data points for total packets sent are highlighted in red. Even though these have different units, it gives us a statistical relative visual of how the time spent by the attacker varies concerning the number of packets sent for the same protocols used.

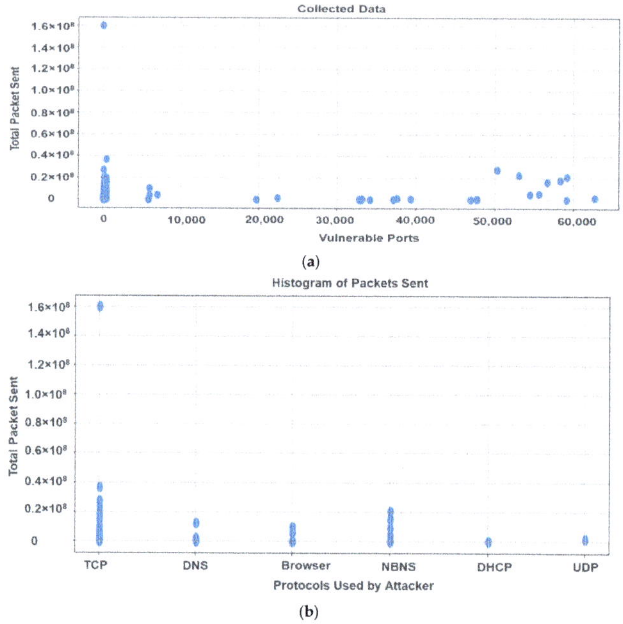

Figure 9. Histogram for (**a**) Total Packets sent vs. Vulnerable Ports, (**b**) Total Packets sent vs. Protocol used by Attackers.

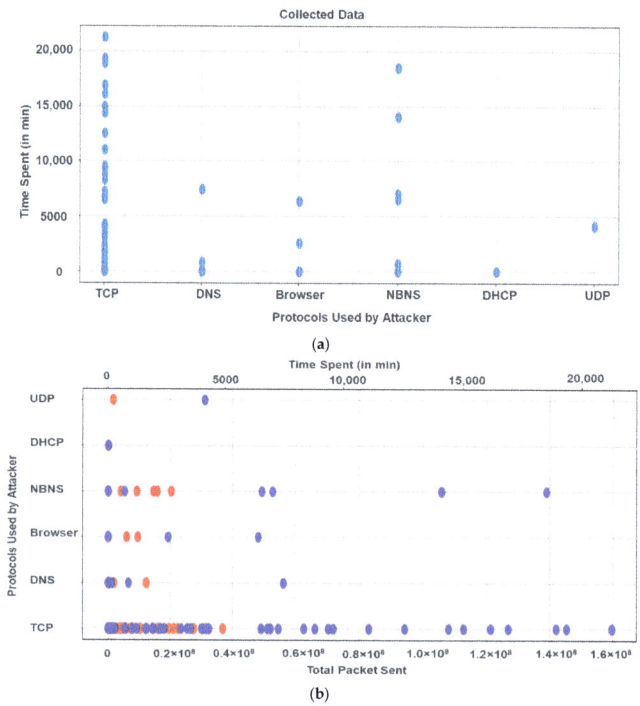

Figure 10. Histogram for (**a**) Time Spent vs. Protocol used by Attackers, (**b**) Protocol used by Attackers vs. Total Packets sent.

5. Conclusions and Future Work

Threats and threat agent's risks are emerging in threat assessment of a network for an organisation and business of the companies. The security risk management practitioners enable a mechanism to explore these risks and enforce their countermeasures based on the threat agent profiling and determining the critical threat intelligence feed to them. This paper presents a semi-automatic model based on the threat assessment of the PCAP files captured by the semi-automatic featured tools during the penetration testing run against the ESXi server of the University of Hertfordshire. The framework captured the data between 2012 and 2019, which illustrates the value of assets stored on the server, and the motivation, opportunity, and capability of the threat agents while accessing the network. We evaluate the situational awareness data through this semi-automatic threat assessment model by exploring the threat profiles for the historically captured data with the aid tools. Furthermore, we provide the threat agent practitioners with an idea of using an automatic model for threat assessment of a network. This research's findings will support decision makers, management, and software developer practitioners regarding the building of threat agent profiling for their historical data. Critical Threat Intelligence feeds for the threat agent's groups might be helpful for the evaluation of new threats found in the network. In the future, we aim to build an automatic machine learning–based threat and vulnerability analysis security reference model as a security risk management tool to evaluate the security needs of networks with sequential requirements of the next to real-time informational environment.

Author Contributions: Conceptualisation, G.S. and S.V.; methodology, G.S.; software, G.S.; validation, G.S., S.V., and C.M.; formal analysis, G.S.; investigation, G.S.; resources, S.V.; data curation, G.S.; writing—original draft preparation, G.S.; writing—review and editing, G.S., S.V., C.M., N.A., and S.K.; visualisation, G.S.; supervision, S.V., C.M. and N.A.; project administration, S.V.; funding acquisition, G.S. All authors have read and agreed to the published version of the manuscript.

Funding: This research received no external funding.

Data Availability Statement: The data presented in this study are available upon request from the corresponding author and are available on GitHub [33].

Acknowledgments: We are grateful for the anonymous reviewers' hard work and comments that allowed us to improve the quality of this paper.

Conflicts of Interest: The authors declare no conflict of interest.

References

1. Iglesias, J.A.; Angelov, P.; Ledezma, A.I.; Sanchis, A. Modelling evolving user behaviours. In Proceedings of the 2009 IEEE Workshop on Evolving and Self-Developing Intelligent Systems, Nashville, TN, USA, 30 March–2 April 2009; pp. 16–23. [CrossRef]
2. Xue, M.; Yuan, C.; Wu, H.; Zhang, Y.; Liu, W. Machine Learning Security: Threats, Countermeasures, and Evaluations. *IEEE Access* **2020**, *8*, 74720–74742. [CrossRef]
3. Jones, A. *Identification of a Method for the Calculation of the Capability of Threat Agents in an Information Environment*; School of Computing, University of Glamorgan: Pontypridd, UK, 2002; pp. 1–134.
4. Mavroeidis, V.; Bromander, S. Cyber Threat Intelligence Model: An Evaluation of Taxonomies, Sharing Standards, and Ontologies within Cyber Threat Intelligence. In Proceedings of the 2017 European Intelligence and Security Informatics Conference (EISIC), Athens, Greece, 11–13 September 2017; pp. 91–98.
5. Atote, B.S.; Saini, T.S.; Bedekar, M.; Zahoor, S. Inferring emotional state of a user by user profiling. In Proceedings of the 2016 2nd International Conference on Contemporary Computing and Informatics (IC3I), Greater Noida, India, 14–17 December 2016; pp. 530–535.
6. Asgari, H.; Haines, S.; Rysavy, O. Identification of Threats and Security Risk Assessments for Recursive Internet Architecture. *IEEE Syst. J.* **2017**, *12*, 2437–2448. [CrossRef]
7. Azaria, A.; Richardson, A.; Kraus, S.; Subrahmanian, V.S. Behavioral Analysis of Insider Threat: A Survey and Bootstrapped Prediction in Imbalanced Data. *IEEE Trans. Comput. Soc. Syst.* **2014**, *1*, 135–155. [CrossRef]
8. Vidalis, S.; Jones, A.; Blyth, A.; Jones, A. Assessing cyber-threats in the information environment. *Netw. Secur.* **2004**, *2004*, 10–16. [CrossRef]
9. Cappelli, D.M.; Moore, A.P.; Trzeciak, R.F. *The CERT Guide to Insider Threats: How to Prevent, Detect, and Respond to Information Technology Crimes (Theft, Sabotage, Fraud)*; Addison-Wesley: Boston, MA, USA, 2012.

10. Susukailo, V.; Opirskyy, I.; Vasylyshyn, S. Analysis of the attack vectors used by threat actors during the pandemic. In Proceedings of the 2020 IEEE 15th International Conference on Computer Sciences and Information Technologies (CSIT), Zbarazh, Ukraine, 23–26 September 2020; Volume 2, pp. 261–264.
11. Wold, S.; Esbensen, K.; Geladi, P. Principal component analysis. *Chemom. Intell. Lab. Syst.* **1987**, *2*, 37–52. [CrossRef]
12. Legg, P.A.; Moffat, N.; Nurse, J.R.; Happa, J.; Agrafiotis, I.; Goldsmith, M.; Creese, S. Towards a conceptual model and reasoning structure for insider threat detection. *J. Wirel. Mob. Netw. Ubiquitous Comput. Dependable Appl.* **2013**, *4*, 20–37.
13. Bishop, M.; Conboy, H.M.; Phan, H.; Simidchieva, B.I.; Avrunin, G.S.; Clarke, L.A.; Osterweil, L.J.; Peisert, S. Insider Threat Identification by Process Analysis. In *2014 IEEE Security and Privacy Workshops*; IEEE: Piscataway, NJ, USA, 2014; pp. 251–264.
14. Morakis, E.; Vidalis, S.; Blyth, A. Measuring vulnerabilities and their exploitation cycle. *Inf. Secur. Tech. Rep.* **2003**, *8*, 45–55. [CrossRef]
15. Vidalis, S.; Jones, A. Threat Agents: What InfoSec officers need to know. *Mediterr. J. Comput. Secur.* **2006**, *1*, 1–12.
16. Sogbesan, A.; Ibidapo, A.; Zavarsky, P.; Ruhl, R.; Lindskog, D. Collusion threat profile analysis: Review and analysis of MERIT model. In Proceedings of the World Congress on Internet Security (WorldCIS-2012), Guelph, ON, Canada, 10–12 June 2012; pp. 212–217.
17. Casillo, M.; Coppola, S.; De Santo, M.; Pascale, F.; Santonicola, E. Embedded intrusion detection system for detecting attacks over CAN-BUS. In Proceedings of the 2019 4th International Conference on System Reliability and Safety (ICSRS), Rome, Italy, 20–22 November 2019; pp. 136–141.
18. Lombardi, M.; Pascale, F.; Santaniello, D. EIDS: Embedded Intrusion Detection System using Machine Learning to Detect Attack over the CAN-BUS. In Proceedings of the 30th European Safety and Reliability Conference and 15th Probabilistic Safety Assessment and Management Conference, Venice, Italy, 21–26 June 2020; p. 2028.
19. Erola, A.; Agrafiotis, I.; Happa, J.; Goldsmith, M.; Creese, S.; Legg, P. RicherPicture: Semi-automated cyber defence using context-aware data analytics. In Proceedings of the 2017 International Conference On Cyber Situational Awareness, Data Analytics and Assessment (Cyber SA), London, UK, 19–20 June 2017; pp. 1–8.
20. Deore, U.D.; Waghmare, V. Cybersecurity automation for controlling distributed data. In Proceedings of the 2016 International Conference on Information Communication and Embedded Systems (ICICES), Chennai, India, 25–26 February 2016; pp. 1–4.
21. Legg, P.A.; Buckley, O.; Goldsmith, M.; Creese, S. Automated Insider Threat Detection System Using User and Role-Based Profile Assessment. *IEEE Syst. J.* **2015**, *11*, 503–512. [CrossRef]
22. Pogrebna, G.; Skilton, M. The Twelve Principles of Safe Places. In *Navigating New Cyber Risks*; Springer: Berlin/Heidelberg, Germany, 2019; pp. 171–197.
23. Iskandar, A.; Virma, E.; Ahmar, A.S. Implementing DMZ in Improving Network Security of Web Testing in STMIK AKBA. *Int. J. Eng. Technol.* **2018**, *7*, 99–104. [CrossRef]
24. Vidalis, S.; Jones, A. Analysing Threat Agents and Their Attributes. In Proceedings of the 4th European Conference on Information Warfare and Security 2005 (ECIW 2005), Glamorgan, UK, 11–12 July 2005.
25. Rubini, R.; Porta, A.; Baselli, G.; Cerutti, S.; Paro, M. Power spectrum analysis of cardiovascular variability monitored by telemetry in conscious unrestrained rats. *J. Auton. Nerv. Syst.* **1993**, *45*, 181–190. [CrossRef]
26. Shin, B.; Lowry, P.B. A review and theoretical explanation of the 'Cyberthreat-Intelligence (CTI) capability' that needs to be fostered in information security practitioners and how this can be accomplished. *Comput. Secur.* **2020**, *92*, 101761. [CrossRef]
27. Chen, R.-C.; Cheng, K.-F.; Chen, Y.-H.; Hsieh, C.-F. Using Rough Set and Support Vector Machine for Network Intrusion Detection System. In Proceedings of the 2009 First Asian Conference on Intelligent Information and Database Systems, Dong Hoi, Vietnam, 1–3 April 2009; pp. 465–470.
28. Available online: https://github.com/Gauravsbin/Excell-sheets-of-pcap-files-and-results-of-ThreatAssessment-analysis (accessed on 12 June 2021).
29. Rynes, A.; Bjornard, T. *Intent, Capability, and Opportunity: A Holistic Approach to Addressing Proliferation as a Risk Management Issue*; Idaho National Laboratory (INL): Idaho Falls, ID, USA, 2011.
30. Rossebo, J.E.; Fransen, F.; Luiijf, E. Including threat actor capability and motivation in risk assessment for Smart GRIDs. In Proceedings of the 2016 Joint Workshop on Cyber-Physical Security and Resilience in Smart Grids (CPSR-SG), Vienna, Austria, 12 April 2016; pp. 1–7.
31. Saygili, G.; Rathje, E.M.; Wang, Y.; El-Kishky, M. Cloud-Based Tools for the Probabilistic Assessment of the Seismic Performance of Slopes. In *Geotechnical Earthquake Engineering and Soil Dynamics V*; American Society of Civil Engineers (ASCE): Houston, TX, USA, 2018; pp. 19–26.
32. Van Veen, H.; Saul, N.; Eargle, D.; Mangham, S. Kepler Mapper: A flexible Python implementation of the Mapper algorithm. *J. Open Source Softw.* **2019**, *4*, 1315. [CrossRef]
33. Narkar, S.; Thomson, B.L.; Fox, P.A. Designing for 2030: The Impact and Potential of Virtual Laboratories. In Proceedings of the American Geophysical Union, Fall Meeting 2020, 1–17 December 2020.

electronics

MDPI

Systematic Review

Data Transformation Schemes for CNN-Based Network Traffic Analysis: A Survey

Jacek Krupski, Waldemar Graniszewski * and Marcin Iwanowski

Institute of Control and Industrial Electronics, Warsaw University of Technology, ul. Koszykowa 75, 00-662 Warszawa, Poland; jacek.krupski@ee.pw.edu.pl (J.K.); marcin.iwanowski@ee.pw.edu.pl (M.I.)
* Correspondence: waldemar.graniszewski@ee.pw.edu.pl

Abstract: The enormous growth of services and data transmitted over the internet, the bloodstream of modern civilization, has caused a remarkable increase in cyber attack threats. This fact has forced the development of methods of preventing attacks. Among them, an important and constantly growing role is that of machine learning (ML) approaches. Convolutional neural networks (CNN) belong to the hottest ML techniques that have gained popularity, thanks to the rapid growth of computing power available. Thus, it is no wonder that these techniques have started to also be applied in the network traffic classification domain. This has resulted in a constant increase in the number of scientific papers describing various approaches to CNN-based traffic analysis. This paper is a survey of them, prepared with particular emphasis on a crucial but often disregarded aspect of this topic—the data transformation schemes. Their importance is a consequence of the fact that network traffic data and machine learning data have totally different structures. The former is a time series of values—consecutive bytes of the datastream. The latter, in turn, are one-, two- or even three-dimensional data samples of fixed lengths/sizes. In this paper, we introduce a taxonomy of data transformation schemes. Next, we use this categorization to describe various CNN-based analytical approaches found in the literature.

Keywords: network traffic analysis; convolutional neural networks; machine learning; network traffic images; visualization of traffic

Citation: Krupski, J.; Graniszewski, W.; Iwanowski, M. Data Transformation Schemes for CNN-Based Network Traffic Analysis: A Survey. *Electronics* **2021**, *10*, 2042. https://doi.org/10.3390/electronics10162042

Academic Editor: Amir Mosavi

Received: 2 July 2021
Accepted: 16 August 2021
Published: 23 August 2021

1. Introduction

1.1. Deep-Learning Approach to Network Traffic Analysis

The rapid growth of computer networks over the last decades [1] has entailed a larger amount of cyber-attacks. In order to minimize the losses, many security methods are in heavy use. Among others, network traffic analysis is in the lead. This day-to-day operation consists of processing typical patterns, such as traffic flow, bandwidth usage or resource access. Together, these patterns identify the normal network behavior, also known as a baseline. Having this baseline in mind, it is possible to interpret abnormal activities, which may indicate an attack.

Deep learning methods have also begun gaining popularity recently. This is mainly caused by the development of computing capabilities based on parallel processors originated from graphics cards. This has resulted in the rapid increase in efficient implementations of computationally demanding complex neural networks and, finally, a remarkable growth in capabilities to solve advanced problems. Among the most successful architectures of this kind are convolutional neural networks (CNNs, conv-nets). They are ideally suited for multidimensional data, originate from image processing but can be successfully applied to other computing domains.

Internet traffic analysis and machine learning are the two worlds that must be connected with one another, especially when applying the latter to the data provided by the former. The originator of the junction of traffic analysis with CNNs is Wang, who, during the innovative presentation at the "Black Hat" conference in 2015 [2], pointed out the similarities between images and TCP flow payloads. Despite the utilization of an autoencoder to identify network traffic, in a later work, Wang signaled the usefulness of CNNs for the same task. To the best of our knowledge, this is the first mention of network traffic identification or malware detection with the advantage of CNNs.

There is a striking change in the number of research papers that are devoted to the analysis of network traffic by CNN models (see Figure 1). To enhance the analysis, we distinguish three possible subjects of the articles: malware detection, traffic classification, and the junction of both. These categories are connected with datasets studied in each paper. The typical indicators of the datasets are the motifs of data, e.g., captured botnets are the foundation of the CTU-13 dataset, so each article utilizing it will belong to the malware detection group.

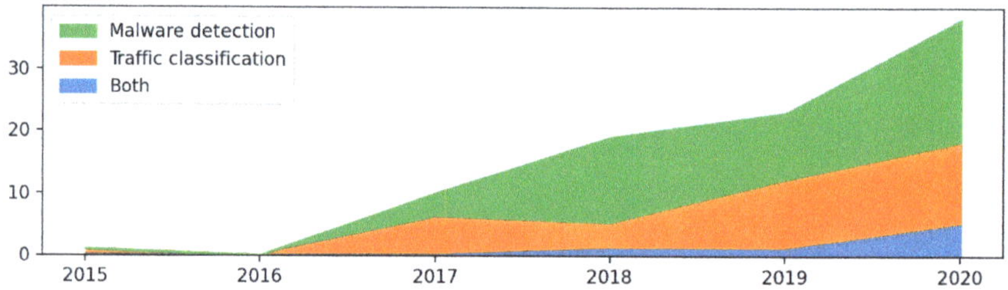

Figure 1. The growth of CNN-based models, which process transformed network traffic in the years 2015–2020.

What is particularly interesting seems to be the overview from the perspective of traffic transformation methods before being given as an entry to the CNNs. Deep learning models require particular data formats that are rarely similar to original computer traffic. In most of the reviewed articles, the traffic data are transformed into the forms needed for the analytical part of the whole workflow. These transformations usually require performing one or more typical actions, e.g., the selection of specific network layers, trimming the data stream, or computing some traffic features. Due to the given architecture of some learning algorithms, these data transformations often demand an increase in the dimensionalities of the traffic data. The network traffic data are a time series, while the CNNs require multidimensional input consisting of equal-length samples. Due to this fact, the original traffic data must always be transformed into a format acceptable by the deep-learning models.

1.2. Our Contributions to the Topic

This survey deals with transformations of the network traffic, which are the input of the deep learning models, with particular attention paid to the CNN models. We have studied many articles and finally chose 136 papers written recently in this field of science. It is essential to highlight that other surveys present these studies from the perspective of cyber-attacks, particular system traffic, or mixed deep-learning models.

We explicitly focus on the network traffic transformations before being given as an entry to the CNN models.

1. This paper proposes the new taxonomy of the network traffic transformations for CNN processing purposes.
2. Additionally, all 136 revised articles are comprehensively investigated and mapped to the adequate transformation algorithms. The survey differentiates three categories of research papers:
    (a) Traffic classification.
    (b) Malware detection.
    (c) The combination of traffic classification and malware detection.

The first contains all articles that focus on encrypted traffic identification. The malware detection category is about finding unwanted traffic. The last includes research about both mentioned categories. These categories are firmly connected with the datasets utilized by each group of authors. Moreover, it is possible to distinguish different themes of the datasets, such as dealing with VPN traffic or exploring features of botnets.

3. This work highlights and describes the utilized datasets and the architecture of each CNN model.

The proposed taxonomy is the first on this topic. While preparing this work, we inherited and developed the concept from the CNN chapter from [3].

As the number of papers in the described field is constantly growing, we decided to review the proposed works and highlight all scientific observations. The detailed comparison in this area can establish current trends as well as enhance network traffic analysis.

1.3. Paper Structure

This paper consists of nine sections. Section 2 touches upon fundamental issues in the discussed scientific field. The following subsections are devoted to main categories of methods that reflect the ways that the network traffic is transformed prior to transferring them into the CNN-based neural models. Section 3 presents approaches based on raw traffic, i.e., network traffic without any filtering. Section 4 is about all transformations working on flows—particles of the entire traffic. Section 5 highlights data manipulations on payloads extracted from the raw traffic. Section 6 focuses on all concepts based on payload that is extracted from flows. Traffic features approaches are the main subject of Section 7. Section 8, in contrast to Section 7, gives a concrete overview of those articles that additionally focus on the feature extraction process. Section 9 concludes the paper.

2. Preliminaries

2.1. Network Monitoring

The internet is based on a protocol suite, which was developed by the Defense Advanced Research Projects Agency (DARPA). The idea of a distributed topology, with a packet switched network is described from the time perspective by its author Baran in [4]. With the rapid growth of the internet at the end of the 20th century, there was also a necessity for network accounting and monitoring. Almost in parallel to network traffic profiling for accounting reasons, frequent and large-scale network attacks have led to an increased need for developing techniques for analyzing network traffic. In the design philosophy of the DARPA internet protocols [5], Clark explained *flow* as being connected with the necessity to treat differently those packets transmitted by intermediary network devices with an appropriate type of service demanded by the endpoints applications. In such a way, particular packets belonging to the same connection can be distinguished. According to the basic principles of packet switching networks, each datagram from a network connection can take different routes. In the beginning, the original ARPANET host-to-host protocol provided flow control based on both bytes and packets. However, later, due to efficiency reason, only the bytes number was used for acknowledgments.

Later, at the beginning of the 1990s, Mills et al. proposed internet accounting [6]. Network accounting introduced packet aggregation based on flows, using packet header information. Then, the idea was developed to use real-time traffic flow measurement (RTFM) [7]. Claffy et al. proposed a methodology for profiling traffic flows on the internet for communication analysis [8].

Then, with the implementation of the NetFlow [9] protocol, the traditional understanding of IP flow was defined as a set of five, up to seven, IP packet attributes flowing in a single direction. When a TCP session is considered, a flow consists of all packets transmitted until this session terminates. NetFlow, among others, uses the following IP packet attributes: IP source address, IP destination address, source port, destination port, layer 3 protocol type, type of service, router or switch interface. All packets with the same earlier-mentioned attributes are grouped into a flow, and then packets and bytes are counted. With the introduction of the IP flow information export (IPFIX) protocol, the number of flow attributes, named IPFIX information elements, increased to several hundred. The RFC7011 explains that [10]: "(...) A Flow is defined as a set of packets or frames passing an Observation Point in the network during a certain time interval. All packets belonging to a particular Flow have a set of common properties. Each property is defined as the result of applying a function to the values of:

1. One or more packet header fields (e.g., destination IP address), transport header fields (e.g., destination port number), or application header fields (e.g., RTP header fields [RFC3550]);
2. One or more characteristics of the packet itself (e.g., number of MPLS labels);
3. One or more of the fields derived from Packet Treatment (e.g., next-hop IP address, the output interface) (...)".

In the studied papers, we have found different usages of the *flow* term with several nouns, such as traffic, packet, data, and IP packets, which can mislead the readers. Therefore, we have decided to unify network traffic terms for this survey as the following definitions:

- Raw traffic —network traffic observed in an observation point, such as a line, to which the probe is attached, an Ethernet-based LAN, or the ports of a switch or router [10].
- Flow (also called traffic flow (e.g., [10]), network connection (e.g., [11]), internet stream (e.g., [12]))—grouped raw network traffic according the same properties, usually 5-tuple: source and destination IP address, source and destination port number, and type of service.
- Session—bi-directional flow. Traffic grouped according to the same properties as a flow, which mimics conversation between the end devices. A session usually requires establishing a TCP connection in the form of a three-way handshake.
- Traffic features [13] (also called *flow features*)—set of features describing the traffic. They can be statistical features of flow data obtained from flow probe, using one of the flow profiling protocols, e.g., IPFIX, or processed using the appropriate software or particular network protocols headers fields [14]. When collected and exported in IPFIX flow records, they are called information elements (IEs) [10]. Some of these features can be exported in IPFIX flow records, using a textual representation of IPFIX [15]. A standard list of IEs is maintained by the internet assigned numbers authority (IANA). Moreover, the internet community can define their new elements, which fulfill the applications' specifications [16]. Hofstede et al. prepared a more detailed specification of flow monitoring with NetFlow and IPFIX [17].
- Payload—transmitted data encapsulated in the particular ISO/OSI model protocol data unit (PDU). The Layer 4 (and above layers) payload (L4+ payload) are the actual upper—layers (L5, L6, L7) data, e.g., HTTP request or response—FTP data. The Layer 3 payload is a segment (TCP) or a datagram (UDP) of the Layer 4 PDU, including the L5-L7 PDUs. The Layer 2 payload (L2 payload) is a packet—usually an IP packet.

Traffic data can be collected from an observation point with a hardware or software solution. Written in C, an open-source library *Libpcap* (see: https://www.tcpdump.org/, accessed on: 2 July 2021) is available for different platforms. This library delivers an application programming interface (API), which can be implemented in capturing software, e.g., tcpdump or Wireshark. Collected traffic data with the libpcap library can be saved in the pcap file format and used to create a dataset for analytics and classification.

Depending on the available datasets (see Section 2.4) and implemented machine learning algorithms, traffic datasets can directly feed the chosen *CNN-based deep learning model* (CDM) or may have to be pre-processed according to several paths, as is presented in the upper part of the workflow diagram in Figure 2. Different possible side paths—raw traffic processing or filtering—are indicated with blue arcs.

The straightforward path is with only trimming or padding block. In the case of using only internet raw traffic for a machine (deep) learning (straight line in Figure 2), there is a necessity to change the length of the original stream data into chunks to prepare these according to the input size of the chosen CDM. If the input dimension of the selected model is smaller than the stream data chunk, the latter has to be trimmed to the size of the CDM's input vector dimension. When the input stream is shorter than is expected by the CDM, the remaining part of the input vector is padded with an arbitrarily selected value—usually with zeros, to fit the suitable CDM's input vector dimension.

Following two side paths—alternatively: flows or sessions—requires grouping traffic data according to the same properties. Then, the flows or sessions' data must be trimmed or padded, again as in raw traffic, to fit the suitable CDM's input vector dimension.

An alternate path for flows or sessions data can lead through selected layers (L2, L3, L4+) payload extraction. In intrusion detection systems (IDS), such processing is called deep packet inspection (DPI). Then again, extracted payloads must be trimmed or padded to fit the suitable CDM's input vector dimension.

Finally, the traffic data samples became an input for machine learning data described in more detail in Section 2.2.

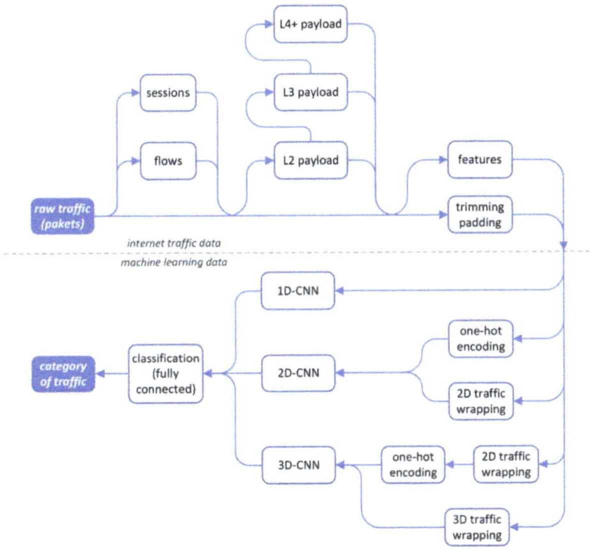

Figure 2. Workflow diagram of network traffic classification using deep convolutional neural networks with various data transformation paths.

2.2. *Convolutional Neural Networks*

Deep neural networks have recently become one of the hottest methodologies applied in machine learning and pattern recognition. They provide machine learning models that surpass previous approaches. Thanks to the rapid development of computing resources and common usage of relatively cheap parallel-computing platforms, previously long-lasting machine learning tasks have become available for everyone. Among the most popular types of deep networks are convolutional neural networks (CNN) [18]. They are based on convolution operators, the weight of which is a subject of learning. Due to the multidimensional nature of convolution, the CNN has gained enormous popularity in image processing and analysis. Their history starts from the LeNet [19] by LeCun et al., which was a breakthrough in image pattern recognition. The CNN-based approach considerably surpassed the previous methods to classify hand-written digits recognition (MINIST datasets). It became possible because the neural network, in this case, is responsible not only for classifying the data samples (as, up to that time, typical neural nets did) but also for extracting features. In particular, the convolutional layers perform this task. In the case of conv-nets, the typical structure of the recognition scheme consists of two parts. The first one is a set of consecutive convolution layers that are stacked alternately with pooling layers. Convolution layers are responsible for extracting data features while pooling layers for reduction of the data size. The combination of feature extraction with size reduction allows for detecting data features at increasing scales. Finally, if necessary, the output of the convolution and pooling layer is flattened to obtain a final feature vector. Its further processing is a typical classification task that is usually based on the structure resembling (or sometimes being equal to) the multilayer perceptron (MLP classifier). Contrary to convolutional layers, in classification layers, all neurons located at a given layer are connected to all in the next one. Because of that, they are called fully connected (FC) or dense layers. The combination of the CNN and FC layers constitute the complete classification framework. In many papers, the name CNN is spread into the complete neural model consisting of both parts, the actual CNN and FC. However, formally speaking, it should rather be used exclusively for the first—feature extraction—part of the model. An example of such a type of network is shown in Figure 3. The diagram shows the LeNet consisting of the two parts mentioned above. The data feature extraction part inputs and the 32 × 32 image, consist of layers—convolution (conv 1), pooling (pool 1), convolution (conv 2), pooling (pool 2)—and outputs the vector of 400 data features. The classification part consists of three fully connected layers: the first with 120 neurons, the second with 84 neurons, and finally, the third with ten neurons. The number of neurons equals the number of output classes, which is equal, in this case, to the number of possible digits that might appear on the input image. For the sake of simplicity, we use shortcuts for the principal layers of the neural model: C—convolutional layer, P—pooling layer and, FC—fully connected layer. The LeNet structure may thus be coded as C | P | C | P | FC | FC | FC.

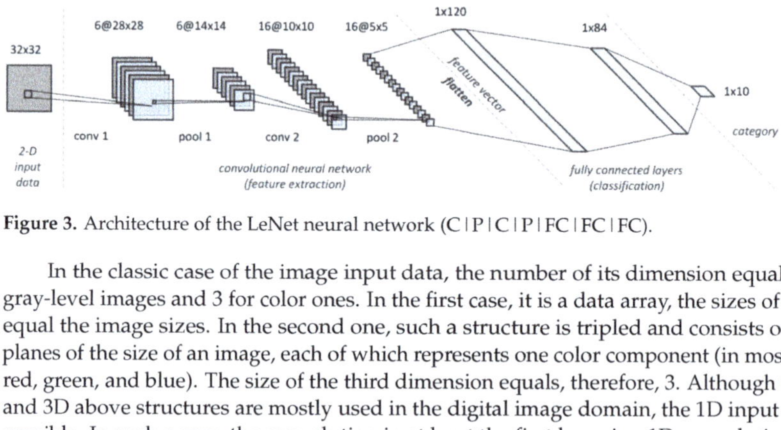

Figure 3. Architecture of the LeNet neural network (C | P | C | P | FC | FC | FC).

In the classic case of the image input data, the number of its dimension equals 2 for gray-level images and 3 for color ones. In the first case, it is a data array, the sizes of which equal the image sizes. In the second one, such a structure is tripled and consists of three planes of the size of an image, each of which represents one color component (in most cases red, green, and blue). The size of the third dimension equals, therefore, 3. Although the 2D and 3D above structures are mostly used in the digital image domain, the 1D input is also possible. In such a case, the convolution in at least the first layer is a 1D convolution.

Following the enormous growth in popularity of the CNN structures, they started to be applied in many domains other than vision systems. One of these domains was the categorization of the IP network traffic. In this case, the input data to be classified are samples of the network traffic. The resulting classes, in turn, are related to the types of traffic.

There are several ways of preparing the traffic data to obtain valuable input for the machine (deep) learning model described in detail in Section 2.1 (see also Figure 2). One of the possible preprocessing methods makes use of the traffic features. These features are, however, different from data features extracted by CNN. The primers are intentionally and carefully selected features of particular meaning: either properties of the traffic (e.g., IP address, port) or some statistics (e.g., number of bytes, packets). The data features, in turn, are automatically selected numbers derived from the original data vector that makes the input of the learning model.

In traffic analysis applications, the input data are one-dimensional time series consisting of consecutive bytes transmitted. Following various dimensionalities of possible inputs of the CNN (1-, 2- or 3D), one may find in the traffic analysis several solutions that either keep the original 1D dimensional nature of the traffic data, or increase the number of dimensions. The 1D CNN solutions consist of 1D convolution filters, at least at the input layer. The 2D solutions add the second dimension by using, in the vast majority of cases, two approaches: traffic wrapping or one-hot encoding. The 3D solution either exploits more sophisticated wrapping or combines both techniques. The schematic diagram showing the data flow in each case is shown in the lower part of Figure 2.

Independently of the method used to add the second dimension of data, the input data for the machine learning model should consist of equal-sized data samples. To obtain such samples, data trimming (for samples originally too long) or padding (for those that are too short) is usually performed (see Section 2.1 for details).

The *traffic wrapping* cuts the data sample consisting of n bytes into n_2 pieces of the same length n_1. Values of n_1 and n_2 are chosen in such a way that $n_1 \cdot n_2 = n$. In the output data 2D array, each data value has not only neighbors that were transmitted just before and just after (these are horizontal neighbors in the 2D array), but also has vertical neighbors that, coming back to the original 1D data sample, are equivalent to the data values that appeared at a certain time before and the same time after the current data value. For example, if the data sample of size n consists of bytes, the t-th byte has two direct horizontal neighbors, $t-1$ and $t+1$, and two direct vertical ones, $t-n_1$ and $t+n_1$. The traffic wrapping is shown in Figure 4. This approach performs in a way that may be called linear stacking and is applied in all but one among the studied approaches. Several atypical approaches to 1D to 2D sample mapping (diagonal, waterfall, spirals) were studied in [20].

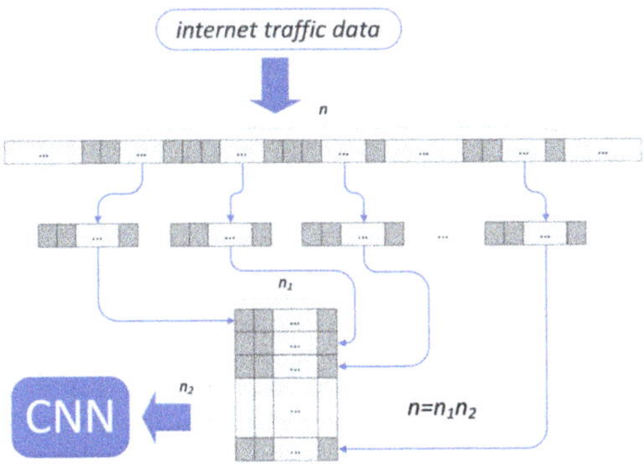

Figure 4. Introducing the second dimension by wrapping the network traffic data.

The second technique of increasing the traffic data dimensions is *one-hot encoding*. This approach replaces numerical integer values by the binary vector such that all but one of its elements equals zero, and the unique element equals one. Such an approach is applicable for numerical variables belonging to a finite set of m possible values (for example, a value of a byte belongs to the set of possible $m = 256$ values). The one-hot-encoder thus inputs an integer of m values and outputs a binary vector of size m, containing value one at the position related to the current input values and zero elsewhere (an alternative solution encodes n-values variable as a binary vector of size $n-1$, where the n-th input value is encoded as an all-zero output). Replacing single values by vectors converts the 1D vector of integers into a 2D binary array—see Figure 5.

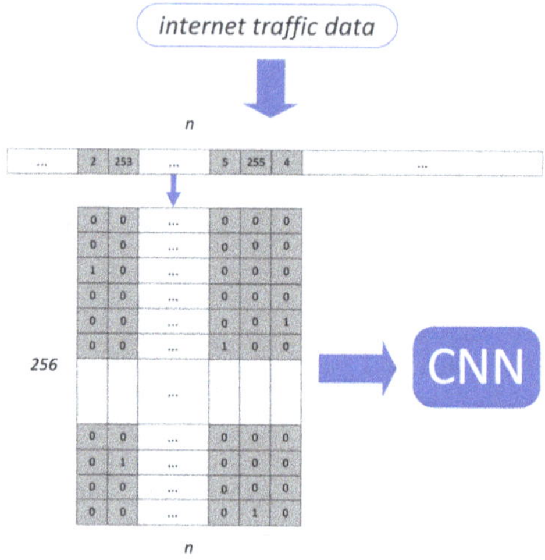

Figure 5. Introducing the second dimension using one-hot encoding.

The origin of the one-hot encoding approach is related to the observation that—in most cases—elements of the traffic data sample (single values) are not ordered, and there is no intrinsic order of values represented by bytes. They usually represent some pieces of information encoded using bytes via standardized codes. They should thus be treated as unordered categorical values, rather than a set of consecutive integers. This property makes them different from, for example, image data, where pixel values are ordered—higher values of pixels represent a higher value of luminance. The property of having ordered values of the input data is essential in neural network learning algorithms, which are inseparable parts of the neural models that use gradient descent approaches to modify network weights iteratively.

The one-hot-encoded vector is the sparse 1D data structure of the length equal to the encoded variable's possible values. Making it shorter is possible, using another classic trick—the embedding technique. It produces a shorter vector of a given length of possibly the same amount of information as the one-hot-encoded input. The vector embedding is performed using a fully connected layer that takes a binary one-hot-encoded vector as the input and produces a shorter embedding vector further processed by the convolution layer(s).

The architecture of the neural models consists of classic convolution, pooling and fully connected layers. It also includes often typical mechanisms found in other deep-neural models, such as regularization (mostly drop-out), preventing overfitting, or softmax output normalization that allows for interpreting the output of the model as probabilities.

In many neural approaches to network traffic analysis, pre-trained neural CNN-based models are used. They gained popularity in the image analysis domain due to their effectiveness and ability to work as backbones in many image analysis fields. In their case, the transfer learning approach in most widely used, where the image pre-trained model is learned to adapt to the network-traffic data. Pre-trained models focus on recognizing single objects located within the image and work usually on images with fixed sizes. To this group belong the following well-known networks: LeNet [19], AlexNet [21], GoogLeNet/Inception [22], DenseNet [23], ResNet [24], VGG [25], XCeption [26], and MobileNet [27].

2.3. Visual Aspects of the Traffic Data

The visualization of the network traffic is one of the classic approaches to traffic monitoring. The most traditional way is visualizing network structure as graphs where nodes and edges represent the network topology. A routing graph is a typical example of such visualization. However, this is just one of the possible network visualizations. Along with developing the internet and constantly increasing abilities to process traffic data, data visualization techniques have always played a significant role in this field. There have been many contributions in this field since the first editions of the Visualization for Cyber Security (VizSec) forum [28,29]. To perform meaningful visualizations, in some cases, authors use data reduction methods, e.g., PCA for dimensionality reduction [30].

Thanks to transforming the 1D time series of the original traffic data into 2- and 3-matrices, one may look at network traffic as digital images [31]. The single elements of the traffic data samples—which, in most cases, are simply bytes—play the role of pixels. The luminance of the pixel refers to the value of a particular element/byte, where higher byte values are represented by lighter pixels. The 1D traffic data converted into higher-dimensional data samples of a fixed length may be displayed as binary, gray level, or color images. In the first case, the input must be binary. One uses this type of traffic-to-image transformation in the case of one-hot-encoded network traffic. In the case of gray-level images, image pixels, one usually applies wrapping techniques. The resulting gray-level image looks like an image of a texture, including either irregular or regular patterns. In rarer cases, the resulting image is a color one, which is the 3D data structure. The third dimension has a fixed size of 3, due to the number of planes referring to three color components. Each of them is a gray-level image with the luminance value associated with the intensity of a particular component.

They interpret the network-traffic samples as images allowed for directly applying the image-processing techniques to this type of, initially, non-image data. They have been used, e.g., for detecting anomalies in internet traffic [32,33].

Because 2- and 3D CNN-based neural models were initially developed to process digital images, image representation of traffic has become an obvious visualization method in the CNN-based neural models. Since the ready-to-use neural backbone models are designed to process the input data of a fixed size, the size of the traffic data sample must become compliant with the input image size.

Because 2- and 3D CNN-based neural models were initially developed to process digital images, image representation of traffic became an obvious visualization method in CNN-based neural models. Since the ready-to-use neural backbone models are designed to process the input data of a fixed size, the size of the traffic data sample must become compliant with the input image size. This fact is noticeable in many network models where the size of the traffic data sample is equal to the size of the input of the neural model initially developed to process images of particular sizes. A typical example of such a strict dependence of the traffic data sample and the input of pre-trained backbone is a sample size that equals 784, which appears in many approaches. They also force the 2D input of the neural model equal to be a square array, where the length of the edge equals 28 ($28 \times 28 = 784$)—see [34]. Such a choice is not motivated by the particular properties of the network traffic, but by the neural LeNet model, which was originally used to recognize hand-written digits on squared bitmaps of size 28×28. Examples of images of network traffic processed using the method [34] are shown in Figure 6. The grayscale images are built from matrices using flow wrapping. The hexadecimal value of black pixels stand for 0x00, and white ones for 0xff. One may see that different samples of the same traffic (rows) look similar, while images derived from different types of internet traffic differ from one another (columns).

Figure 6. Traffic visualizations of trojan Zeus (the first column), Skype (the second column), Outlook (the third column), backdoor Htbot (fourth column) and botnet Virut (fifth column). Images were created by the authors with the advantage of the tool introduced in [34] on the USTC-TFC2016 dataset.

2.4. The Datasets

The crucial role in all machine learning methods is that of the datasets. They are necessary to perform the learning process of classifiers. They also help compare various approaches. In the case of traffic classification, several open datasets are commonly used in papers under study. The datasets include various types of traffic data: raw traffic, flows and features. Short characteristics of the most frequently employed in the studied papers are listed in Table 1. Figure 7 shows the popularity of particular datasets in the investigated papers.

The group of 28 articles use less popular datasets (Figure 7). These datasets in alphabetic order are as follows:

Table 1. The summary of the most popular datasets used in the studied papers—sorted by the year of creation.

Dataset	Applied in	Format	Size [GB]	Year
KDD Cup 1999	9 articles: [35–43]	features	0.74	1998
NSL-KDD	7 articles: [38,43–48]	features	0.04	2009
ISCX-IDS-2012	6 articles: [49–54]	flow, raw packets	8.42	2012
CTU-13	5 articles: [55–59]	features, raw packets	74.27	2013
UNSW-NB15	5 articles: [31,42,60–62]	features, flow, raw packets	0.55	2015
ISCX VPN-nonVPN	19 articles: [12,49,51,57,63–77]	features, raw packets	28	2016
USTC-TFC2016	11 articles: [34,66,68,69,78–84]	raw packets	3.71	2017
ISCX-IDS-2017	5 articles: [42,49,54,85,86]	features, flow, raw packets	51.1	2018

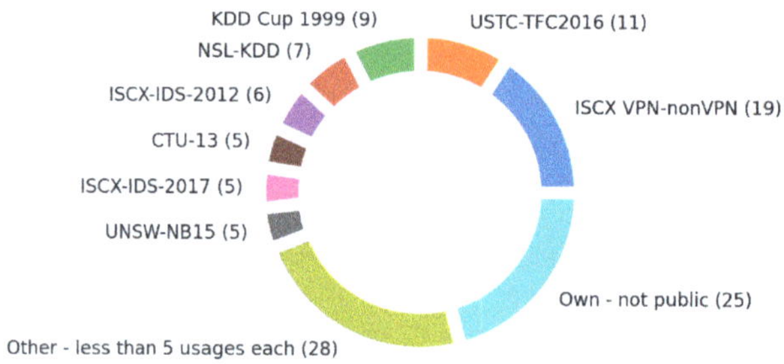

Figure 7. Popularity of the datasets within the reviewed articles. Digits in the brackets stand for the number of all occurrences of each dataset.

BoT-IoT, CAN 2017, CIC-AAGM2017, CIRA-CIC-DoHBrw-2020, CSE-CIC-IDS2018, CTU-Malware, CTU-Mixed, DARPA 1998, DARPA 1999, EDU1, ISCX-Bot-2014, ISCX Tor-nonTor, Malware Capture Facility Project (malware), MAWILab, Mirai-RGU, NIMS, NLANR AMP, NLANR MAWI, SCU-RNE, UPC Broadband Traffic Research group's dataset, VAST 2013 challenge and WRCCDC.

All the datasets contain a certain number of labeled traffic samples. Labels refer to the traffic classes. Classes always belong to one of two groups. In most cases, these groups are malware and benign traffic. One dataset, VPN-nonVPN, contains classes grouped according to the VPN connections within the frames of which the traffic was registered. For details regarding classes, see Table 2.

There are many datasets used in scientific papers for network monitoring and classification. They usually consist of real or simulated data. Some of them are described only in publications but are not available for other researchers for methods evaluation. In this survey, we have selected and compared only those datasets that were used in the research described in the studied papers. A comprehensive analysis that highlights datasets utilized for IDS concepts purposes is in [87]. The paper touched upon the question of pcap and NetFlow differences. It analyzed common datasets concerning the wanted traffic occurrence (not malware), the data format, anonymity, volume of the traffic, type of traffic, labeling, etc. It is crucial to point out that some described datasets are publicly available.

Table 2. The traffic details of the most popular datasets. Datasets are sorted by the number of occurrences in articles.

Dataset	Type of Data	Traffic Details
ISCX VPN-nonVPN	encrypted	14 classes: Browsing, VPN-Browsing, Email, VPN-Email, Chat, VPN-Chat, Streaming, VPN-Streaming, File Transfer, VPN-File Transfer, VoIP, VPN-VoIP, P2P and VPN-P2P [63].
USTC-TFC2016	malware	20 classes: 10 malware and 10 benign traffic. Malware: Cridex (a worm), Geodo (a trojan), Htbot (a backdoor), Miuref (a trojan), Neris (a botnet), Nsis-ay (a botnet), Shifu (a trojan), Tinba (a trojan), Virut (a botnet), Zeus (a trojan). Benign traffic: BitTorrent, FTP, Facetime, Gmail, MySQL, Outlook, SMB, Skype, Weibo, WorldOfWarcraft [34].
KDD Cup 1999	malware	41 traffic features and 22 attacks. The 4 attack categories are: Dos, R2L, U2R and Probing [43].
CTU-13	malware	7 botnets: Neris, Rbot, Virut, Menti, Sogou, Murlo, NSIS.ay in 9 different characteristics: IRC, SPAM, ClickFraud, Port Scan, DDos, FastFlux, P2P, HTTP and compiled and controlled by the researchers [88].
ISCX-IDS-2012	malware	4 attack scenarios: Infiltrating the network from the inside, HTTP DoS, DDoS using an IRC botnet and SSH brute force [87].
NSL-KDD	malware	41 features and 1 label. The 3 groups of features are basic features, content features; and traffic features. Possible attack labels are: DoS, probe, U2R and R2L [45].
ISCX-IDS-2017	malware	16 types of attacks: Brute Force (FTP-Patator, SSH-Patator), DoS/DDoS (DoS slowloris, DoS Slowhttptest, DoS Hulk, DoS GoldenEye, DDoS LOIT), Web Attacks (Brute Force, XSS, SQL Injection), Infiltrations (Dropbox download, Cool disk – MAC), Bugs/Exploits (Heartbleed, Meta exploit Win Vista), Botnet ARES and Port Scan [89].
UNSW-NB15	malware	9 types of attacks: Fuzzers, Analysis, Backdoors, DoS, Exploits, Generic, Reconnaissance, Shellcode and Worms [90].

Having a given dataset, in the case of network traffic classification, one follows the classic machine learning workflow. The dataset is divided into training and testing sets. The former is used to train the neural model, while the latter is used to test it. However, this workflow is preceded by transforming the network traffic data into data samples that the neural model may use. Finally, the evaluation of the results is usually performed using typical measures: precision, recall, accuracy, and F1-score.

2.5. Other Surveys

Because the classification of computer network traffic seems to be a leading trend in the latest research, many surveys have been published that widely discuss this topic. However, each paper examines the scientific problem from a different perspective.

Identifying malware by machine learning techniques was widely investigated in [91]. The paper focused on different types of malware analysis. In addition, one can find a brief description of common malware types. Then, feature selection, classification, and clustering for malware detection were discussed. Finally, the authors mentioned trends of malware development. The survey did not focus on network traffic.

Another paper [92], was written on the subject of traffic classification for quality of service purposes. The article examined many machine learning methods and their advantages for anomaly and intrusion detection. It is important to highlight that the survey also discussed the practicality of the methods. Unfortunately, when it comes to CNNs, there was only one paragraph fully devoted to the history of CNNs.

Unwanted network traffic detection in the Internet of Medical Things (IoMT) was extensively discussed in [93]. Researchers analyzed types of malware attacks, architectures of the IoT environment and taxonomy of security IoT protocols. The latter focused on key management, authentication, access control and intrusion detection. The article stated that future research will be based, among others, on blockchain usage and cross-platform detection.

Work in the topic of Android malware classification was categorized in [94]. In the paper, a novel taxonomy of android malware families was introduced. The interesting part of the paper is a list of Android malware datasets and the surveyed articles' limitations. The paper finished with future directions.

A comprehensive review of malware analysis tools that detect and analyze malware executables is given in [95]. Except for reverse engineering tools as well as memory forensics, packet analysis, detection tools, online scanners and sandboxes were elaborated.

Deep learning techniques were introduced as those that can quickly solve complex problems [96]. The article highlighted the following architectures of deep neural networks (DNNs): feed-forward neural network (FNN), convolutional neural network (CNN), recurrent neural network (RNN) and generative adversarial network (GAN). One section touched on the deep learning private data frameworks. The deep learning threats and attacks, as well as defense techniques, were also examined.

A survey [1] for collecting articles that propose deep learning-based modes to find intrusion in the network data was introduced. In the paper, one can find the taxonomy of deep learning models. In the list of research papers on supervised instance classification models for intrusion detection, there is a brief mention of [34]. The authors, among others, concluded that advances in deep learning methods are noticeable. On top of that, they said that it is often impossible to reproduce some deep learning models, due to the lack of adequate information. The authors also proposed a novel classification of four network traffic datasets.

The utilization of deep learning methods for the purposes of cybersecurity was examined in [97]. The paper singled out types of machine learning, types of deep learning and algorithms for both. Then, deep learning platforms were examined. Finally, the article outlined network attacks. CNN usage by [34] was only mentioned. The detection of cyberattacks to the IoT infrastructure with the advantage of deep learning articles was widely discussed in [98]. The researchers reviewed the IoT architecture, reference models and IoT protocols. Then, they introduced threats against IoT systems and continued with intrusion detection systems (IDS). An interesting IDS taxonomy was also described in the paper. In the CNN section, they mentioned a few articles, but only [34] is related to security, based on network traffic. The next survey dealt with the development and detection trends of unwanted software [99]. In addition, the authors focused on those areas that were omitted by other surveys, e.g., advances in the creation of new types of malware.

The detection of intrusions throughout analysis of images generated from network traffic was outlined in [100]. The paper distinguished classical and neural networks methods. The authors dwelt on deep learning models of the convolutional neural network (CNN), long short-term memory (LSTM), support vector machine (SVM) and hybrid ones. When it comes to only CNN models, they detailed the works of [20,34,56,85]. This review points out that [101] was one of the first concepts of converting network traffic to images. The paper aroused our interest. The work proposed the creation of two-dimensional images that consist of 4 bytes in an IP address structure. The matrix then shows the intensity of traffic in the image representation. Nevertheless, this idea does not refer explicitly to CNN, and is not covered in further sections of this survey.

A systematic literature review highlighted interesting trends in the IoT infrastructure [102]. The researchers concluded that the majority of attacks take place in the network layer. There is a mention of the most popular datasets as well as common attacks.

Network traffic classification algorithms, for instance, based on the port number, statistical characteristics, host behaviors and deep learning, was considered in [3]. The last category encapsulated the following models: stack autoencoder (SAE), CNN, LSTM and deep belief networks (DBN). The CNN section described only three methods, where the CNN input was transformed beforehand—From the one-dimensional data to different one-dimensional data, to two-dimensional data or to three-dimensional data.

While preparing this survey, we decided to develop the concept of describing nothing but the CNN models' usage for traffic classification and malware detection purposes. Contrary to [3], our article consists of 91 papers, i.e., all papers on this topic written until 2021. The conclusion of the CNN chapter in [3] is that the transformation from the 1st dimension to the 3rd dimension is better than other transformations. We believe that it is hard to hypothesize with only a few examples. On top of that, the compared examples utilized a variety of methods. Therefore, the proposed survey inherits and enhances the classification of different forms of transformations.

3. Raw Traffic

The first group of transformation methods works on the captured packets as they come in—the raw traffic. This type of data seems to be the most direct input of the CNN-based deep learning model (CDM), as its considerable merit is the lack of necessity of the prepossessing phase. However, only four research groups decided to base their work on this type of data while crafting the CNN input. These are 1D transformation [78] and 2D input concepts [35,103] (see Table 3). In addition, ref. [79] proposed both a 1D approach and a 2D one.

The one-dimensional entry to CNN is a vector created from raw traffic packets. While transforming packets, Marín et al. proposed the removal of only two attributes of traffic from protocol data units (PDU), i.e., MAC and IP addresses [78]. After that, a fixed size of 1300 bytes is set. It means that all longer packets are trimmed, while smaller packets are zero-padded. In the end, each packet is labeled to be either benign or malware. Finally, vectors are given to the 6-Layer CNN, which is tested on the USTC-TFC2016 dataset. This is an unwanted traffic detection approach.

The idea of wrapping raw traffic packets into the matrix was proposed by Ko et al. to test an 11-Layer CDM [103]. Traffic originated from the EDU1 dataset. This research proposed 200 × 200 bytes images. This is a traffic classification approach.

The scientific concept of raw traffic packets wrapping was further studied by Jia et al. [35]. The traffic images were based on the DARPA 1999. The authors unified the packets length so that each reached 784 bytes. Then, they wrapped the vector to create a matrix of 28 × 28 bytes size. The paper provided the images as an entry for LeNet [19]. The work's aim was to enhance malware detection—in particular, the detection of intrusions.

The following paper, written by Zhang et al., used two different versions of CNN input [79]. The 1300 bytes size vectors are given to two 10-Layer CDM. 2D CNN obtains a traffic matrix, whose size is not revealed. It is important to highlight that vector as an entry to 1D CNN achieves better results than the matrix given to the 2D CDM. In the proposed approach, the packets are left unchanged. The deep learning model works on raw traffic from the USTC-TFC2016 dataset to detect malware. The paper proposed two types of transformations.

Table 3. The summary of articles that are in the raw traffic transformation group.

Article	Input Dimension	Layers	Dataset	Year
[78]	1D	C I P I C I P I FC I FC	USTC-TFC2016	2018
[103]	2D (200 × 200)	C I C I P I C I P I C I P I F I FC I FC I FC	EDU1	2019
[35]	2D (28 × 28)	LeNet [19]	DARPA 1999	2020
[79]	1D, 2D	C I C I P I C I C I P I F I FC I FC I FC	USTC-TFC2016	2019

4. Flows

Flows, grouped raw network traffic according to the same properties, form the second part of network traffic that could be processed while preparing CNN entries. Some papers work on datasets that already consist of flows, whereas others order the raw traffic to pick up all packets belonging to each flow.

4.1. One Dimensional CNN Input

A big group of research articles processed, in CNN-based tools, vectors built from flows. The transformations are basically differentiated in two areas: sizes of input vectors and data manipulations [49,64,66,80–82,104,105] (Table 4).

An extended version of [78] was widely elaborated by Casas et al. for flow vectors in network security, i.e., malware detection [104]. The authors decided to use only two packets and the first 100 bytes of each. This approach was based on statistical calculations. The 3-Layer CDM was tested on the MAWILab dataset.

The same research group, Marín et al., continued to investigate malware detection [105]. The authors checked the same, previously proposed CNN tool with vectors crafted from flows of the CTU-Malware dataset. Then, the tests were extended with USTC-TFC2016 in the next two papers [80,81]. In each, the 3-Layer CNN obtained the flow vector as an input. The articles also tested different machine learning models, also not related exclusively to CNNs.

The objective of traffic classification enhancement was set out by Song et al. in [64]. The authors utilized 8-Layer CDM, which contains an embedding layer (EMB). Pcaps from ISCX VPN-nonVPN were then transformed to flow vectors to test the CNN models. Traffic prepossessing was done according to the idea of Wang Wei et al. [34]. Wang's concept is described in the next subsection. After the normalization process, the one-hot encoding method was used for each byte in the vector, which enlarges up to 255 different values of bytes. The matrix consists of concatenated vectors. To increase the speed and effectiveness of the process, all one-hot-encoded vectors of the matrix are converted into low-dimensional dense vectors.

Hwang et al. widely tested different sizes of CNN input vectors [82]. An example of the vector is 2 by 50 bytes, which means two flows and 50 bytes of each. The researchers used pcaps from the USTC-TFC2016, the Mirai-RGU and their own datasets. The introduced 11-Layer CDM to deal with malware detection, especially anomaly detection problems.

The proposed concept of Chen et al. can determine whether traffic belongs to any of the known classes [66]. The idea requires unchanged flows that form vectors, which later become the input of the 16-Layer CDM. This function enhances the capability of the detection of yet-unknown traffic. The tool was tested on two datasets: USTC-TFC2016 and ISCX VPN-nonVPN. This article is an example of both approaches: traffic classification as well as malware detection.

Flows were also used by Chen et al. to form a 1D-CNN input [49]. Vectors of the sizes of 784 bytes were created due to the idea of [63]. Three different datasets—ISCX VPN-nonVPN, ISCX-IDS-2012 and ISCX-IDS-2017—were used to check the proposed 5-Layer CDM's effectiveness. The CDM classifies traffic.

4.2. Two-Dimensional CNN Input

Contrary to the previously described approaches, the two-dimensional CDM input requires increasing the dimensionality of the traffic data. Flow wrapping seems to be one of the leading trends of traffic manipulations within discussed CNN entries [34,50,55–58,60,61,68,106] (Table 4). The practical concept, which started in 2017, wraps the network traffic data into the matrix [34].

The very first article that fulfilled the concept of [2] is the paper written by Wang Wei et al. [34]. This paper seems to be the first practical approach to utilize CNNs to process network traffic. Ref. [34] uses 6-Layer CNN to detect malware in the USTC-TFC2016 dataset. The authors decided to give the CDM a matrix of 28 bytes per 28 bytes. The process of creating the image (matrix) is the following: raw traffic packets are aggregated into flows or sessions, and then data are anonymized. The next step is to trim the flows to 784 bytes and 'wrap' the vector so then one has a matrix of 28 × 28 bytes, visualized as a gray level image. The authors decided to share the tool used to create the matrix. While aggregating the packets into flow, one can choose one of the four versions of the process:

- Trim according to flows with all network layers;
- Trim according to flow with only Layer 7 (L4+);
- Trim according to session with all network layers;
- Trim according to sessions with only Layer 7 (L4+).

The choice of the only L4+ could have been placed in Section 6 of our survey. Nevertheless, the remaining two aggregating methods are also widely discussed in the paper, and this indicator makes the paper ideal for this section.

Moskalenko and Moskalenko proposed a typical flow wrapping to check 2-Layer CNN for malware detection [55]. To create the matrix, raw packets are aggregated into flows. Then, 784 bytes of sequential flows are taken to create a wrapped vector—the matrix of 28 × 28 pixels. The last step is to normalize the matrix values (pixels' brightnesses) in the range [0,1]. The CDM is tested by pcaps from the CTU-Mixed and CTU-13 datasets.

Flow wrapping is also utilized to detect malware—more specifically, botnets by Taheri et al. [56]. The flows from the CTU-13 dataset are transformed into grayscale images of 28 bytes × 28 bytes and delivered to the entry of the DenseNet CNN [23]. It is important to underline that all layers of flows are utilized.

Zhou et al. delivered the following sizes of session images to the entry of the 5-Layer CDM: 16 × 16, 20 × 20, 28 × 28, 32 × 32 [106]. The CDM detecting botnets was tested on the ISCX-Bot-2014 dataset. The raw traffic packets from the dataset were aggregated into flows.

The transformation concept of [34] was utilized in the malware detection article of Wang et al., which introduced the 5-Layer CDM [60]. The tool tests were conducted on captured packets from the UNSW-NB15 dataset. From the raw traffic, sessions were cropped. Finally, the CNN input was a 28 by 28 bytes matrix.

The same transformation to 32 by 32 bytes images was used in the research in which malware detection was a theme [57]. Huang et al., in their article, tested the 7-Layer CDM's quality against sessions with all layers from the CTU-13 and ISCX VPN-nonVPN datasets.

The novel transformation of flow via one-hot encoding was proposed by Wang et al. to test 5-Layer CDM (named HAST-I) [50]. The CNN tool was introduced to detect malware. The network traffic came from the DARPA 1998, and ISCX-IDS-2012 datasets. In the beginning, raw packets were aggregated into flows. Then, during thorough tests, flows were trimmed to either 600 or 800 bytes. Later, one-hot encoding transformed each byte in the vector into a vector. All the vectors were transpositioned and then concatenated so that a matrix was formed. The smaller, 256 × 600 bytes image achieved the best classification results for the ISCX-IDS-2012 dataset, while the 256 × 800 bytes were ideal for DARPA 1998.

A few different traffic transformations for malware detection purposes were based on three CDMs [61]. Out of the proposed tools, only one was exclusively CNN. The different CDMs of Millar et al. were given three different types of entries: 50 byte flow vector, 24 traffic features and flow wrapping. The first two methods tested non-CNN based models, whereas the last type of entry was a 2D flow image, which was given to the CNN model. In these flows' images, each pixel represents a byte of data in the network. A row of the image stands for the next packet in the flow. In each field, the value means the packet filling. The CNN model was tested on UNSW-NB15.

Moskalenko et al. investigated flow wrapping inherited from [34] to detect malware [58]. It used LeNet [19]. The tests input was taken from two datasets: CTU-Mixed and CTU-13.

A simple flow wrapping method to 28 × 28 bytes matrices was used during the image generating process [68]. Li et al. proposed traffic classification for 9-Layer CDM, which was tested on data from the ISCX VPN-nonVPN and USTC-TFC2016 datasets.

4.3. Various Dimensionalities

A few papers verified the various dimensions of CNN inputs built from flows [12,63,65,67,69,83,107,108] (see Table 4).

Flow vectors are the input of the 6-Layer CDM proposed by Wang et al. [63]. Data are transformed as in [34] until the 784-byte vectors are formed. In this approach, the CDM deals with the ISCX VPN-nonVPN dataset. Additionally, the researchers mentioned that the proposed method is compared with 2D transformation. The interesting outcome achieved by the authors was that the 2D approach achieved worse results than the 1D approach.

A thought-provoking transformation of network traffic into the third dimension to classify traffic was proposed by Ran et al. The researchers utilized the 8-Layer CDM [83], and then tested it on pcaps from USTC-TFC2016. The 3D model was built in four steps. The first one identified flows within packets. The next step extracted a chosen number of bytes from each flow. The third step concerned trimming all packets to one fixed size. Longer packets were trimmed, whereas shorter ones were padded with zeros. Then, each packet was transformed into a 2D matrix with the usage of one-hot encoding. To create a 3D image, all 2D images of the same packets had to be put together.

Flow vectors were also used as a CNN entry in the research articles of [107,108]. Aceto et al. utilized three types of CNN entries. Two methods, based on forming 784-byte vectors, were taken from [63]. The third method proposed a matrix of traffic features as a CNN entry. We describe this 2D concept in detail in Section 8. The articles utilized 6-Layer CDMs from [34,63]. Flows for this traffic classification approach were taken from the authors' own dataset.

While discussing in detail the wrapping flows, one should mention the approach of Cui et al., which improved it slightly, with the advantage of the sessions' weights [67]. On top of that, 6-Layer CDM of [63] was checked by traffic that originates from ISCX VPN-nonVPN. Additionally, during flow transformation, unrelated SNMP, DNS and ARP sessions were diluted, whereas valid sessions' weights were increased. The paper's aim was to classify traffic. It is important to underline that the paper introduced a 5-Layer CDM, CapsNet. The core part of the paper, which is a 2D transformation model, achieved a better outcome than the 1D classification.

The next paper of He and Li distinguished two types of traffic from the ISCX VPN-nonVPN dataset and touched upon flow wrapping [65]. The raw traffic packets were aggregated into sessions. Then, for non-VPN traffic, the first 90 non-zero payloads of flows were taken. In the second group, the VPN traffic one, the first 20 non-zero payloads were chosen for further processing. In both groups, the tool, introduced in [34], was used. Additionally, all DNS and NetBIOS names packets were erased. The authors decided to remove also the three-way handshake packets. Then, traffic images of 28 × 28 byte size were provided to the 5-Layer CDM for traffic classification. The paper also proposed a 1D model, which works on 784-byte vectors, and compared its results with the CNN of [70].

Yet another work on the topic of traffic classification slightly modified flow wrapping [12]. The experiment transformed the traffic of the first 20 packets of each flow to not only 28 by 28 bytes images, but also other square images. These values tested by

Pacheco et al. were not explicitly specified. In this paper, CNNs from [34,63] were tested on traffic captured during research as well as on the VPN-NonVPN dataset.

Recently, the concept of wrapping flows to create CNN input images was also utilized by Chen et al. for malware detection purposes [69]. The tests were conducted on pcaps originated from the following datasets: VPN-nonVPN and USTC-TFC2016.

Table 4. All research works of the flow transformation method

Article	Input Dimension	Layers	Dataset	Year
[104]	1D	C I FC I FC	MAWILab	2019
[105]	1D	same as [104]	CTU-Malware	2019
[80,81]	1D	same as [104]	USTC-TFC2016 and CTU-Malware	2019 and 2020
[64]	1D	EMB I C I C I C I C I P I FC I FC	ISCX VPN-nonVPN	2019
[82]	1D	C I P I C I P I FC I FC I FC I FC I FC I FC I FC	USTC-TFC2016 and Mirai-RGU and own	2020
[66]	1D	C I C I P I C I C I P I C I C I C I P I C I C I C I FC I FC I FC	USTC-TFC2016 and ISCX VPN-nonVPN	2020
[49]	1D	C I P I C I P I FC	ISCX VPN-nonVPN, ISCX-IDS-2012 and ISCX-IDS-2017	2020
[34]	2D (28 × 28)	C I P I C I P I FC I FC	USTC-TFC2016	2017
[55]	2D (28 × 28)	C I C	CTU-Mixed and CTU-13	2018
[56]	2D (28 × 28)	DenseNet [23]	CTU-13	2018
[106]	2D (16 × 16, 20 × 20, 28 × 28 and 32 × 32)	C I P I C I P I FC	ISCX-Bot-2014	2018
[60]	2D (28 × 28)	C I P I C I P I FC	UNSW-NB15	2018
[57]	2D (32 × 32)	C I P I C I P I C I C I FC	CTU-13 and ISCX VPN-nonVPN	2018
[50]	2D (256 × 600 and 256 × 800)	C I P I C I P I FC	DARPA 1998 and ISCX-IDS-2012	2018
[58]	2D (28 × 28)	LeNet [19]	CTU-Mixed and CTU-13	2020
[68]	2D (28 × 28)	C I P I C I P I C I P I C I P I FC	ISCX VPN-nonVPN and USTC-TFC2016	2020
[63]	1D, 2D (probably 28 × 28)	C I P I C I P I FC I FC	ISCX VPN-nonVPN	2017
[83]	1D, 2D(NS),3D	C I C I P I C I C I P I FC I FC	USTC-TFC2016	2018
[107,108]	1D, 2D (28 × 28)	same as [34,63]	Own	2018
[67]	1D, 2D (28 × 28)	same as [63]	ISCX VPN-nonVPN	2019
[65]	1D, 2D (NS)	C I P I C I P I FC	ISCX VPN-nonVPN	2020
[12]	1D, 2D (28 × 28 and other square images)	same as [34,63]	own	2020
[69]	1D, 2D (32 × 32)	C I P I C I P I C I FC	ISCX VPN-nonVPN and USTC-TFC2016	2020

5. Payload Extracted from Raw Traffic

The next possible transformation is based on the extraction of chosen payloads from the raw traffic. For example, according to Figure 2, one can remove the headers of Layers 2, 3 and 4. The most popular idea is to remove the L4 header and form the CDM entry from only the L4+ payload.

5.1. One-Dimensional CNN Input

This section outlines those papers that provided the CNN model a vector, formed with the advantage of payloads and header manipulations [70–73] (Table 5). In this section, extractions are made from raw traffic packets.

An interesting concept of Lotfollahi et al. is based on 1D vectors, which reach the length of 1500 bytes [70]. In the first step, headers of L2 are removed. On top of that, there are some changes in the L4 layer. According to the design, normally shorter than TCP, UDP packets are padded with zeros to reach the TCP length. The next step is to remove not only the entire three-way handshake communication, but also the DNS queries. Finally, the vectors reach the length of 1500 bytes, in which each byte value is normalized. The paper used the ISCX VPN-nonVPN dataset to test the 6-Layer CDM. This is typical traffic classification work.

The same transformation idea, like in the previous work, was inherited to classify traffic [71]. The authors, Akbari and Tahoun, utilized a federated learning model, called the model-averaging technique, and created 3-Layer and 5-Layer CDMs. The publication's models were tested on the USTC-TFC2016 dataset.

5.2. Two-Dimensional CNN Input

The concept of creating traffic images from the extracted payload of raw traffic packets is a next form of 2D-CNN input. This transformation was carried out in the following group of scientific investigations [73,109–112] (see Table 5).

In the research of He and Shi, images were generated with the advantage of wrapping the payload of raw traffic [109]. It seems that the authors chose the L4+ payload, so they removed the L2, L3 and L4 headers. The researchers aimed to identify traffic, especially SSH applications. The used 5-Layer CDM was tested on traffic from the article's own dataset. The authors informed that the CNN input is a 28 by 28 bytes image.

Li et al. removed the L2 headers and modified the L4+ headers to form the CNN input [110]. The modification in L4 means unifying the length of TCP and UDP headers. On top of that, all duplicated and empty packets (with no payload) were erased. In this research paper, the transformation of packets to 30 × 30 byte matrices was utilized in order to classify traffic for virtualization purposes with the 5-Layer CNN. Tests were conducted on traffic captured by the authors.

The payload of L4+ was extracted from the raw traffic packets to create a 2D image [111]. The creation of the image required taking 10,000 packets from each application traffic captured in the UPC Broadband Traffic research group. Then, payloads of each packet were divided by four to constitute one pixel of a future image. The sizes of all application's traffic were readjusted to the following: 36, 64, 256 and 1024 pixels. In the case of a smaller number of payload samples, the images were padded with zeros. The paper of Lim et al. used 4-Layer CDM and also ResNet to classify the captured traffic.

A similar concept of choosing only L4+ payload while creating images was applied by Xue et al. [112]. The transformation's last step was to wrap vectors in order to create 2D images. The paper utilized six different CNN networks: ResNet [24], VGG16, VGG19 [25], Inception V3 [22], Xception [26] and MobileNet [27]. Their task was to work on traffic classification issues. Models were tested on traffic captured within the authors' research.

5.3. Various Dimensionalities

Papers in this section compare a few transformations methods (see Table 5).

Only the transport layer's payload (L4+) was taken from the raw traffic to form the CNN input [72]. The authors Xu et al. tested four sizes of input data: 400, 625, 784 and 900 bytes, which were later left as a vector or transformed to an image. Vectors are the input to the first 8-Layer CDM. Moreover, the entry of the second 12-Layer CDM is a square matrix. Four variants were investigated: 20 by 20, 25 by 25, 28 by 28 and 30 by 30 bytes. Traffic classification tests were extensively conducted with the advantage of pcap files of the ISCX VPN-nonVPN dataset. Due to the dataset choice, this was a typical traffic classification approach. The CNNs that work on vectors outperformed those models that deal with matrices.

The paper of Zhang et al. applied three versions of transformations, i.e., to the vector (1D), to the matrix (2D) and to the cubic form (3D) for traffic classification purposes [73]. The one-dimensional CNN input is a 1456 byte vector that consists of the raw traffic payload of L4+. The second dimension was implemented by wrapping a different initial vector (1521 bytes) into a 39 by 39 byte matrix. The third dimension was also created by wrapping. An initial vector (1452 bytes) was changed into 22 by 22 by 3 bytes RGB colored cubic forms. The network traffic was taken from the ISCX VPN-nonVPN dataset as well as their own dataset. For this transformation, the paper used 5-Layer CDM. The results of the experiments indicate the matrix as the input for achieving the highest classification results.

Table 5. Works that extract payload of raw traffic.

Article	Input Dimension-Payload	Layers	Dataset	Year
[70]	1D -L3	C I C I P I FC I FC I FC	ISCX VPN-nonVPN	2019
[71]	1D -L3	C I C I P I FC I FC and C I P I FC	USTC-TFC2016	2019
[109]	2D (28 × 28) -L4+	C I P I C I P I FC	Own	2018
[110]	2D (30 × 30) -L4+	C I C I C I C I FC	Own	2019
[111]	2D (6 × 6, 8 × 8, 16 × 16 and 32 × 32 [pixels]) -L4+	C I C I P I FC and ResNet [24]	Own	2019
[112]	2D (128 × 128) -L4+	ResNet [24], VGG16, VGG19 [25], Inception V3 [22], Xception [26] and MobileNet [27]	The dataset of the UPC Broadband Traffic Research Group	2020
[72]	1D, 2D (20 × 20, 25 × 25, 28 × 28 and 30 × 30) -L4+	C I C I P I C I C I P I FC I FC	ISCX VPN-nonVPN	2020
[73]	1D, 2D (39 × 39), 3D -L4+	C I C I FC I FC I FC	ISCX VPN-nonVPN and own	2020

6. Payload Extracted from Flows

This section is entirely devoted to the transformations of raw traffic, which extract payloads from grouped packets, i.e., flows. This section discusses 13 research papers.

6.1. One Dimensional CNN Input

Another group of articles proposed giving the CNN model an extracted payload of flows [51,52,113,114] (see Table 6). All papers worked on L4+ payloads, which means that headers from L2, L3 and L4 were decapsulated.

Zeng et al. created 900 byte vectors from flows and used them to form a 5-Layer CDM entry [51,52]. While creating the vectors, TCP and UDP headers were removed. In [51] the malware detection model's performance was checked with data from ISCX VPN-nonVPN and ISCX-IDS-2012. The latter paper detected malware in the vehicular ad hoc network (VANET) by testing the CNN model on the network traffic of the ISCX-IDS-2012 dataset and their own simulated dataset, NS-3 VANET. The datasets contained pcap files from which flows were aggregated. In both papers, the main concept was a hybrid deep learning model, which obtains 30 byte by 30 byte images. These flow images were built according to the concept of [63]. As the hybrid models do not fulfill the requirements of this CNN-based survey, the two papers [51,52] are not described in the Section 4.

The next paper of Wang et al. also introduced a CNN input vector, that is, the L4+ payload [113]. The deep learning model's entry reached the size of 200 bytes. The work introduced a few models for traffic classification. The solel CNN was App-CNN, which is a 5-Layer CDM. Flows were taken from the researchers' own dataset.

Similarly, in the approach of Wang et al., CNN's entry is a fix-length vector, only consisting of the flow payloads from L4+ [114]. Then, only the top hundreds of flow bytes are stuck into the vector. Three different deep learning models, in which one of them is solely a CNN, were investigated. The 5-Layer CDM classifies traffic. Additionally, it was tested on the authors' own dataset.

6.2. Two-Dimensional CNN Input

Some papers decided to process the payload of grouped raw traffic—flows [20,74,84,115–119] (Table 6). After the extraction step, which is the common part for all papers, the changes within these concepts arise. The biggest differences are mainly the layer choice as well as the selection of headers for removal.

A matrix of 32 bytes × 32 bytes was proposed to examine the 6-Layer CDM of Ma and Qin in the work [115]. The input was formed from 1024 bytes of the L4+ payload. The flows were caught by the authors. According to the paper, the first 1024 bytes contain crucial information.

In the next paper, Zhao and Chen used the L4+ payload while transforming network traffic [116]. On top of that, much larger unidirectional flow images of 87 bytes per 87 bytes were used to classify the traffic of smartphone applications. The researchers tested the

5-Layer CDM model on their own dataset. During the preprocessing phase, all tiny flows with less than two packets were removed. After that, five flow vectors, 1500 bytes each, were converted into a 2D image of the mentioned size.

Wrapping the L2 payload of flows to create an image was the dimension transformation used by Zhang et al. [84]. The paper dealt with the malware detection problem with the advantage of different CNN models: LeNet [19], AlexNet [21] and VGGNet [25]. Tests were accomplished on the USTC-TFC2016 dataset. Each input image had 28 by 28 bytes.

Removal of the L4 header, and so choosing the L4+ payload of flow, was applied by Zhou and Cui [74]. Additionally, the authors examined the usefulness of Alexnet [21] to classify traffic from the ISCX VPN-nonVPN dataset. The usage of the datasets means that the authors were dealing with the encrypted payloads.

The next article, written by Feng et al., inherited the idea of [34] of wrapping only the L4+ payload of flow and widely utilized it for traffic classification purposes [117]. The paper used 6-Layer CDM. The tests of the model were based on flows coming from the DARPA 1998 dataset.

A different idea of choosing the L3 payload was tested by Zhao et al. While transforming flows from the Malware Capture Facility Project (malware samples) and their own dataset (benign samples), researchers decided to focus on the first 32 packets of each flow, and the first 512 bytes of each packet [119]. Then, chosen data werer saved as a matrix of 32 by 512 bytes. Later, after the normalization process, the final matrix was reshaped to a 128 by 128 byte size. If anything was smaller than the desired size, they were padded with zeros. The matrices were given as an input to the proposed 7-Layer CDM network. The paper also utilized an interesting metric regularization term, which enforced the model to learn more discriminative features. This feature impacted the classification so that the results were more precise.

A novel approach of the L2 payload of flow transformation was utilized by Saleh and Ji. The authors constituted images by one of the five possible mappings of flow vector (1D) into a 2D matrix. Prior to that, pcaps from the authors' own dataset were aggregated into flows [20] for the purpose of network traffic classification. Then, all invalid connections were removed. Matrices of 17 by 17 bytes size or 25 by 25 bytes size were an entry to the VGG-16 CNN [25], the 16-Layer CNN model. The authors proposed the following mappings: linear, diagonal, waterfall, center spiral and edge spiral. The first method is frankly wrapping flows. The diagonal mapping starts by placing bytes from the top left corner and then arranges them diagonally. The waterfall method is said to imitate nature: a water stream pulling into a cliff. Here, the first byte is also in the top left corner. The second byte moves along the diagonal, the third one to the left side, the fourth up and again along the diagonal and so on. The center spiral starts from the central position and locates the next bytes around the previous ones. The last mapping is the center spiral in the reverse order. Despite attempts of various mappings, the classic, linear one—the flow wrapping—achieved the best results.

6.3. Various Dimensionalities

Research works described in this section deal with two various dimensionalities of CNN input (Table 6).

Android traffic was transformed into images [118] according to the method of removing 24 bytes, i.e., the header of L4. The authors, Yunjie et al., decided to enlarge images, as they used 1024 bytes. Thus, the images achieved 32 by 32 byte sizes. The interesting part of the algorithm was the step where third party traffic was removed. The paper adds to a growing corpus of malware detection research. The 7-Layer CDM was used to find unwanted traffic within the CIC-AAGM2017 dataset. On top of that, the authors dealt with two various dimensions of CNN input. The 2D method outperformed the 1D concept.

In the following approach, the one dimension is changed into three dimensions to better detect unwanted traffic [31]. Consequently, the CNN input is three dimensional. The paper of Millar et al. proposed a segmented CDM of 1D- and 2D-CNNs. Additionally, the

1D-CNN and separable 2D-CNN models were introduced. Their quality was tested on 3D flow images generated from the UNSW-NB15 dataset. The 1D CDM was given a 2D flow image. At the beginning of image creation, 97 bytes of flow were chosen. A total of 47 bytes were taken from the flow's header, whereas the remaining 50 were from the payload. Then, an additional nine flows were added, so the 2nd dimension was achieved by the flow wrapping concept. The third dimension was built with the advantage of one-hot encoding. While comparing the separate models of the 1D- and 2D-CNNs, one can see that the application of the one-dimensional transformation resulted in higher effectiveness.

Table 6. Papers that belong to the group extracted payload—flows.

Article	Input Dimension-Payload	Layers	Dataset	Year
[51]	1D -L4+	C I P I C I P I FC	ISCX VPN-nonVPN and ISCX-IDS-2012	2019
[52]	1D -L4+	C I P I C I P I FC	ISCX-IDS-2012 and own	2019
[113]	1D -L4+	C I P I C I P I FC	Own	2020
[114]	1D -L4+	C I P I C I P I FC	Own	2020
[115]	2D (32 × 32) -L4+	C I C I C I P I FC I FC	Own	2017
[116]	2D (87 × 87) -L4+	C I P I C I P I FC	Own	2018
[84]	2D (28 × 28) -L2	LeNet [19], AlexNet [21] and VGGNNet [25]	USTC-TFC2016	2020
[74]	2D (28 × 28) -L4+	Alexnet [21]	ISCX VPN-nonVPN	2020
[117]	2D (28 × 28) -L4+	C I P I C I P I FC I FC	DARPA 1998	2020
[119]	2D (128 × 128) -L3	C I P I C I P I FC I FC I FC	Own (benign traffic) and Malware Capture Facility Project (malware traffic)	2020
[20]	2D (17 × 17, 25 × 25 and 49 × 49) -L2	VGG [25]	Own	2020
[118]	1D, 2D (32 × 32) -L4+	C I P I C I P I FC I FC	CIC-AAGM2017	2020
[31]	2D (97 × 10), 3D -L4+	C I C I P I C I P I FC I FC I FC and segmented CNN: C I C I P I C I P I FC with C I C I P I C I P I FC I FC	UNSW-NB15	2019

7. Traffic Features

The papers collected in this chapter proposed a transformation of the features of the network traffic to the CNN entry. The difference between this chapter's concept and the next one is that here, the research groups utilized only those datasets that consist of traffic features (e.g., KDD Cup 1999). In contrast, in the next chapter, the papers not only created interesting CNN entries, but also proposed feature extraction techniques.

7.1. *One-Dimensional CNN Input*

The transformation of chosen network traffic features into a vector that later becomes the CNN deep learning model input is a core part of Refs. [36–38,120–122]. These papers used network traffic datasets with explicitly traffic features, or extracted them from flow, pcap based datasets. On top of that, four works combined traffic features with the traffic payload [53,75,123,124].

A simple vector of features was given as an entry to different CDMs, which were used to detect unwanted traffic, e.g., intrusions [36]. The solely CNNs which were used were 3-Layer CNN, 4-Layer CNN and 5-Layer CNN. The authors, Vinayakumar et al., chose the KDD Cup 1999 dataset to test the proposed models.

The same transformation was used by Vinayakumar et al. in a work that focused on SSH traffic identification [120]. The paper concept was ten different deep learning models. The most interesting are two CNN models, i.e., 3-Layer and 6-Layer. The vector consisted of flow features, for instance, protocol, duration of flow, maximum packet, etc. The article made use of publicly available datasets: NLANR AMP, NLANR MAWI and NIMS.

The CNN model is given a vector, which consists of Can 2017 dataset features, which were collected from in-vehicle on-board diagnostics [121]. The article of Lokman et al. considered malware and intrusion detections with the advantage of 4-Layer CDM.

The 6-Layer CDM, to detect unwanted traffic, was also tested with a vector of network traffic features [37]. The traffic samples in Manimaran et al. research were taken from the KDD Cup 1999 dataset.

Another paper, written by Liu and Zhang also proposed 1D input of traffic features to improve malware detection [38]. Here, the 5-Layer CDM was tested on data from the KDD Cup 1999 and NSL-KDD datasets.

The same transformation was performed by Susilo and Sari on the BoT-IoT dataset [122]. It appears that the 5-Layer CDM input was the vector of features. The paper showed the malware detection approach.

The discussed transformation approach was extended by combining ten network features with additional traffic payloads [53]. The researchers, Cui et al., decided to test GoogLeNet [22] on the ISCX-IDS-2012 dataset. This work widely investigated malware as well as intrusion detection.

A combination of network traffic features with flow payloads was classified by a few AI models [123]. The paper of Zhao et al. used 6-Layer CDM ([63]) and other classical methods, e.g., random forest. The CNN was given a vector with 29 attributes, where 12 were statistical features, 16 byte values, and the last one was a port number. The statistical features were the payload size (5 features) and the packet length (7 features). The byte values were 16 bytes of the payload. The model was tested on the researchers' own dataset, which consisted of flows.

The trend of combining traffic payload with its statistical features continued in the article of Dong et al. [75]. Firstly, all unneeded packets, such as DHCP and NetBios, were removed from the pcap files. The second step was to aggregate raw traffic packets with respect to the sessions. After removing all retransmission flows and those related to a particular application, each packet was trimmed to the set size. Then everything was joined into one vector. The last step was the normalization of the vector's data. In this paper, two 6-Layer CNNs from different articles [63,70], were utilized. Both CDMs aimed to classify encrypted traffic. The input of CNNs was crafted from the ISCX VPN-nonVPN dataset.

The idea of Yang et al. was to create the CNN input in four steps: payload extraction, inter-arrival time calculation, truncating/padding process and normalization process [124]. The 8-Layer CDM tested this kind of an payload and time feature input. Flows were originated from the WRCCDC dataset. This article is an example of a traffic classification approach.

7.2. Two-Dimensional CNN Input

The next method of CNN input transformation is vector of features wrapping [39–48,76,85,125]. This idea changes the form of input data representation from a vector to a matrix, similar to that done with flows. The combination of both traffic features and payload was also proposed in [126].

Vector of features wrapping was first introduced in the work of Liu et al., which was focused on malware detection and intrusion detection purposes [39]. The paper proposed 32 by 32 byte matrices to be given as an entry of LeNet [19]. CNN was tested on KDD Cup 1999, which consisted of feature vectors. To create feature wrapping images, the authors chose 1024 bytes from feature vectors, which were later transformed into images.

A novel transformation of network traffic was proposed by Liu et al. in their work focused on malware detection and the intrusion detection challenge [40]. For this task, the paper used two CNNs: ResNet 50 [24] and GoogLeNet [22]. Network traffic was taken from the NSL-KDD dataset. The paper introduced an innovative method to create input for CNN images. Firstly, all symbolic features from the dataset, i.e., protocol type, flag and service, were converted into binary vectors (one-hot encoded). All continuous features were normalized to scale [0–1]. After that, the authors discretized the scaled continuous value into ten intervals. The next step was to use one-hot encoding again. This time, the method ordered intervals into binary vectors. The vector with 484 features was then changed into a greyscale image. Eight bytes were changed into one pixel. Finally, the data became an image of 8 bytes by 8 bytes in size. If necessary, the images were padded with zeros. It is important to draw attention to the fact that the dataset consisted of vectors of 41 network traffic features. To sum up, vectors of 41 traffic features were transformed into 2D images.

Replicating vectors of features as a 11-Layer CDM entry was proposed by Naseer and Saleem in their work which dwelt on traffic classification malware detection, mainly intrusion detection [41]. The tool was tested on transformed features vectors from the KDD Cup 1999 dataset. The vector contained 41 features. Three symbolic features: 'protocol_type', 'service' and 'flag' were converted to become a quantitative date. Then, whole vectors were replicated three times, and five chosen features were concatenated. These actions created 128 features vectors. Later, the vectors were again replicated (probably eight times) to create an image 32 bytes by 32 bytes—2D matrices. These matrices then became greyscale images, which were the CNN tool entry.

Malware and intrusion detection, more precisely anomaly detection, were closely investigated [42]. Kim et al. utilized the GoogleLeNet CNN model and tested its usefulness for the topic with the advantage of three datasets: KDD Cup 1999, UNSW-NB15, and ISCX-IDS-2017. This means that they dealt with vectors of network traffic features, flows and raw packets. While processing the dataset, the authors normalized numerical data with the min–max normalization algorithm. Then they transformed categorical features into numerical ones with the advantage of one-hot encoding. Later, all data were encoded to a greyscale vector and reorganized into a greyscale image. Finally, the created images were of the following sizes:

- 12 by 12 bytes images for KDD Cup 1999.
- 14 by 14 bytes images for UNSW-NB15.
- 9 by 9 bytes images for ISCX-IDS-2017.

A novel transformation of the feature vector into an image was widely examined Mohammadpour et al. [44]. The first step was taken to convert nominal attributes into discrete attributes with the advantage of one-hot encoding. This action established the number of attributes to 122. Then, the authors removed one of the 122 features. The remaining features were normalized in the range of [0, 1] by max–min normalization. Finally, the 121 feature vector was wrapped to a 2D matrix. The paper used 7-Layer CDM to deal with the NSL-KDD dataset traffic. The authors' aim in this paper was to develop intrusion as well as malware detection issues.

The same transformation of a feature vector into a 2D matrix was introduced in the paper of Wang et al., which was fully devoted to the detection of unwanted traffic in the network [43]. The authors checked the usefulness of the proposed 9-Layer CDM and LeNet [19], on vectors of network features from the KDD Cup 1999 and the NSL-KDD datasets.

Unchanged transformation from [44] was used to test the 4-Layer CDM of Hu et al. The introduced tool had to detect malware as well as intrusions in wireless networks. In the CDM, there is a split convolution module (SPC), which is a special layer to minimize the problem of an unbalanced dataset [46]. The paper made use of the NSL-KDD dataset.

The researchers Li et al. decided to utilize randomly repeating features to enhance traffic images [47]. The paper focused on 9 by 9 bytes, 9 by 10 bytes, 10 by 10 bytes and 11 by 11 bytes matrices. The authors decided to find malware, especially intrusions in the network, with 7-Layer CDM. The idea was tested with the advantage of the NSL-KDD dataset.

Network traffic transformation proposed by Mohammadpour et al. [44] was continued [85]. This time, the authors detected malware and intrusions with 4-Layer CDM on the traffic from the ISCX-IDS-2017 dataset. The model consists of a layer known as a mean convolutional layer (MC). This layer enhances classification so that all anomaly samples are separated during computing. Moreover, this helps in learning the prediction error filters, which can generate low-level abnormal features.

The same idea of 2D transformation was utilized [125], where Zhang proposed a 6-Layer CDM to deal with malware detection. The vector of features images was 32 × 32 bytes. They were formed from the KDD Cup 1999 dataset.

To detect malware as well as detect intrusions, Pham et al. utilized two methods of network traffic transformations [76]. The first one, based on histogram creation, was inherited from [77]. The second one, for the purpose of image creation, multiplied the packet's length by the normalized delivery time. This was done in order to differentiate two packets of the same length, collected at different times. Thanks to multiplication, the same length packets were stuck in different parts of the image, not disturbing the sequence pattern. The next step was to reduce the multiplication outcome to the image size in order to achieve data within the image's size—the so-called modulo operation. The researchers created 30 by 30 pixel images for the CSE-CIC-IDS2018 dataset traffic and a 300 by 300 pixels matrix for ISCX VPN-nonVPN. The used CDM was a 9-Layer one.

A novel sliding window based approach was introduced in [126] for traffic classification. Li et al. used 7-Layer CDM, which was later evaluated by flows from their own dataset. The CNN input was an image created in a few steps. At first, the flow traffic was divided into segments that corresponded to particular applications activities. Then, each segmented traffic stream was represented by a matrix and a vector. The matrix consisted of a number of packets received in the chosen time unit. The vector held frequency-domain features of the traffic.

The same network data transformation, as in [40], was applied by Su et al. The authors utilized the neuro evolution of augmenting topologies algorithm to find the optimal CNN architecture [48]. As there was not one chosen CNN for malware detection purposes, this paper will not be covered in the summary table at the end of this section. Tests of different CNNs were conducted on the NSL-KDD dataset.

7.3. Three-Dimensional CNN Input

A few articles proposed CLM models that require a 3D entry [127–130] (see Table 7).

Probability distributions of the network flow sequence to images were converted [128]. To fulfill the task, reproducing kernel Hilbert space (RKHS) embeddings were used by Chen et al. This method is said to create a neat image representation of a (conditional) distribution. Network flows were originated from the researchers own dataset. The article aimed to develop traffic classification methods with the advantage of a 7-Layer CDM.

There is a traffic classification in terms of QoS and a security approach in which CNN input is an RGBA image [127]. The article used four predefined CNNs: LeNet [19], AlexNet [21], ConvNet and GoogleNet [22]. The network traffic in the form of pcaps was taken from the researchers' own dataset. Raw traffic packets were firstly aggregated into flows. Then, four features—size (s), interarrival_time (t), protocol (p) and direction (d)—were taken. Merged together, the following vector of the packet's feature was formed: [s, t, p, d]. Later, vectors with the packets' features formed a flow matrix, so that each matrix element was a vector. Salman et al. highlighted that a feature vector of four elements can become an RGBA pixel [127]. They followed this idea and created RGBA images. The size of each was firmly connected to the mode of the model: offline vs. online. The online mode worked on smaller images with 16 packets of the flow, whereas the offline was capable of processing 28 packets of the flow.

Volumetric colored images that represent the amount of the data captured within a chosen time was also utilized as a CNN entry [129]. The concept assumed a colored input of 656 by 874 pixels. This input was built from the dataset of De Schepper et al. and tested 8-Layer CDM in terms of traffic classification accuracy.

The concept of building a 3D entry from a features vector was used by Arivudainambi et al. for malware detection [130]. With the advantage of PCA compressions, seven attributes were minimized to only two crucial ones. Then, the CDM model was given an entry from the traffic captured by the authors. Details of the CNN architecture were not revealed.

In contrast to the previous articles, one work tested various dimensions within the discussed transformation approach [45]. In the article of Wu et al., 11 by 11 byte images were given as an entry of a 5-Layer CDM. The classification tool was that of [44], which was tested on network traffic features from the NSL-KDD dataset. CNN input images were created as in [44]. This is a malware detection approach. The paper compared the 2D transformation and classification results with those of 1D. Having analyzed these results, one can see that the feature wrapping concept is a better method for classification.

Table 7. The summary of all feature-based articles.

Article	Input Dimension	Layers	Dataset
[36]	1D	C I P I FC, C I C I P I FC and C I C I C I P I FC	KDD Cup 1999
[120]	1D	C I P I FC and C I C I C I C I P I FC	NLANR AMP, NLANR MAWI and NIMS
[121]	1D	EMB I C I P I FC	Can 2017
[37]	1D	C I P I C I P I FC I FC	KDD Cup 1999
[38]	1D	C I P I C I P I FC	
[122]	1D	EMB I C I P I FC I FC	BoT-IoT
[53]	1D	GoogLeNet [22]	ISCX-IDS-2012
[123]	1D	same as [63]	Own
[75]	1D	[63,70]	ISCX VPN-nonVPN
[124]	1D	C I C I P I C I C I P I FC I FC	WRCCDC
[39]	2D (32 × 32)	LeNet [19]	KDD Cup 1999
[40]	2D (8 × 8)	ResNet [24] and GoogLeNet [22]	KDD Cup 1999
[41]	2D	C I P I C I P I C I P I C I P I F I FC I FC	KDD Cup 1999
[42]	2D	GoogLeNet [22]	KDD Cup 1999, UNSW-NB15 and ISCX-IDS-2017
[44]	2D (11 × 11)	C I P I C I P I FC I FC I FC	NSL-KDD
[43]	2D (probably 11 × 11)	LeNet [19] and C I C I C I C I C I P I FC I FC I FC	KDD Cup 1999 and NSL-KDD
[46]	2D	C I SPC I SPC I FC	NSL-KDD
[47]	2D (9 × 10, 11 × 11, 9 × 9 and 10 × 10)	C I P I C I P I FC I FC I FC	NSL-KDD
[85]	2D (11 × 11)	MC I C I C I FC	ISCX-IDS-2017
[125]	2D (32 × 32)	C I P I C I P I C I FC	KDD Cup 1999
[76]	2D (30 × 30 and 300 × 300 [pixels])	C I P I C I P I C I P I C I FC I FC	CSE-CIC-IDS 2018 and ISCX VPN-nonVPN
[126]	2D	C I P I C I P I C I P I FC	own
[128]	3D	C I P I C I P I FC I FC I FC	own
[127]	3D	LeNet [19], AlexNet [21], ConvNet, GoogleNet [22] and ResNet [24]	own
[129]	3D	C I P I C I P I C I P I FC I FC	own
[45]	1D, 2D (11 × 11)	C I P I C I P I FC	NSL-KDD

8. Extracted Features

When compared to the previous section, this category of traffic transformations focused not only on classification but also on feature extraction. Here, the used datasets were mainly flow or packet-based. Therefore, after the extraction process, the CNN models dealt with traffic features.

8.1. One-dimensional CNN input

These following batch of papers widely elaborated feature extraction approaches to form input vectors [54,59,62,131,132] (see Table 8).

The different CNN model works on the basis of CTU-Malware, UNSW-NB15 and SCU-RNE datasets [62]. Shao et al. also proposed a novel method to extract features by a 4-Layer CDM. The idea learns the representation of a time series input data at each network model layer with the advantage of a hierarchical transformation of a CNN. The researchers compared the extraction tool with other feedforward networks. Further, the method avoids explicit feature extraction. The research paper proposed a 5-Layer CNN classifier to detect malware within computer network traffic.

MontazeriShatoori et al. created in a novel way CNN input vectors from pcaps and flows originated from the CIRA-CIC-DoHBrw-2020 dataset [131]. While preparing an input vector for CNN, all statistical features of flows were chosen from raw packets, i.e., the number of flow bytes sent, the rate of flow bytes sent, the number of flow bytes received, the rate of flow bytes received, packet length (e.g., mean, variance), packet time (e.g., mean, variance), request/response time difference (e.g., mean, variance). It is important to highlight that the authors decided to share their statistical feature extractor tool 'DoHMeter' publicly. They used CNN and other hybrid deep learning models to detect malware within DNS over HTTPS tunnels. The details of the CNN were not described.

Yet another publication written by Kolcun et al. utilized vector of features as the CNN entry to deal with the traffic classification challenge in IoT [132]. There are a few models proposed, but only one meets the requirements of this review, the 4-Layer CDM. The model's input is a vector of features that originates from the authors' own dataset. They are 19 chosen features, among others: source and destination ports, a number of received bytes, mean size of packets, a variance of the packets' sizes and duration of the stream.

Fourteen features from the CTU-13 dataset were extracted to form a 5-Layer CNN input [59]. These features included the traffic flow start time, protocol, the total number of packets or average packet rate, etc. This is a concept of detecting malware in the network traffic, especially botnets.

Differentiation of the features of flow packets due to time arrivals was proposed by Doriguzzi-Corin et al. to create network traffic vectors [54]. Once all flows in a particular time window were chosen, 11 special features were extracted. Longer flows were truncated. Then, these feature vectors were normalized and, if needed, zero-padded. The last step was devoted to labeling. The authors gave the vectors as an entry to 3-Layer CDM. The research was based on the traffic taken from the following datasets: ISCX-IDS-2012, ISCX-IDS-2017 and CSE-CIC-IDS2018.

8.2. Two-dimensional CNN input

The following eight articles created matrices of wrapped traffic features. These features were first extracted from the chosen datasets [77,86,107,108,133–136] (Table 8).

The first 2D concept of extracting traffic features from captured flows was proposed by Lopez-Martin et al. [133]. The authors took advantage of the 6-Layer CDM to classify the traffic. The deep learning model was given matrices containing six flow features, i.e., source port, destination port, the number of bytes in payload, TCP window size, interarrival time, and packet's direction in each row. The six features were taken from the 20 packets. Thus, the 2D input size was 20 by 6. Flows were originated from the authors own dataset, from Spanish research centers.

Two papers [107,108] utilized a few concepts of traffic transformations. The flow wrapping approaches are widely discussed in Section 4. On top of that, Aceto et al. followed the idea of [133] and also provided matrices of extracted features to classification models. The CDM used in this category of traffic manipulations was a 6-Layer CDM. This is a typical traffic classification study. The traffic was taken from the author's own capture.

Images based on the arrival time of packets are the input of the deep learning model of Yang et al. in another traffic classification research article [134]. Scientific work uses AlexNet CNN [21]. The tool was tested on the author's own dataset, where flows were captured. CNN's input was 10 by 10 bytes matrices. These 2D images are generated from the inter-arrival time of the first 50 packets of a session or their lengths. For packet lengths, the 1500 byte maximum transmission unit (MTU), and for an inter-arrival time, the 1200 milliseconds, constitute states which later form matrices.

The same concept of CNN entry was further tested by Hussein et al. on a few models: LeNet [19], AlexNet [21], ConvNet and GoogleNet [22,135]. Vectors were crafted from traffic features, which originated from the authors' own dataset. During processing, input data were transformed into images with a size of 16 × 16 bytes. The goal of the paper was to detect malware or find intrusions.

CNNs' input was created by concatenating matrices [136]. The article tested malware detection on a few different models. Among others, two were exclusively CNN based: 5-Layer CDM and ResNet [24]. The test was based on flows from VAST 2013 challenge collections. When it comes to the deep learning models' inputs, the authors created interesting correlation matrices on the numeric features of flows. While doing this, they omitted categorical data. This means that each numerical feature of the flow had a correlation matrix. Then all matrices for all traffic features were concatenated. It is important to highlight that each matrix was surrounded by a chosen value of top features. The image was called SC matrices. This is an outstanding concept of Liu et al. of the discussed topic when compared to other proposed ideas. LeNet [19] was used to enhance network traffic classification field [77]. The tests of the model were carried out on pcaps from the ISCX VPN-nonVPN and ISCX Tor-nonTor datasets. Raw traffic packets were processed beforehand. The first step of the process was to aggregate packets into flows. Then flows were divided into 60-s blocks. The next step was time normalization: the opening time was zero, and the final time was 1500. That means that 60 s is now 1500. Later, all pairs of IP datagram sizes and arrival times of the flow were registered in the 2D histogram. Each cell in the histogram contains the number of received packets in a particular time and of a particular size. Histograms are 1500 by 1500 bytes size and are named Flowpic. Shapira and Shavitt provide Flowpic as an input of CNN [77]. This is an interesting concept, dealing with the topic of transformations from different perspective of input data.

A thought-provoking article of Zhang et al. dwells on the traffic classification scientific problem and amends the concept of [50]. The changes included extracting features from the raw traffic [86]. The paper assumed that each flow consistsed of five packets, which, according to the authors' suggestion, were the most important ones. This assumption reduced redundant features from the top network layer and proposed more compact flows. The authors summarized these changes with the statement that more flows can be processed, and the introduction of zero elements was more firmly reduced. The image was 16 by 16 bytes in size. The proposed CNN model was a segmented CDM. The top branch of the model was responsible for image segmentation tasks that handle pixel-level classifications. The bottom branch main task was to deal with abnormal traffic that was imbalanced. The model was tested on ISCX-IDS-2017.

Table 8. Extracting features papers.

Article	Input Dimension	Layers	Dataset	Year
[62]	1D	C I C I P I FC	CTU-Malware, UNSW-NB15 and SCU-RNE	2019
[132]	1D	C I C I P I FC	Own	2020
[59]	1D	C I C I P I FC I FC	CTU-13	2020
[54]	1D	C I P I FC	ISCX-IDS-2012, ISCX-IDS-2017 and CSE-CIC-IDS2018	2020
[133]	2D (20 × 6 [features])	C I P I C I P I FC I FC	Own	2017
[107,108]	2D (20 × 6 [features])	same as [133]	Own	2020
[134]	2D (66 × 10, 81 × 10 and 76 × 10 [pixels])	AlexNet	Own	2018
[135]	2D (16 × 16)	LeNet [19], AlexNet [21], ConvNet and GoogleNet [22]	Own	2019
[136]	2D (3 × 30)	C I C I C I P I FC and ResNet [24]	VAST 2013 challenge	2019
[77]	2D (1500 × 1500)	LeNet [19]	ISCX VPN-nonVPN and ISCX Tor-nonTor	2019
[86]	2D (16 × 16)	Segmented CNN: I C I C I C I C I C I P I FC and C I P I C I P I C I P I FC	ISCX-IDS-2017	2019

9. Summary and Conclusions

There have been many scientific publications on CNN-based deep learning models (CDMs) for traffic classification and malware detection since 2015, as indicated in Figure 1. The aim of this survey was to study different dataset transformations described in the selected papers, using different criteria. The following aspects were considered:

- Network traffic data (raw traffic, flow, L2, L3, L4+ payload, traffic feature as shown in Table 9.
- Network traffic data transformation to the form of the input required by CDM.

- Different structure of CNN layers and models.
- Dimensionality of the CDM's input data.
- Current trends in CDMs for network traffic classification.

The type of network traffic data as an input for CDM is one of the crucial elements. Network traffic data used in the studied papers were acquired from different sources: test-beds, real traffic, or datasets prepared for and shared to the scientific community. Acquisition and the preprocessing of network traffic are an essential part of data analysis. The two most popular datasets within the elaborated topic are ISCX VPN-nonVPN (19 articles) and USTC-TFC2016 (11 articles). On top of that, many scientists did not share their datasets (25 articles).

As shown in Table 9, the numerousness of papers in each category highlights the paths followed by researchers. The most popular categories are manipulations of flows and traffic features. Using raw traffic so data do not need preprocessing is the least popular category. Under that reasoning, feature vectors as well flows were widely taken from utilized datasets.

Table 9. The summary of the most common transformation methods of the CNNs' inputs.

Transformation	Articles' No.	Articles
Raw traffic (1D)	2	[78,79]
Raw traffic (2D)	2	[35,103]
Flows (1D)	13	[49,63–67,80–82,104,105,107,108]
Flows (2D)	14	[12,34,50,55–58,60,61,65,67–69,106]
Flows (3D)	1	[83]
Extracted payload—raw traffic (1D)	4	[70–73]
Extracted payload—raw traffic (2D)	6	[72,73,109–112]
Extracted payload—raw traffic (3D)	1	[73]
Extracted payload—flows (1D)	4	[51,52,113,114]
Extracted payload—flows (2D)	9	[20,31,74,84,115–119]
Extracted payload—flows (3D)	1	[31]
Feature-based approaches (1D)	10	[36–38,53,75,120–124]
Feature-based approaches (2D)	13	[39–48,76,85,125]
Feature-based approaches (3D)	4	[127–130]
Extracting Features (1D)	5	[54,59,62,131,132]
Extracting Features (2D)	8	[77,86,107,108,133–136]

Analyzing CNN layers and models, LeNet was the most common CDM. Moreover, some papers amended their architecture with one or more additional layers. This was caused by the usefulness and practicality of the model in other scientific areas, such as data science and image recognition. Then, there is only a need to adjust the input data (network traffic), so it fits the requirements of the trained LeNet on, for instance, the MNIST dataset. This aspect is called transfer learning.

This survey is CNN based, so the majority of papers decided to form a 2D input to the deep learning model. Vectors as CNN entries were not so frequently used. Methods that proposed a 3D input to the 3D-CNN, dyed red, were in the minority (see Figure 8).

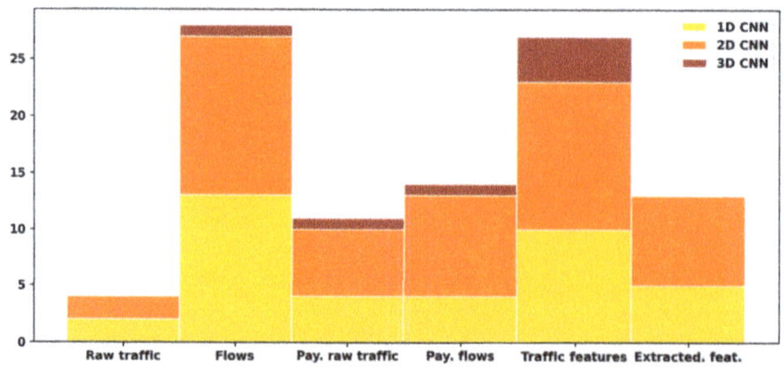

Figure 8. The popularity of discussed transformation methods, with the CNN architecture.

Regarding the comparison of dimensions, as the input data for CDM, we observed the following trends. Among various dimensions, 2D was the most common approach. The majority of articles added the second dimension with the advantage of wrapping. In the group of 2D methods, the entry size of 28 by 28 bytes was the leading trend. This concept may have been taken from the CNN structure used for the MNIST dataset.

As presented in Table 10, a noticeable batch of research papers utilized two or more dimensions of the CNN entries. We found out that seven papers gave proof of high results obtained by lower dimensions. This conclusion was not only unexpected, but also relevant for further studies (see Table 10). On top of that, manipulations on flows were the most common ones within the papers that compared various dimensions of CNN entry.

Table 10. The comparison of papers using more than one transformation model for CNN entry purposes.

Article	1D	2D	3D	Transformation
[79]	✓(best)	✓	—	Raw traffic
[63]	✓(best)	✓	—	Flows
[83]	✓	✓	✓(best)	Flows
[107]	✓	✓(best)	—	Flows (1D, 2D) & Extracting Features (2D)
[108]	✓	✓(best)	—	Flows (1D, 2D) & Extracting Features (2D)
[67]	✓	✓(best)	—	Flows
[65]	✓(best)	✓	—	Flows
[12]	✓(best)	✓	—	Flows
[69]	✓	✓(best)	—	Flows
[72]	✓(best)	✓	—	Extracted payload—raw traffic
[73]	✓	✓(best)	✓	Extracted payload—raw traffic
[118]	✓	✓(best)	—	Extracted payload—flows
[31]	-	✓(best)	✓	Extracted payload—flows
[45]	✓	✓(best)	-	Feature-based approaches

In some of the studied papers, researchers also used other CDMs to analyze network traffic. For example, the following methods were applied to the study of network traffic analysis: classical methods (tree-based, K-nearest neighbor, naive Bayes, logistic regression, support vector machine and semi-supervised) and neuronal methods (recurrent, multilayer perceptron, autoencoder and hybrid models).

Considering the constant increase in the number of papers on CNN-based models for computer network traffic analysis, one may conclude that this approach is becoming one of the classic approaches to traffic classification. One may also predict that the growth in the number of applications will continuously improve both the efficiency and detection/classification speed.

Funding: The research was funded by POB Cybersecurity and Data Analysis of Warsaw University of Technology within the Excellence Initiative: Research University (IDUB) program.

Conflicts of Interest: The authors declare no conflict of interest.

References

1. Gamage, S.; Samarabandu, J. Deep learning methods in network intrusion detection: A survey and an objective comparison. *J. Netw. Comput. Appl.* **2020**, *169*, 102767. [CrossRef]
2. Wang, Z. The applications of deep learning on traffic identification. *BlackHat USA* **2015**, *24*, 1–10.
3. Li, J.; Pan, Z. Network Traffic Classification Based on Deep Learning. *KSII Trans. Internet Inf. Syst.* **2020**, *14*, 062021.
4. Baran, P. The beginnings of packet switching: Some underlying concepts. *IEEE Commun. Mag.* **2002**, *40*, 42–48. [CrossRef]
5. Clark, D. The Design Philosophy of the DARPA Internet Protocols. *SIGCOMM Comput. Commun. Rev.* **1988**, *18*, 106–114. [CrossRef]
6. Mills, C.; Hirsh, D.; Ruth, G. *Internet Accounting: Background*; Internet Requests for Comments; RFC Editor, 1991. Available online: https://ieeexplore.ieee.org/abstract/document/920864/ (accessed on 19 July 2021).
7. Brownlee, N. *RTFM: Applicability Statement*; Internet Requests for Comments; RFC Editor, 1999. Available online: https://www.hjp.at/doc/rfc/rfc2721.html (accessed on 24 July 2021).

8. Claffy, K.; Braun, H.; Polyzos, G. A parameterizable methodology for Internet traffic flow profiling. *IEEE J. Sel. Areas Commun.* **1995**, *13*, 1481–1494. [CrossRef]
9. Claise, B. *Cisco Systems NetFlow Services Export Version 9*; RFC 3954; RFC Editor, 2004. Available online: https://datatracker.ietf.org/doc/html/rfc3954.html (accessed on 9 June 2021). [CrossRef]
10. Aitken, P.; Claise, B.; Trammell, B. *Specification of the IP Flow Information Export (IPFIX) Protocol for the Exchange of Flow Information*; RFC 7011; RFC Editor, 2013. Available online: https://datatracker.ietf.org/doc/html/rfc7011 (accessed on 13 June 2021). [CrossRef]
11. Folino, F.; Folino, G.; Guarascio, M.; Pisani, F.; Pontieri, L. On learning effective ensembles of deep neural networks for intrusion detection. *Inf. Fusion* **2021**, *72*, 48–69. [CrossRef]
12. Pacheco, F.; Exposito, E.; Gineste, M. A framework to classify heterogeneous Internet traffic with Machine Learning and Deep Learning techniques for satellite communications. *Comput. Netw.* **2020**, *173*, 107213. [CrossRef]
13. Zhao, J.; Jing, X.; Yan, Z.; Pedrycz, W. Network traffic classification for data fusion: A survey. *Inf. Fusion* **2021**, *72*, 22–47. [CrossRef]
14. Moore, A.; Zuev, D.; Crogan, M. Discriminators for Use in Flow-Based Classification. Ph.D. Thesis, The Queen Mary University of London, London, UK, 2005
15. Trammell, B. *Textual Representation of IP Flow Information Export (IPFIX) Abstract Data Types*; RFC 7373; RFC Editor, 2014. Available online: https://www.hjp.at/doc/rfc/rfc7373.html (accessed on 3 July 2021). [CrossRef]
16. Claise, B.; Trammell, B. *Information Model for IP Flow Information Export (IPFIX)*; RFC 7012; RFC Editor, 2013. Available online: https://www.hjp.at/doc/rfc/rfc5102.html (accessed on 3 July 2021). [CrossRef]
17. Hofstede, R.; Čeleda, P.; Trammell, B.; Drago, I.; Sadre, R.; Sperotto, A.; Pras, A. Flow Monitoring Explained: From Packet Capture to Data Analysis With NetFlow and IPFIX. *IEEE Commun. Surv. Tutor.* **2014**, *16*, 2037–2064. [CrossRef]
18. LeCun, Y.; Bengio, Y.; Hinton, G. Deep learning. *Nature* **2015**, *521*, 436–444. [CrossRef]
19. LeCun, Y.; Boser, B.; Denker, J.S.; Henderson, D.; Howard, R.E.; Hubbard, W.; Jackel, L.D. Backpropagation Applied to Handwritten Zip Code Recognition. *Neural Comput.* **1989**, *1*, 541–551. [CrossRef]
20. Saleh, I.; Ji, H. Network Traffic Images: A Deep Learning Approach to the Challenge of Internet Traffic Classification. In Proceedings of the 10th Annual Computing and Communication Workshop and Conference (CCWC), Las Vegas, NV, USA, 6–8 January 2020; pp. 0329–0334. [CrossRef]
21. Krizhevsky, A.; Sutskever, I.; Hinton, G.E. ImageNet Classification with Deep Convolutional Neural Networks. *Commun. ACM* **2017**, *60*, 84–90. [CrossRef]
22. Szegedy, C.; Liu, W.; Jia, Y.; Sermanet, P.; Reed, S.; Anguelov, D.; Erhan, D.; Vanhoucke, V.; Rabinovich, A. Going Deeper with Convolutions. *arXiv* **2014**, arXiv:cs.CV/1409.4842.
23. Huang, G.; Liu, Z.; van der Maaten, L.; Weinberger, K.Q. Densely Connected Convolutional Networks. *arXiv* **2018**, arXiv:cs.CV/1608.06993.
24. He, K.; Zhang, X.; Ren, S.; Sun, J. Deep Residual Learning for Image Recognition. *arXiv* **2015**, arXiv:cs.CV/1512.03385.
25. Simonyan, K.; Zisserman, A. Very Deep Convolutional Networks for Large-Scale Image Recognition. *arXiv* **2015**, arXiv:cs.CV/1409.1556.
26. Chollet, F. Xception: Deep Learning with Depthwise Separable Convolutions. In Proceedings of the 2017 IEEE Conference on Computer Vision and Pattern Recognition (CVPR), Honolulu, HI, USA, 21–26 July 2017; pp. 1800–1807. [CrossRef]
27. Howard, A.G.; Zhu, M.; Chen, B.; Kalenichenko, D.; Wang, W.; Weyand, T.; Andreetto, M.; Adam, H. MobileNets: Efficient Convolutional Neural Networks for Mobile Vision Applications. *arXiv* **2017**, arXiv:cs.CV/1409.1556.
28. Nataraj, L.; Karthikeyan, S.; Jacob, G.; Manjunath, B.S. Malware Images: Visualization and Automatic Classification. In *Proceedings of the 8th International Symposium on Visualization for Cyber Security*; Association for Computing Machinery: New York, NY, USA, 2011; VizSec '11. [CrossRef]
29. Guimarães, V.T.; Freitas, C.M.D.S.; Sadre, R.; Tarouco, L.M.R.; Granville, L.Z. A Survey on Information Visualization for Network and Service Management. *IEEE Commun. Surv. Tutorials* **2016**, *18*, 285–323. [CrossRef]
30. Tan, Z.; Jamdagni, A.; He, X.; Nanda, P.; Liu, R.P.; Hu, J. Detection of Denial-of-Service Attacks Based on Computer Vision Techniques. *IEEE Trans. Comput.* **2015**, *64*, 2519–2533. [CrossRef]
31. Millar, K.; Cheng, A.; Chew, H.G.; Lim, C.C. Using convolutional neural networks for classifying malicious network traffic. In *Deep Learning Applications for Cyber Security*; Springer: Berlin/Heidelberg, Germany, 2019; pp. 103–126.
32. Fontugne, R.; Hirotsu, T.; Fukuda, K. An Image Processing Approach to Traffic Anomaly Detection. In *Proceedings of the 4th Asian Conference on Internet Engineering (AINTEC '08)*; Association for Computing Machinery: New York, NY, USA, 2008; pp. 17–26. [CrossRef]
33. Kim, S.; Reddy, A. Image-Based Anomaly Detection Technique: Algorithm, Implementation and Effectiveness. *IEEE J. Sel. Areas Commun.* **2006**, *24*, 1942–1954. [CrossRef]
34. Wang, W.; Zhu, M.; Zeng, X.; Ye, X.; Sheng, Y. Malware traffic classification using convolutional neural network for representation learning. In Proceedings of the 2017 International Conference on Information Networking (ICOIN), Da Nang, Vietnam, 11–13 January 2017; pp. 712–717. [CrossRef]

35. Jia, W.; Liu, Y.; Liu, Y.; Wang, J. Detection Mechanism Against DDoS Attacks based on Convolutional Neural Network in SINET. In Proceedings of the 2020 IEEE 4th Information Technology, Networking, Electronic and Automation Control Conference (ITNEC), Chongqing, China, 12–14 June 2020; Volume 1, pp. 1144–1148.
36. Vinayakumar, R.; Soman, K.; Poornachandran, P. Applying convolutional neural network for network intrusion detection. In Proceedings of the 2017 International Conference on Advances in Computing, Communications and Informatics (ICACCI), Udupi, India, 13–16 September 2017; pp. 1222–1228.
37. Manimaran, A.; Chandramohan, D.; Shrinivas, S.; Arulkumar, N. A comprehensive novel model for network speech anomaly detection system using deep learning approach. *Int. J. Speech Technol.* **2020**, *23*, 305–313. [CrossRef]
38. Liu, G.; Zhang, J. CNID: Research of Network Intrusion Detection Based on Convolutional Neural Network. *Discret. Dyn. Nat. Soc.* **2020**, *2020*, 4705982. [CrossRef]
39. Liu, Y.; Liu, S.; Zhao, X. Intrusion detection algorithm based on convolutional neural network. *DEStech Trans. Eng. Technol. Res.* **2017**, *10*, 9–13. [CrossRef]
40. Li, Z.; Qin, Z.; Huang, K.; Yang, X.; Ye, S. Intrusion detection using convolutional neural networks for representation learning. In *International Conference on Neural Information Processing*; Springer: Berlin/Heidelberg, Germany, 2017; pp. 858–866.
41. Naseer, S.; Saleem, Y. Enhanced Network Intrusion Detection using Deep Convolutional Neural Networks. *TIIS* **2018**, *12*, 5159–5178.
42. Kim, T.; Suh, S.C.; Kim, H.; Kim, J.; Kim, J. An encoding technique for CNN-based network anomaly detection. In Proceedings of the 2018 IEEE International Conference on Big Data (Big Data), Seattle, WA, USA, 10–13 December 2018; pp. 2960–2965.
43. Wang, X.; Yin, S.; Li, H.; Wang, J.; Teng, L. A Network Intrusion Detection Method Based on Deep Multi-scale Convolutional Neural Network. *Int. J. Wirel. Inf. Netw.* **2020**, *27*, 503–517. [CrossRef]
44. Mohammadpour, L.; Ling, T.C.; Liew, C.S.; Chong, C.Y. A convolutional neural network for network intrusion detection system. *Proc. Asia Pac. Adv. Netw.* **2018**, *46*, 50–55.
45. Wu, K.; Chen, Z.; Li, W. A Novel Intrusion Detection Model for a Massive Network Using Convolutional Neural Networks. *IEEE Access* **2018**, *6*, 50850–50859. [CrossRef]
46. Hu, Z.; Wang, L.; Qi, L.; Li, Y.; Yang, W. A Novel Wireless Network Intrusion Detection Method Based on Adaptive Synthetic Sampling and an Improved Convolutional Neural Network. *IEEE Access* **2020**, *8*, 195741–195751. [CrossRef]
47. Li, Y.; Xu, Y.; Liu, Z.; Hou, H.; Zheng, Y.; Xin, Y.; Zhao, Y.; Cui, L. Robust detection for network intrusion of industrial IoT based on multi-CNN fusion. *Measurement* **2020**, *154*, 107450. [CrossRef]
48. Su, B.; Li, R.; Zhang, H. Evolving Deep Convolutional Neural Network for Intrusion Detection Based on NEAT. In Proceedings of the 2020 23rd International Symposium on Wireless Personal Multimedia Communications (WPMC), Okayama, Japan, 19–26 October 2020; pp. 1–6.
49. Chen, M.; Wang, X.; He, M.; Jin, L.; Javeed, K.; Wang, X. A Network Traffic Classification Model Based on Metric Learning. *CMC Comput. Mater. Contin.* **2020**, *64*, 941–959.
50. Wang, W.; Sheng, Y.; Wang, J.; Zeng, X.; Ye, X.; Huang, Y.; Zhu, M. HAST-IDS: Learning Hierarchical Spatial-Temporal Features Using Deep Neural Networks to Improve Intrusion Detection. *IEEE Access* **2018**, *6*, 1792–1806. [CrossRef]
51. Zeng, Y.; Gu, H.; Wei, W.; Guo, Y. Deep-Full-Range : A Deep Learning Based Network Encrypted Traffic Classification and Intrusion Detection Framework. *IEEE Access* **2019**, *7*, 45182–45190. [CrossRef]
52. Zeng, Y.; Qiu, M.; Zhu, D.; Xue, Z.; Xiong, J.; Liu, M. DeepVCM: A Deep Learning Based Intrusion Detection Method in VANET. In Proceedings of the 2019 IEEE 5th Intl Conference on Big Data Security on Cloud (BigDataSecurity), IEEE Intl Conference on High Performance and Smart Computing, (HPSC) and IEEE Intl Conference on Intelligent Data and Security (IDS), Washington, DC, USA, 27–29 May 2019; pp. 288–293. [CrossRef]
53. Cui, J.; Long, J.; Min, E.; Liu, Q.; Li, Q. Comparative study of CNN and RNN for deep learning based intrusion detection system. In *International Conference on Cloud Computing and Security*; Springer: Berlin/Heidelberg, Germany, 2018; pp. 159–170.
54. Doriguzzi-Corin, R.; Millar, S.; Scott-Hayward, S.; Martinez-del Rincon, J.; Siracusa, D. LUCID: A practical, lightweight deep learning solution for DDoS attack detection. *IEEE Trans. Netw. Serv. Manag.* **2020**, *17*, 876–889. [CrossRef]
55. Moskalenko, V.; Moskalenko, A. Growing Convolutional Neural Network For Malware Traffic Detection. In Proceedings of the 2018 International Conference on Information and Telecommunication Technologies and Radio Electronics (UkrMiCo), Odessa, Ukraine, 10–14 September 2018; pp. 1–5. [CrossRef]
56. Taheri, S.; Salem, M.; Yuan, J.S. Leveraging image representation of network traffic data and transfer learning in botnet detection. *Big Data Cogn. Comput.* **2018**, *2*, 37. [CrossRef]
57. Huang, H.; Deng, H.; Chen, J.; Han, L.; Wang, W. Automatic Multi-task Learning System for Abnormal Network Traffic Detection. *Int. J. Emerg. Technol. Learn.* **2018**, *13*, 4–20. [CrossRef]
58. Moskalenko, A.; Moskalenko, V.; Shaiekhov, A.; Zaretskyi, M. Multi-layer model and training method for information-extreme malware traffic detector. In Proceedings of the Third International Workshop on Computer Modeling and Intelligent Systems (CMIS-2020), Zaporizhzhia, Ukraine, 27 April–1 May 2020; CEUR-WS.org: Aachen, Germany, 2020; pp. 288–299.
59. Nugraha, B.; Nambiar, A.; Bauschert, T. Performance Evaluation of Botnet Detection using Deep Learning Techniques. In Proceedings of the 2020 11th International Conference on Network of the Future (NoF), Bordeaux, France, 12–14 October 2020; pp. 141–149.

60. Wang, Y.; An, J.; Huang, W. Using CNN-based representation learning method for malicious traffic identification. In Proceedings of the 2018 IEEE/ACIS 17th International Conference on Computer and Information Science (ICIS), Singapore, 6–8 June 2018; pp. 400–404.
61. Millar, K.; Cheng, A.; Chew, H.G.; Lim, C.C. Deep learning for classifying malicious network traffic. In *Pacific-Asia Conference on Knowledge Discovery and Data Mining*; Springer: Berlin/Heidelberg, Germany, 2018; pp. 156–161.
62. Shao, G.; Chen, X.; Zeng, X.; Wang, L. Deep Learning Hierarchical Representation From Heterogeneous Flow-Level Communication Data. *IEEE Trans. Inf. Forensics Secur.* **2019**, *15*, 1525–1540. [CrossRef]
63. Wang, W.; Zhu, M.; Wang, J.; Zeng, X.; Yang, Z. End-to-end encrypted traffic classification with one-dimensional convolution neural networks. In Proceedings of the 2017 IEEE International Conference on Intelligence and Security Informatics (ISI), Beijing, China, 22–24 July 2017; pp. 43–48. [CrossRef]
64. Song, M.; Ran, J.; Li, S. Encrypted Traffic Classification Based on Text Convolution Neural Networks. In Proceedings of the 2019 IEEE 7th International Conference on Computer Science and Network Technology (ICCSNT), Dalian, China, 19–20 October 2019; pp. 432–436. [CrossRef]
65. He, Y.; Li, W. Image-based Encrypted Traffic Classification with Convolution Neural Networks. In Proceedings of the 2020 IEEE Fifth International Conference on Data Science in Cyberspace (DSC), Hong Kong, China, 27–30 July 2020; pp. 271–278. [CrossRef]
66. Chen, Y.; Li, Z.; Shi, J.; Gou, G.; Liu, C.; Xiong, G. Not Afraid of the Unseen: A Siamese Network based Scheme for Unknown Traffic Discovery. In Proceedings of the 2020 IEEE Symposium on Computers and Communications (ISCC), Rennes, France, 7–10 July 2020; pp. 1–7. [CrossRef]
67. Cui, S.; Jiang, B.; Cai, Z.; Lu, Z.; Liu, S.; Liu, J. A Session-Packets-Based Encrypted Traffic Classification Using Capsule Neural Networks. In Proceedings of the 2019 IEEE 21st International Conference on High Performance Computing and Communications; IEEE 17th International Conference on Smart City; IEEE 5th International Conference on Data Science and Systems (HPCC/SmartCity/DSS), Zhangjiajie, China, 10–12 August 2019; pp. 429–436. [CrossRef]
68. Li, W.; Zhang, X.Y.; Shi, H.; Liu, F.; Ma, Y.; Li, Z. A Glimpse of the Whole: Path Optimization Prototypical Network for Few-Shot Encrypted Traffic Classification. *arXiv* **2020**, arXiv:2010.13285.
69. Chen, L.; Jiang, Y.; Kuang, X.; Xu, A. Deep Learning Detection Method of Encrypted Malicious Traffic for Power Grid. In Proceedings of the 2020 IEEE International Conference on Energy Internet (ICEI), Sydney, NSW, Australia, 24–28 August 2020; pp. 86–91.
70. Lotfollahi, M.; Siavoshani, M.J.; Zade, R.S.H.; Saberian, M. Deep packet: A novel approach for encrypted traffic classification using deep learning. *Soft Comput.* **2020**, *24*, 1999–2012. [CrossRef]
71. Akbari, I.; Tahoun, E. PrivPkt: Privacy Preserving Collaborative Encrypted Traffic Classification. 2019. Available online: http://www.informationweek.com/news/201202317 (accessed on 29 June 2021).
72. Xu, L.; Zhou, X.; Ren, Y.; Qin, Y. A Traffic Classification Method Based on Packet Transport Layer Payload by Ensemble Learning. In Proceedings of the 2019 IEEE Symposium on Computers and Communications (ISCC), Barcelona, Spain, 29 June–3 July 2019; pp. 1–6. [CrossRef]
73. Zhang, J.; Li, F.; Ye, F.; Wu, H. Autonomous Unknown-Application Filtering and Labeling for DL-based Traffic Classifier Update. In Proceedings of the IEEE INFOCOM 2020-IEEE Conference on Computer Communications, Toronto, ON, Canada, 6–9 July 2020; pp. 397–405.
74. Zhou, Y.; Cui, J. Research and Improvement of Encrypted Traffic Classification Based on Convolutional Neural Network. In Proceedings of the 2020 IEEE 8th International Conference on Computer Science and Network Technology (ICCSNT), Dalian, China, 20–22 November 2020; pp. 150–154. [CrossRef]
75. Dong, C.; Zhang, C.; Lu, Z.; Liu, B.; Jiang, B. CETAnalytics: Comprehensive effective traffic information analytics for encrypted traffic classification. *Comput. Netw.* **2020**, *176*, 107258. [CrossRef]
76. Pham, V.; Seo, E.; Chung, T.M. Lightweight Convolutional Neural Network Based Intrusion Detection System. *J. Commun.* **2020**, *15*, 808–817. [CrossRef]
77. Shapira, T.; Shavitt, Y. FlowPic: Encrypted Internet Traffic Classification is as Easy as Image Recognition. In Proceedings of the IEEE Conference on Computer Communications Workshops (INFOCOM WKSHPS), Paris, France, 29 April–2 May 2019; pp. 680–687. [CrossRef]
78. Marín, G.; Casas, P.; Capdehourat, G. Rawpower: Deep learning based anomaly detection from raw network traffic measurements. In Proceedings of the ACM SIGCOMM 2018 Conference on Posters and Demos, Budapest, Hungary, 20–25 August 2018; pp. 75–77.
79. Zhang, W.; Wang, J.; Chen, S.; Qi, H.; Li, K. A Framework for Resource-aware Online Traffic Classification Using CNN. In Proceedings of the 14th International Conference on Future Internet Technologies, Phuket, Thailand, 7–9 August 2019; pp. 1–6.
80. Marín, G.; Casas, P.; Capdehourat, G. Deep in the Dark-Deep Learning-Based Malware Traffic Detection Without Expert Knowledge. In Proceedings of the 2019 IEEE Security and Privacy Workshops (SPW), San Francisco, CA, USA, 19–23 May 2019; pp. 36–42.
81. Marín, G.; Caasas, P.; Capdehourat, G. Deepmal-deep learning models for malware traffic detection and classification. In *Data Science–Analytics and Applications*; Springer: Berlin/Heidelberg, Germany, 2021; pp. 105–112.
82. Hwang, R.; Peng, M.; Huang, C.; Lin, P.; Nguyen, V. An Unsupervised Deep Learning Model for Early Network Traffic Anomaly Detection. *IEEE Access* **2020**, *8*, 30387–30399. [CrossRef]

83. Ran, J.; Chen, Y.; Li, S. Three-dimensional convolutional neural network based traffic classification for wireless communications. In Proceedings of the 2018 IEEE Global Conference on Signal and Information Processing (GlobalSIP), Anaheim, CA, USA, 26–29 November 2018; pp. 624–627.
84. Zhang, L.; Li, B.; Liu, Y.; Zhao, X.; Wang, Y.; Wu, J. FPGA Acceleration of CNNs-Based Malware Traffic Classification. *Electronics* **2020**, *9*, 1631. [CrossRef]
85. Mohammadpour, L.; Ling, T.; Liew, C.; Aryanfar, A. A Mean Convolutional Layer for Intrusion Detection System. *Secur. Commun. Netw.* **2020**, *2020*, 8891185. [CrossRef]
86. Zhang, Y.; Chen, X.; Guo, D.; Song, M.; Teng, Y.; Wang, X. PCCN: Parallel Cross Convolutional Neural Network for Abnormal Network Traffic Flows Detection in Multi-Class Imbalanced Network Traffic Flows. *IEEE Access* **2019**, *7*, 119904–119916. [CrossRef]
87. Ring, M.; Wunderlich, S.; Scheuring, D.; Landes, D.; Hotho, A. A survey of network-based intrusion detection data sets. *Comput. Secur.* **2019**, *86*, 147–167. [CrossRef]
88. Garcia, S.; Grill, M.; Stiborek, J.; Zunino, A. An empirical comparison of botnet detection methods. *Comput. Secur.* **2014**, *45*, 100–123. [CrossRef]
89. Sharafaldin, I.; Lashkari, A.H.; Ghorbani, A.A. Toward generating a new intrusion detection dataset and intrusion traffic characterization. *ICISSp* **2018**, *1*, 108–116.
90. Moustafa, N.; Slay, J. UNSW-NB15: A comprehensive data set for network intrusion detection systems (UNSW-NB15 network data set). In Proceedings of the 2015 Military Communications and Information Systems Conference (MilCIS), Canberra, ACT, Australia, 10–12 November 2015 ; pp. 1–6.
91. Ye, Y.; Li, T.; Adjeroh, D.; Iyengar, S.S. A survey on malware detection using data mining techniques. *ACM Comput. Surv. (CSUR)* **2017**, *50*, 1–40. [CrossRef]
92. Boutaba, R.; Salahuddin, M.A.; Limam, N.; Ayoubi, S.; Shahriar, N.; Estrada-Solano, F.; Caicedo, O.M. A comprehensive survey on machine learning for networking: Evolution, applications and research opportunities. *J. Internet Serv. Appl.* **2018**, *9*, 1–99. [CrossRef]
93. Wazid, M.; Das, A.K.; Rodrigues, J.J.; Shetty, S.; Park, Y. IoMT malware detection approaches: Analysis and research challenges. *IEEE Access* **2019**, *7*, 182459–182476. [CrossRef]
94. Alswaina, F.; Elleithy, K. Android malware family classification and analysis: Current status and future directions. *Electronics* **2020**, *9*, 942. [CrossRef]
95. Talukder, S. Tools and techniques for malware detection and analysis. *arXiv* **2020**, arXiv:2002.06819.
96. Tariq, M.I.; Memon, N.A.; Ahmed, S.; Tayyaba, S.; Mushtaq, M.T.; Mian, N.A.; Imran, M.; Ashraf, M.W. A Review of Deep Learning Security and Privacy Defensive Techniques. *Mob. Inf. Syst.* **2020**, *2020*, 6535834. [CrossRef]
97. Geetha, R.; Thilagam, T. A review on the effectiveness of machine learning and deep learning algorithms for cyber security. *Arch. Comput. Methods Eng.* **2020**, *28*, 2861–2879. [CrossRef]
98. Asharf, J.; Moustafa, N.; Khurshid, H.; Debie, E.; Haider, W.; Wahab, A. A review of intrusion detection systems using machine and deep learning in internet of things: Challenges, solutions and future directions. *Electronics* **2020**, *9*, 1177. [CrossRef]
99. Caviglione, L.; Choraś, M.; Corona, I.; Janicki, A.; Mazurczyk, W.; Pawlicki, M.; Wasielewska, K. Tight Arms Race: Overview of Current Malware Threats and Trends in Their Detection. *IEEE Access* **2020**, *9*, 5371–5396. [CrossRef]
100. Konopa, M.; Fesl, J.; Janeček, J. Promising new Techniques for Computer Network Traffic Classification: A Survey. In Proceedings of the 2020 10th International Conference on Advanced Computer Information Technologies (ACIT), Deggendorf, Germany, 16–18 September 2020; pp. 418–421.
101. Kim, S.S.; Reddy, A.N. Modeling network traffic as images. In Proceedings of the IEEE International Conference on Communications, Seoul, Korea, 16–20 May 2005; Volume 1, pp. 168–172.
102. Bahaa, A.; Abdelaziz, A.; Sayed, A.; Elfangary, L.; Fahmy, H. Monitoring Real Time Security Attacks for IoT Systems Using DevSecOps: A Systematic Literature Review. *Information* **2021**, *12*, 154. [CrossRef]
103. Ko, T.; Raza, S.M.; Binh, D.T.; Kim, M.; Choo, H. Network prediction with traffic gradient classification using convolutional neural networks. In Proceedings of the 2020 14th International Conference on Ubiquitous Information Management and Communication (IMCOM), Taichung, Taiwan, 3–5 January 2020; pp. 1–4.
104. Casas, P.; Marín, G.; Capdehourat, G.; Korczynski, M. MLSEC-Benchmarking Shallow and Deep Machine Learning Models for Network Security. In Proceedings of the 2019 IEEE Security and Privacy Workshops (SPW), San Francisco, CA, USA, 19–23 May 2019; pp. 230–235.
105. Marín, G.; Casas, P.; Capdehourat, G. DeepSec meets RawPower-Deep Learning for Detection of Network Attacks Using Raw Representations. *ACM SIGMETRICS Perform. Eval. Rev.* **2019**, *46*, 147–150. [CrossRef]
106. Zhou, Z.; Yao, L.; Li, J.; Hu, B.; Wang, C.; Wang, Z. Classification of botnet families based on features self-learning under Network Traffic Censorship. In Proceedings of the 2018 Third International Conference on Security of Smart Cities, Industrial Control System and Communications (SSIC), Shanghai, China, 18–19 October 2018; pp. 1–7. [CrossRef]
107. Aceto, G.; Ciuonzo, D.; Montieri, A.; Pescapé, A. Mobile encrypted traffic classification using deep learning. In Proceedings of the 2018 Network traffic measurement and analysis conference (TMA), Vienna, Austria, 26–29 June 2018; pp. 1–8.
108. Aceto, G.; Ciuonzo, D.; Montieri, A.; Pescapé, A. Mobile encrypted traffic classification using deep learning: Experimental evaluation, lessons learned, and challenges. *IEEE Trans. Netw. Serv. Manag.* **2019**, *16*, 445–458. [CrossRef]

109. He, L.; Shi, Y. Identification of SSH Applications Based on Convolutional Neural Network. In Proceedings of the 2018 International Conference on Internet and e-Business, Singapore, 25–27 April 2018; pp. 198–201.
110. Li, L.; Ota, K.; Dong, M. DeepNFV: A Lightweight Framework for Intelligent Edge Network Functions Virtualization. *IEEE Netw.* **2019**, *33*, 136–141. [CrossRef]
111. Lim, H.K.; Kim, J.B.; Heo, J.S.; Kim, K.; Hong, Y.G.; Han, Y.H. Packet-based network traffic classification using deep learning. In Proceedings of the 2019 International Conference on Artificial Intelligence in Information and Communication (ICAIIC), Okinawa, Japan, 11–13 February 2019; pp. 046–051.
112. Xue, J.; Chen, Y.; Li, O.; Li, F. Classification and identification of unknown network protocols based on CNN and T-SNE. *J. Phys. Conf. Ser.* **2020**, *1617*, 012071. [CrossRef]
113. Wang, X.; Chen, S.; Su, J. Automatic Mobile App Identification From Encrypted Traffic With Hybrid Neural Networks. *IEEE Access* **2020**, *8*, 182065–182077. [CrossRef]
114. Wang, X.; Chen, S.; Su, J. Real network traffic collection and deep learning for mobile app identification. *Wirel. Commun. Mob. Comput.* **2020**, *2020*, 4707909. [CrossRef]
115. Ma, R.; Qin, S. Identification of unknown protocol traffic based on deep learning. In Proceedings of the 2017 3rd IEEE International Conference on Computer and Communications (ICCC), Chengdu, China, 13–16 December 2017; pp. 1195–1198.
116. Zhao, S.; Chen, S. Smartphone Application Identification by Convolutional Neural Network. In *International Conference on Machine Learning and Intelligent Communications*; Springer: Berlin/Heidelberg, Germany, 2018; pp. 105–114.
117. Feng, W.; Hong, Z.; Wu, L.; Fu, M.; Li, Y.; Lin, P. Network protocol recognition based on convolutional neural network. *China Commun.* **2020**, *17*, 125–139. [CrossRef]
118. Yujie, P.; Weina, N.; Xiaosong, Z.; Jie, Z.; Wu, H.; Ruidong, C. End-To-End Android Malware Classification Based on Pure Traffic Images. In Proceedings of the 2020 17th International Computer Conference on Wavelet Active Media Technology and Information Processing (ICCWAMTIP), Chengdu, China, 18–20 December 2020; pp. 240–245.
119. Zhao, L.; Cai, L.; Yu, A.; Xu, Z.; Meng, D. A novel network traffic classification approach via discriminative feature learning. In Proceedings of the 35th Annual ACM Symposium on Applied Computing, Brno, Czech Republic, 30 March–3 April 2020; pp. 1026–1033.
120. Vinayakumar, R.; Soman, K.; Poornachandran, P. Secure shell (ssh) traffic analysis with flow based features using shallow and deep networks. In Proceedings of the 2017 International Conference on Advances in Computing, Communications and Informatics (ICACCI), Udupi, India, 13–16 September 2017; pp. 2026–2032.
121. Lokman, S.F.; Othman, A.T.B.; Abu-Bakar, M.H. Optimised Structure of Convolutional Neural Networks for Controller Area Network Classification. In Proceedings of the 2018 14th International Conference on Natural Computation, Fuzzy Systems and Knowledge Discovery (ICNC-FSKD), Huangshan, China, 28–30 July 2018; pp. 475–481.
122. Susilo, B.; Sari, R.F. Intrusion Detection in IoT Networks Using Deep Learning Algorithm. *Information* **2020**, *11*, 279. [CrossRef]
123. Zhao, S.; Chen, S.; Sun, Y.; Cai, Z.; Su, J. Identifying known and unknown mobile application traffic using a multilevel classifier. *Secur. Commun. Netw.* **2019**, *2019*, 9595081. [CrossRef]
124. Yang, K.; Xu, L.; Xu, Y.; Chao, J. Encrypted Application Classification with Convolutional Neural Network. In Proceedings of the 2020 IFIP Networking Conference (Networking), Paris, France, 22–26 June 2020; pp. 499–503.
125. Zheng, W.F. Intrusion detection based on convolutional neural network. In Proceedings of the 2020 International Conference on Computer Engineering and Application (ICCEA), Guangzhou, China, 18–20 March 2020; pp. 273–277.
126. Li, D.; Li, W.; Wang, X.; Nguyen, C.T.; Lu, S. ActiveTracker: Uncovering the Trajectory of App Activities over Encrypted Internet Traffic Streams. In Proceedings of the 2019 16th Annual IEEE International Conference on Sensing, Communication, and Networking (SECON), Boston, MA, USA, 10–13 June 2019; pp. 1–9.
127. Salman, O.; Elhajj, I.H.; Chehab, A.; Kayssi, A. A Multi-level Internet Traffic Classifier Using Deep Learning. In Proceedings of the 2018 9th International Conference on the Network of the Future (NOF), Poznan, Poland, 19–21 November 2018; pp. 68–75. [CrossRef]
128. Chen, Z.; He, K.; Li, J.; Geng, Y. Seq2img: A sequence-to-image based approach towards ip traffic classification using convolutional neural networks. In Proceedings of the 2017 IEEE International Conference on Big Data (big data), Boston, MA, USA, 11–14 December 2017; pp. 1271–1276.
129. De Schepper, T.; Camelo, M.; Famaey, J.; Latré, S. Traffic classification at the radio spectrum level using deep learning models trained with synthetic data. *Int. J. Netw. Manag.* **2020**, *30*, e2100. [CrossRef]
130. Arivudainambi, D.; Varun Kumar, K.A.; Sibi Chakkaravarthy, S.; Visu, P. Malware traffic classification using principal component analysis and artificial neural network for extreme surveillance. *Comput. Commun.* **2019**, *147*, 50–57.
131. MontazeriShatoori, M.; Davidson, L.; Kaur, G.; Habibi Lashkari, A. Detection of DoH Tunnels using Time-series Classification of Encrypted Traffic. In Proceedings of the 2020 IEEE Intl Conf on Dependable, Autonomic and Secure Computing, Intl Conf on Pervasive Intelligence and Computing, Intl Conf on Cloud and Big Data Computing, Intl Conf on Cyber Science and Technology Congress (DASC/PiCom/CBDCom/CyberSciTech), Calgary, AB, Canada, 17–22 August 2020; pp. 63–70. [CrossRef]
132. Kolcun, R.; Popescu, D.A.; Safronov, V.; Yadav, P.; Mandalari, A.M.; Xie, Y.; Mortier, R.; Haddadi, H. The Case for Retraining of ML Models for IoT Device Identification at the Edge. *arXiv* **2020**, arXiv:2011.08605.
133. Lopez-Martin, M.; Carro, B.; Sanchez-Esguevillas, A.; Lloret, J. Network traffic classifier with convolutional and recurrent neural networks for Internet of Things. *IEEE Access* **2017**, *5*, 18042–18050. [CrossRef]

134. Yang, Y.; Kang, C.; Gou, G.; Li, Z.; Xiong, G. TLS/SSL encrypted traffic classification with autoencoder and convolutional neural network. In Proceedings of the 2018 IEEE 20th International Conference on High Performance Computing and Communications; IEEE 16th International Conference on Smart City; IEEE 4th International Conference on Data Science and Systems (HPCC/SmartCity/DSS), Exeter, UK, 28–30 June 2018; pp. 362–369.
135. Hussein, A.; Salman, O.; Chehab, A.; Elhajj, I.; Kayssi, A. Machine Learning for Network Resiliency and Consistency. In Proceedings of the 2019 Sixth International Conference on Software Defined Systems (SDS), Rome, Italy, 10–13 June 2019; pp. 146–153. [CrossRef]
136. Liu, X.; Tang, Z.; Yang, B. Predicting Network Attacks with CNN by Constructing Images from NetFlow Data. In Proceedings of the 2019 IEEE 5th Intl Conference on Big Data Security on Cloud (BigDataSecurity), IEEE Intl Conference on High Performance and Smart Computing, (HPSC) and IEEE Intl Conference on Intelligent Data and Security (IDS), Washington, DC, USA, 27–29 May 2019; pp. 61–66. [CrossRef]

electronics

MDPI

Article

A Method for Fast Selection of Machine-Learning Classifiers for Spam Filtering

Sylwia Rapacz †, Piotr Chołda † and Marek Natkaniec *,†

Faculty of Computer Science, Electronics and Telecommunications, Institute of Telecommunications, AGH University of Science and Technology, Al. Mickiewicza 30, 30-059 Kraków, Poland; rapaczsylwia@gmail.com (S.R.); cholda@agh.edu.pl (P.C.)
* Correspondence: natkanie@agh.edu.pl; Tel.: +48-12-617-4040
† These authors contributed equally to this work.

Abstract: The paper elaborates on how text analysis influences classification—a key part of the spam-filtering process. The authors propose a multistage meta-algorithm for checking classifier performance. As a result, the algorithm allows for the fast selection of the best-performing classifiers as well as for the analysis of higher-dimensionality data. The last aspect is especially important when analyzing large datasets. The approach of cross-validation between different datasets for supervised learning is applied in the meta-algorithm. Three machine-learning methods allowing a user to classify e-mails as desirable (ham) or potentially harmful (spam) messages were compared in the paper to illustrate the operation of the meta-algorithm. The used methods are simple, but as the results showed, they are powerful enough. We use the following classifiers: *k*-nearest neighbours (*k*-NNs), support vector machines (SVM), and the naïve Bayes classifier (NB). The conducted research gave us the conclusion that multinomial naïve Bayes classifier can be an excellent weapon in the fight against the constantly increasing amount of spam messages. It was also confirmed that the proposed solution gives very accurate results.

Keywords: classifiers; e-mail; ham; machine learning; spam

Citation: Rapacz, S.; Chołda, P.; Natkaniec, M. A Method for Fast Selection of Machine-Learning Classifiers for Spam Filtering. *Electronics* **2021**, *10*, 2083. https://doi.org/10.3390/electronics10172083

Academic Editors: Amir Mosavi and Juan M. Corchado

Received: 22 June 2021
Accepted: 25 August 2021
Published: 27 August 2021

1. Introduction

The spam problem is an ongoing issue: in 2018 14.5 billion spam e-mails were sent per day [1]. According to the Internet Security Threat Report [2] released in 2019 by Symantec, spam levels for their customers increased in 2018. What draws the attention is that small enterprises were attacked more often than large companies, and e-mail malware reached stable levels. Therefore, there is a need to tailor even simple tools for detection and filtering of spam in all organizations.

For the sake of the presented study, we follow the definition by Emilio Ferrara, stating that this is any "attempt to abuse, or manipulate, a techno-social system by producing and injecting unsolicited and/or undesired content aimed at steering the behavior of humans or the system itself, at the direct or indirect, immediate or long-term advantage of the spammer(s)" [3]. Here, we focus on so-called junk e-mails. These are unwanted messages sent at large scale by e-mail. The term spam refers to the undesired (or even harmful) e-mails, while ham is used to indicate the valid and important messages desired by the recipient. Additionally, we assume the scenario where junk e-mails are sent by botnets and they are not aimed at specific users (contrary to, e.g., spear phishing).

This paper proposes a method for identification of the best-performing machine-learning-based classifiers and selection of the one with the leading parameters. The proposed solution solves the problem of fast recognition of the most interesting parameters. This allows for quick analysis of data of higher dimensionality. This is especially important if large datasets are to be analyzed and we want to assure the proper scalability of our system. In our paper, we also show how to find a database to train a machine-learning model used for spam detection (defined here as a binary classifier), to process the text so that the data can be fed to a machine-learning model and how to implement a selected machine-learning model-based classifier. We also propose a method that allows for cross-validation between different datasets in the training and test phases. The obtained results show that our solution gives accurate results consistent with other literature studies and outperforms the reported results in some cases. To the best of our knowledge, our paper is the first which discusses the efficiency of SVM, MNB, *k*-NN algorithms for such comprehensive datasets as almost the whole Enron (4 datasets) and Lingspam databases. Moreover, it uses an unusual cross-validation concept by mixing and applying different datasets for training and test purposes. Such an approach is extremely rare in the literature. Finally, it presents a multistage algorithm for fast and precise selection of machine-learning classifiers for spam filtering. It allows for quick selection of interesting parameters, which is essential for working with large datasets. The quality of the results is proven by a big numerical example given for the method validation.

The structure of the paper is as follows. The review of spam filters based on different machine-learning tools with typical performance metrics and several publicly available datasets is presented in Section 2. In Section 3, the materials and methods are discussed. The assumptions, useful databases of spam messages, text-preprocessing aspects (including tokenization, conversion, removal of punctuation marks, stemming/lemmatization, and dictionary construction) as well as the considered supervised learning solutions are described. The performance of the selected methods is evaluated on four large datasets in Section 4. The dataset structures created with the unique approach of assuring cross-validation between different datasets in training and test phases are analyzed first. Next, the text preprocessing impact on the used dictionary is studied. An innovative multistage meta-algorithm for checking the classifier performance is described in action and validated. The final summary is given in Section 5.

2. Related Work

The increasing number of spam e-mails has created a strong need to develop more reliable and efficient anti-spam filters, including ones based on machine-learning tools. They are efficient, since they only require the preparation of a set of training samples, i.e., pre-classified e-mails [4]. In recent years, various machine-learning methods have been successfully used to effectively detect and filter unwanted messages. The following classification methods are most commonly used for spam filtering: Support Vector Machine (SVM), Naïve Bayes classifier (NB), *k*-Nearest Neighbours (*k*-NN), Artificial Neutral Network (ANN), Decision Tree (DT), Random Forest (RF), Logistic Regression (LR). Below, we present some results reported in the literature. Note that some of the metrics results are compared with our method during the validation of our approach. The values are given at the end of the numerical study in separate table.

The applicability of using different machine-learning methods to recognize spam e-mails was analyzed in [5]. The SpamAssassin dataset, which contains 6000 e-mails with the spam rate 37.04% used in all experiments. Sharma and Arora in [6] analyzed Bayes Net (BN), Logic Boost (LB), RT, JRip (JR), J48-based DTs, Multilayer Perceptron (MP), Kstar (KS), RF, and Random Committee (RC) machine-learning algorithms. The dataset with 4601 instances and 55 spam base attributes downloaded from UCI Machine-Learning Repository were used in the performed research. Harisinghaney et al. [7] applied the following three different algorithms: *k*-NN, NB, and DBSCAN-based clustering. The performance for the four metrics accuracy, precision, sensitivity, and specificity were calculated and compared. Unfortunately, contrary to our approach, only a small set of the Enron Corpus dataset was used in the analysis (2500 mails for training and another 2500 mails for testing from 200,399 messages of the cleaned Enron Corpus). In [8] a comprehensive study of machine-learning mechanisms for spam mail detection such as NB, SVM, and *k*-NN combined with NB is presented. The TREC 2007 public corpus with 12 attributes and 4899 messages as the spam base dataset was used for performance evaluation. The accuracy and F-measure were calculated and compared for all algorithms. The authors in [9] prepared a special dataset called SHED: Spam Ham E-mail Dataset. They collected 6002 e-mails (4490 spam and 1512 ham e-mails) and extracted from them various features. The performance of different classification approaches (NB, BN, AdaBoost, and RF) was evaluated using four metrics: accuracy, precision, recall, and time taken to build the model. In [10] the NB, SVM and hybrid solutions were studied using Lingspam dataset. The authors observed that the SVM algorithm in most cases offers high precision and recall, while NB offers faster classification speed. They also require fewer training samples. The authors in [11] showed how to develop a high-performance and low-computation method for classifying spam e-mails. The UCI SpamBase dataset was used with a total of 4601 data instances for experimentation. The following classifiers were evaluated and compared: RF, ANN, Logistic, SVM, Random Tree, *k*-NN, Decision Table, BN, NB, and neural networks applying Radial Basis Functions (RBF). Seven metrics were used to evaluate the performance of the classifiers. In [12], another comparison between different machine-learning classifiers was presented. The classifiers analyzed in this paper include SVM, NB, and J48. The dataset used in this research was enron1 from the Enron collection of e-mails. It contained 3762 spam messages and 5172 ham messages. The performance analysis of seven machine-learning techniques for e-mail spam classification was analyzed in [13]. The following techniques were compared: NB, SVM, *k*-NN, RF, Bagging, Boosting (AdaBoost), and Ensemble Classifier. The evaluation was performed on the e-mail spam dataset from UCI Machine-Learning Repository and Kaggle website. In [14], the problem of spam review detection is addressed. The authors proposed in their system deep-learning methods: Multilayer Perceptron (MLP), Convolutional Neural Network (CNN), and a variant of Recurrent Neural Network (RNN) based on Long Short-Term Memory (LSTM) cells. They also applied traditional classifiers such as NB, *k*-NN and SVM. They worked on Ott and Yelp Datasets in their study. The presented results showed that considering accuracy, both SVM and NB classifiers performed almost same. The problem of spam and malware elimination from e-mails was discussed in [15]. The authors analyzed and compared ten classification techniques: *k*-NN, SVM, DT, RF, AdaBoost, Extra Tree (ET), Gaussian Naïve Bayes (GNB), Multinomial Naïve Bayes (MNB), Bernoulli Naïve Bayes (BNB), and Gradient Boosting (GB). These algorithms were trained on previously labeled data from the shortened Enron and CMU datasets (26,000 spam and 19,000 ham e-mails) and the accuracy of each classifier was computed. The SVM obtained the best results. We would like to emphasise that—although we also compare some classifiers—our main aim is to propose a general meta-algorithm to deal with various classifiers. This differs us from works such as [15].

Guarav et al. [16] examined the efficiency of NB, DT, and RF algorithms used in the classification process. The experiments were carried out on three different types of datasets: Lingspam, Enron and PU. In the comparative study, the authors showed that the accuracy level for all algorithms highly depended on a specific dataset. In [17] the four classifiers: NB, DT, Ensemble Boosting and Ensemble Hybrid Boosting (EHB) were analyzed and compared. The authors used UCI Machine-Learning Repository as a spam dataset. The mentioned dataset has 4601 instances, 57 attributes, and a single output which allows classification of e-mail as spam or ham. A large group of machine-learning techniques for e-mail spam classification was also analyzed and presented in [18]. The authors studied the efficiency of the following algorithms: SVM, *k*-NN, NB, DT, RF, AdaBoost and Bagging. They used e-mail data sets from different websites, such as Kaggle, along with some datasets created on their own. A spam e-mail dataset from Kaggle was used for training. The performed research showed that the NB gave the best results, but expressed a limitation due to class-conditional independence. Gibson et al. [19] analyzed machine-learning algorithms that are optimized with bio-inspired methods. They implemented Multinominal Naïve Bayes (MNB), SVM, RF, DT, and Multilayer Perceptron algorithms which were tested on seven different e-mail datasets: Lingspam, PUA, PU1, PU2, PU3, Enron, and SpamAssassin. The bio-inspired algorithms such as Particle Swarm Optimization (PSO) and Genetic Algorithm

(GA) were added for performance optimization of classifiers. The GA worked well for RF and DT, whereas PSO worked well for MNB. The authors proved that MNB with GA performed the best overall. In [20], three techniques, namely NB, *k*-NN and SVM, were studied on a prepared dataset. The corpus consists of 16,843 messages, 11,291 of which are marked as spam (from the Babletext web site) and 5552 are labeled as ham (from the SpamAssassin web site). The best accuracy was obtained for NB. The authors in [21] compared: Logistic Regression (LR), DT, NB, *k*-NN, and SVM as the classifiers. The assumed dataset was a spam database taken from UCI Machine-Learning Repository. The RD and *k*-NN obtained the same performance; however, *k*-NN algorithm requires more time to build the model. The accuracy of both algorithms exceeded 99%. Saidini et al. [22] explored the use of a semantic-based classification approach to improve the accuracy of spam detection. The NB, *k*-NN, DT, AdaBoost, and RF machine-learning classifiers were compared in terms of accuracy, recall, precision, and F-measure. The test dataset was collected from several public sources: Enron, Lingspam and some specialized forums. To extend the evaluation part, the authors also used another dataset, called CSDMC2010. They noted that NB and SVM performed better than the other tested classifiers. The categorization by domain significantly improved the spam detection process. The best results were obtained using AdaBoost, NB, and RF classifiers, where the accuracy achieved more than 98% in most of domains. In [23], the authors implemented MNB, RF, *k*-NN, GB, as well as RNN and MLP for deep-learning implementation. The dataset with 4601 instances (1813 spam and 2788 non-spam messages) from the UCI Machine-Learning Repository was applied for analysis. Rastenis et al. [24] proposed an automated spam and phishing e-mail classification solution, which is based on e-mail message body text automated classification. It also solves the problem of correct classification of e-mails written in different languages. They compared NB, General Linearized Model (GLM), Fast Large Margin (FLM), DT, RF, GB, and SVM on Nazario, SpamAssassin, and Vilnius Technical University datasets. Records from different datasets were mixed into one reduced dataset (700 spam and 700 phishing e-mails).

Although we focus here on the usage of many classifications simultaneously, it can be mentioned that a large part of the literature is devoted to the analysis of one type of model to classify e-mails (e.g., [25]) or the potential attacks on classification tasks (such as for instance in [26]). Additionally, it is necessary to remember that some works report that although it is evident that algorithms that perform well in the spam classification (e.g., NB), in other contexts they offer poor performance (e.g., [27,28]). Therefore, the model should always be aligned with a specific problem and data type.

3. Materials and Methods

3.1. Assumptions

E-mail spam filtering is a compound task, and in general we follow the methods elaborated before, where [29] is the main source of inspiration for us. The main goal of this paper is to explore one of its key areas, i.e., machine-learning-based classification, to help with the initial decision if a given e-mail message is indeed spam or ham. The element that enables this research is a dataset selected as a pool for training. The dataset is a collection of real e-mail examples. Access to a useful dataset is not a trivial issue, since typically in the academical world it is not possible to obtain e-mails for scientific research. Additionally, it is necessary to gain access to the database where the messages are already labeled as spam or ham.

Here, we propose a multistage meta-algorithm that allows us to select the best hyper-parameters for various classification algorithms and then compare their performance to decide on which one to use. The meta-algorithm is presented in Figure 1. Please note that the classification algorithms shown are only used as illustration. The following stages of the meta-algorithm are as follows:

1. Selection of a database.
2. Text analysis.

3. Spam detection: cross-validation on different datasets.
4. Final selection.

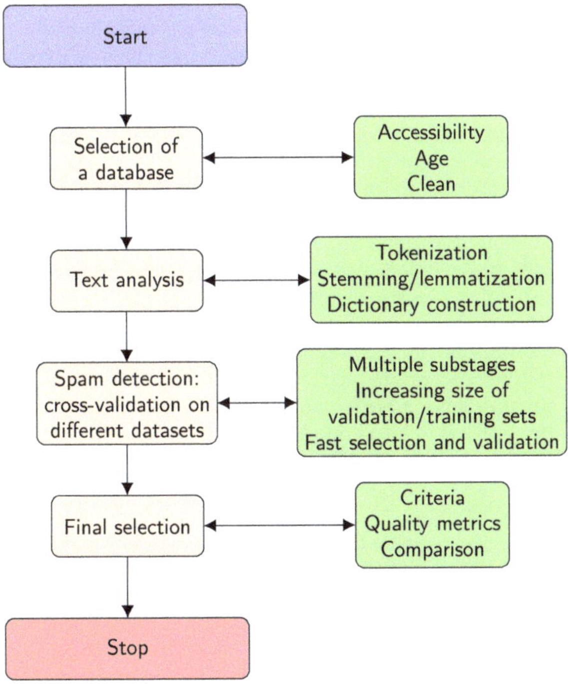

Figure 1. The proposed multistage meta-algorithm for performance check of the spam detection algorithms.

These elements are presented in the subsequent part of the paper. As for now, we can emphasise that our approach deals mainly with the impact the text preprocessing has on the classification process and then analyzes illustratively some of the machine-learning methods performance in this difficult task. Our solution consists of two parts. The first one focuses on the text documents (e-mails) analysis and preprocessing (points 1 and 2 above), so that the documents can be represented as an input for the methods used afterwards. The second one (points 3 and 4 above) implements the classifiers and provides the tools to evaluate them. First, we present the selection of the database (assumptions in Section 3.2 and their concretization in Section 4.1) to obtain the samples to train, adjust, validate, and test any model. Second, we elaborate on how to process the dataset to make it usable for various models and valuable enough to provide meaningful data. As in many cases, data processing (along with feature selection) is important since the quality strongly depends on it. The assumptions behind the text analysis are discussed in Section 3.3, while the details related to concrete data are shown in Section 4.2. Third, the main part of the method is performed in a few substages (five in our example case), and assures the proper scalability of the system. It consists mainly in the preselection of the classifiers and adjustment of their hyperparameters. The concept lays in the fact that the largest number of tests is conducted on the smallest dataset. This approach allows us to obtain the most interesting parameters relatively quickly, and then proceed to check them on data of higher dimensionality. The exemplary classifiers are shortly refreshed in Section 3.4. We emphasise that these models are used only to illustrate our method. All substages are thoroughly shown in the numerical example (Section 4.3). Fourth, as concerns the final selection, we just present the comparison of the output in Section 4.4. The selection should be performed based on a specific application or user's needs, and we do not settle these concerns here.

As can be seen, the proposed meta-algorithm does not solve any specific machine-learning problem, but is a kind of super-algorithm able to select the best algorithms to solve classification problems. As concerns the complexity of the meta-algorithm, we can see that it does not involve any loops or recurrences, so it is purely linear and, therefore, its scalability is very good. In fact, the only elements that can increase the complexity are related to its elements. Potentially problematic stages are related to text analysis, but is it necessary to mention that tokenization, lemmatization, stemming, etc. operate linearly from the viewpoint of the dataset size and its efficiency is mainly related to the search mechanisms involved. As we are using the mechanisms built in the popular machine-learning package, we do not consider their internal complexity. Clearly, a problematic part of the calculations can be also related to the models themselves. Although it is known that the pessimistic complexity of the used classification algorithms (k-NNs, NBs, SVMs) is in general polynomial (no larger than cubic—even in the case of naive implementations), we additionally purposely limit the calculation time by cutting the hyperparameter and training processing times by fast skipping of the models with poor performance based on the training sets with increasing size and complexity. In practice, our experiments were done on a standard desktop PC and the processing time has not exceeded standard times reported in the literature.

3.2. Databases

The first issue to solve while dealing with e-mail spam filtering is to find a dataset needed to train and test the models. It is extremely difficult to find a useful dataset of this kind. Although the total number of e-mails sent/received worldwide in 2019 was expected to reach 293.6 billion [30] per day, the access to the data is hindered due to privacy issues. We had to use publicly accessible data that are free and open to the whole world, which diminishes the set of potential candidates. Additionally, we were interested in databases conforming the following properties: (a) accessible: public and free to download for academic purposes; (b) relatively new: the old databases are not useful since the spamming environment is extremely dynamic; (c) virus-free. During the research, a few sources were selected. Their short descriptions are given below.

- Enron Corpus: chosen to be a foundation for this paper. The corpus is described in detail in the following.
- Lingspam: a part of the database (962 e-mails) was preprocessed and used by Gregory Piatetsky-Shapiro and Matthew Mayo in their implementation of e-mail spam filtering [29]. The dataset was also downloaded and used in our experiments.
- SpamAssassin (SA): a public corpus which was last updated in 2006 [31]. SA is an open-source anti-spam platform [32], filtering e-mail and blocking spam. The tests are carried out on e-mail headers and bodies.
- Honeypot: the last event entered in the website is up to date. Honeypot gathers statistics about harvesters, spam servers, dictionary attackers, and comment spammers. The owners claim that they "periodically collate the e-mail messages they receive and share the resulting corpus with anti-spam developers and researchers" [33]. Unfortunately, they do not provide any ham e-mails.
- MailBait: fills the inbox with e-mails by signing up the provided address for mailing lists and newsletters [34]. It is not anonymised (browser data and IP pass through) and it does not provide ham.

The Enron Corpus [35] was collected at Enron Corporation in 2002, during the investigation after the bankruptcy of the company. The original set was generated by 158 employees and consists of more than 600,000 e-mails. This database has already been used in the studies on machine-learning-based spam detection [36]. The corpus consists of two subdirectories: the 'raw' one (original messages with no modifications) and the 'preprocessed' one (where the messages in non-Latin encoding, virus-infected e-mails and ham sent by the owners to themselves were removed).

3.3. Processing of the Data

Text preprocessing plays a crucial role in spam filtering [24,37]. For any spam detection model to be effective, the content of the e-mails should be normalized and represented as feature vectors. The starting point is the tokenization of the raw text data. Then there are several steps shown in Figure 2 to obtain the data in the form that is ready to be analyzed by the model.

RAW TEXT (e-mails collection)
↓
Tokenization
↓
Conversion to lowercase
↓
Removal of punctuation marks, digits and stop words
↓
Stemming/lemmatization
↓
Dictionary
↓
FEATURE VECTORS (text representation)

Figure 2. Text preprocessing steps.

Tokenization technique allows us to split the content of the e-mails into basic processing units that are called tokens or features. Given that the paper deals with text data, the tokens are simply separate words. For instance, the tokenized sentence "Subject: christmas tree farm pictures" is a collection of strings: "Subject", ":", "christmas", "tree", "farm" and "pictures". The next step involves converting all tokens to lowercase. As a result of this simple operation, the number of words taken into account is significantly reduced. Instead of treating "Example", "example" and "EXAMPLE" as three different words, after converting them to lowercase, we make sure that the program will count them as one ("example"). Punctuation marks, digits, and stop words are all common in both spam and ham e-mails and do not add any value to text analysis. Since we implement our solution in Python, we refer to tools related to this programming language. There are several libraries and functions that may be applied to eliminate the mentioned language elements not essential from the spam detection viewpoint. Below is the list of functionalities chosen by us.

- Python method string.isalpha() checks whether the characters in the string are alphabetic or not. If the character is a digit, the method returns False.
- The method string.punctuation() allows removal of common punctuation marks, such as commas, periods, semicolons, etc.
- Natural Language Toolkit (NLTK) offers a module containing a list of stop words that are the most common words in a language. The examples of stop words are short words, for example: "the", "is", "at", "which", or "on" [38]. The universal list of stop words does not exist, any set can be adopted depending on the purpose.

Next, stemming reduces the morphological variants of the word to its base (stem). The algorithms enabling that operation are often called stemmers. In Python, that may be implemented with the use of NLTK [39]. For English language, there exist two stemmers: PorterStemmer and LancasterStemmer. For the purpose of this paper, the PorterStemmer (PS) was chosen and tested with the designed models because of its simplicity and the speed of its operation. PS is dated to 1979 and often generates stems that are not authentic

English words. It results from the fact that it is based on suffix stripping (examples shown in Table 1). Instead of considering linguistics to build the stem, it applies a set of algorithmic rules that decide if it is reasonable to remove the suffix or not.

Table 1. Examples of stemming with PS.

Word Before	Word After
Cats	Cat
Trouble	Troubl
Troubling	Troubl
Troubled	Troubl

Other option, known as lemmatization, is a more complex approach to searching a word's stem. In this case, the root word is referred to as a lemma. First, the algorithm identifies the part of the speech of a word; and then, based on this information, it applies appropriate normalization. As in the stemming case, lemmatization mechanisms are also provided by NLTK [39]. WordNet Lemmatizer (WNL) generates lemmas by searching for them in the WordNet Database. Examples are shown in Table 2. In the research reported here, text preprocessing was supported by the most basic lemmatization version in specific test cases. However, the method works most efficiently when one defines the context by assigning the value to pos parameter (for instance by giving it the value v—verb). Testing with the pos value defined is outside of the scope of this paper, but its usefulness may be noticed after the analysis of the impact the pos = v has on the verbs shown in Table 2.

Table 2. Examples of lemmatization with WNL.

Value pos Undefined	
Word Before	**Word After**
He	He
Was	Wa
Has	Ha
Playing	Playing
pos = v	
Word Before	**Word After**
He	He
Was	Be
Has	Have
Playing	Play

One may ask which one is better: stemming or lemmatization? The answer is that it depends on the program and the requirements that one is working with. If speed is a priority, then it is more beneficial to use stemming. When language is crucial for the application's purpose, lemmatizing should be a choice as it is more precise.

In e-mail spam filtering, the goal of building the dictionary structure (key-value with unique keys) consists of assessing the word's weight and importance given all available text documents. First, word occurrences are calculated. In the case of the application presented here, words are limited to strings of the length between 3 and 20 characters. Single letters and extremely short/long strings do not add value to the paper (they are common for both ham and spam).

First, we create two separate dictionaries (spamWords and hamWords). The function responsible for the dictionary generation returns the number n n (defined during the tests) of the most common words for each of them. Next, another function builds dictionaries which include common words (subtractFromSpam, subtractFromHam). Based on these structures, three others are defined:

1. spamDictionary = spamWords − subtractFromSpam;
2. hamDictionary = hamWords − subtractFromHam;
3. finalDictionary = spamDictionary + hamDictionary.

According to the informal research carried out by Dave C. Trudgian [40], the unbalanced distribution of spam and ham most common words significantly affects the models' accuracy. The results were improved when the final dictionary included more spam's most common words than ham's most common words. Table 3 presents the ratios implemented in the application described in this paper.

Table 3. Implemented most common words ratios (spam:ham).

No.	Spam	Ham	Total
1	150	50	200
2	900	600	1500
3	2000	1000	3000

Employing machine-learning methods to classify an e-mail as spam or ham requires representation of the text in a specific form. Given the chosen classifiers (described in Section 3.4 below), the structures they need are feature vectors. Signal-to-noise ratio (SNR) may be used to facilitate the understanding of the feature engineering concept. Although the exact definition varies depending on the function in spam detection, its basic idea is straightforward. SNR is the ratio of the input considered relevant to insignificant data. In spam classification, a signal might be a typical word occurring in spam messages, and is a noise word that is common for the given language and occurs in both spam and ham e-mails (for example, one of the stop words) [41]. If the separation of the signal from the noise is done badly, the noise can blur the true meaning of the signal. There are many feature elimination techniques that might help us to identify the critical features, as well as decide which ones should be removed. The methods used in this paper have already been shown once (Figure 2). The objective of every single stage in the process of building the dictionary is to reduce the number of irrelevant words. That is why the function responsible for the dictionary creation and the one that converts e-mails into feature vectors, start with the same lines of code, from the process of content tokenization to stemming/lemmatization.

The function that extracts features generates a feature matrix as an output. For each e-mail, it creates a vector (the array data type in Python) of the dictionary's length, filled with 0 s. After going through all preprocessing stages, it compares the e-mail's content with the dictionary (word by word). If a word from the e-mail occurs in the dictionary, 1 is added to the vector's elements. As a result, we obtain a feature matrix in which the number of e-mails is the number of rows and the dictionary's length describes the number of columns.

3.4. Methods

The solutions discussed in this paper are based on supervised learning, since they apply training sets with the target labels annotated. The generated dictionary is a mixture of labeled words that are assigned to one of the two target categories: spam or ham. The models make their predictions based on the dictionary's content. One can imagine that a question is posed to a program: if this e-mail consists of these words, is it spam or ham? The model responds to this unknown question by comparing it to the similar questions and answers (labels) it was given at the starting point.

The process of labeling (generating a dictionary in the case of the described application) is carried out with the use of a training set. A test set is used to measure the program's performance during the last step of the experiment.

Classification, interesting in the context of this paper, is one of the prevailing supervised machine-learning tasks. Its goal is to predict discrete values (might be categories, classes, or labels) for new examples (that had not been seen by the program before) from one or more features. The set of classes is finite and there are several types of learning. Spam filtering is a two-class learning (also referred to as binary classification) [41]. The program (or its part) performing a classification task is called a classifier. In this paper, the classifiers were implemented with scikit-learn (sklearn), which is a free machine-learning library for the Python programming language.

The training phase is aimed at minimizing the errors, but it is important to remember that no model is perfect. Here, we use a set of typical measures defined in the context of a confusion matrix: true negatives (TN), false positives (FP), false negatives (FN), and true positives (TP). Out of the four, the most undesirable outcome in the case of spam filtering is a false positive as it may result in losing a portion of critical information. Several parameters which allow evaluation of the classifiers are built based on the values that make up the confusion matrix: accuracy, sensitivity, specificity, positive predictive value (PPV), and negative predictive value (NPV). Accuracy was the main indicator of the classifier performance in the tests carried out in this research. In the most interesting cases, all five parameters were calculated for each tested classifier.

Below, we present three machine-learning algorithms we are comparing on the task of spam detection.

Despite its simplicity, k-nearest neighbours (k-NN) proved to be successful in a great number of supervised machine-learning tasks [42]. k-NN perform the classification of the new point (in the multidimensional space, where each point is a vector representing a sample being a single e-mail), based on k elements in its nearest distance. k-NN is sometimes called a "lazy learner", which means that it does not need to learn, but waits for classification until the very last moment. Gathering and labeling data could be referred to as a training phase. Once it is ready, the training stage is also completed. However, this fact leads to a time-consuming testing phase, during which the pairwise distances are calculated and compared.

Supervised neighbour-based learning methods are provided by the sklearn.neigbours library. k-NN may be implemented with the use of KNeighboursClassifier and the specific line of code responsible for the model definition is (when $k = 5$):

model = KNeighboursClassifier (nneighbours = 5)

When a new query point is given, KNeighboursClassifier carries out learning based on its k nearest points (*n_neighbours*). The distance function applied by us is simply the standard Euclidean distance.

When the corpus we are working with is large, there may be hundreds of thousands features in the dictionary. If we convert the text documents (for instance e-mails) into feature vectors, each of them will then have hundreds of thousands of components and most of them will be zero. Such vectors are referred to as sparse. High-dimensional data are problematic for all machine-learning tasks due to the well-known curse of dimensionality [43].This is due to the higher demand for memory and computation compared to low-dimensional vectors. This difficulty may be overcome with scipy Python library using data types that can pull nonzero elements out of the sparse vectors. The second aspect is related to the fact that with the high dimensionality comes a threat of the insufficient number of documents in the training set. It is necessary to make sure that there are enough training instances to cover all features. Otherwise, the algorithm operation may result in overfitting, where the quality results are satisfactory for the training set of samples, but not for the testing set (and the following usage cases).

Support vector machines (SVM) [44] are most typically used in classification applications, although their usefulness is broader (e.g., outlier detection). If given a labeled dataset, SVM finds a classification (separation) hyperplane by searching for the maximum distance between data points (vectors representing samples) belonging to different classes. There exist two types of SVM models: hard-margin (each point needs to be classified accurately) and soft-margin (incorrect classification is also acceptable). Contrary to k-NN classifier, it is beneficial for the SVM to operate in high dimensions [45]. By increasing the number of features, data points tend to be more efficiently separated. The points that are closest to the classification hyperplane are called support vectors. A hyperplane is also referred to as a decision boundary and separates elements belonging to different categories. The gap between the two hyperplanes drawn on support vectors is called a margin. The bigger the margin, the better.

In the application built for the purposes of this paper, two support vector classification (SVC()) based models, NuSVC() and LinearSVC(), were implemented with sklearn; with all parameters taking default values.

The family of naïve Bayes (NB) classifiers is based on the Bayes theorem that bounds absolute and conditional probabilities. In the case of machine learning and spam recognition, the probabilities can be associated with the relative frequencies of word appearance in messages (i.e., relative frequency counting of words). The second concept is the so-called naïve assumption that all features are independent of each other given the output (a class to which they belong). Although this assumption of independence rarely holds true, naïve Bayes classifier can perform a very successful classification, even if the training data does not provide many examples. Moreover, the classifiers that belong to NB family are known to be fast and simple.

The variant tested for the purpose of this paper is provided by sklearn. Multinomial naïve Bayes classifier MultinomialNB() applies the NB algorithm to multinomially distributed data [46]. It is also the most common option used in text classification. The data are represented in the form of word vector counts.

4. Numerical Results with Validation

The results were obtained based on our proprietary-software solution developed in Python 3.7.3.

4.1. Datasets Structure

The classifiers were tested on four datasets of various sizes. Three of them (composed of four datasets: enron1, enron2, enron4, enron5) are the extracts of the Enron Corpus [35]. In this phase, we propose to introduce cross-validation between different datasets (enron 1 and 4 as well as enron 2 and 5) in the training and test phases. These datasets' structure is described in detail in Tables 4–6. The fourth dataset (Table 7) is the exact copy of the part of the Lingspam corpus, used by Gregory Piatetsky-Shapiro and Matthew Mayo as a foundation for the paper described in [29]. The variety of the datasets provides the opportunity to carry out broad research.

Table 4. Dataset 1 structure.

Training (≈73%)		**Test (≈27%)**	
enron1		enron2	
Ham	Spam	Ham	Spam
351	351	130	130
702		260	
962			

Table 5. Dataset 2 structure.

Training (≈63%)		**Test (≈37%)**	
enron1		enron2	
Ham	Spam	Ham	Spam
3672	1500	1493	1493
5172		2986	
8158			

Table 6. Dataset 3 structure.

Training (≈67%)				Test (≈33%)			
enron1	enron4	enron1	enron4	enron2	enron5	enron2	enron5
Ham		Spam		Ham		Spam	
3672	1500	1500	4492	1464	1293	1464	1290
5172		5992		2757		2754	
11,164				5511			
16,675							

Table 7. Dataset 4 structure.

Training (≈73%)		Test (≈27%)	
Lingspam			
Ham	Spam	Ham	Spam
351	351	130	130
702		260	
962			

4.2. Text-Preprocessing Impact on the Dictionary

Although the purpose of using the basic tex preprocessing methods (tokenization, etc. see Section 3.3) is straightforward and easy to explain, things become complicated regarding stemming and lemmatization. This chapter shows the differences in the ten most common words in the dictionary when none of the two methods is applied and when stemming or lemmatization is implemented. The test was repeated for each dataset and the results are shown in Tables 8–11.

Table 8. Ten most common features for dataset 1.

Basic Methods		Stemming		Lemmatization	
Word	Occurrences	Word	Occurrences	Word	Occurrences
enron	462	enron	462	enron	462
nbsp	310	meter	329	meter	329
meter	298	nbsp	310	nbsp	310
pills	267	pill	279	pill	279
http	264	**deal**	**269**	**deal**	**270**
subject	229	http	264	http	264
deal	**201**	subject	229	subject	230
thanks	195	**need**	**203**	thanks	195
height	**179**	thank	202	**volume**	**188**
width	**171**	**volum**	**188**	**need**	**183**

For dataset 1, both stemming and lemmatization caused the number of occurrences of the word **deal** increased by almost 70. Moreover, the words **need** and **volum(e)** appeared in the table, pushing the words **height** and **width** out (Table 8). For dataset 2, implementing either stemming or lemmatization resulted in the increase of the number of occurrences of the word **deal** by almost 700. Furthermore, the word **volum(e)** appeared in top 10, pushing the word **forwarded** out (Table 9). Table 10 presents the results for dataset 3, which is the biggest one (includes 16,675 e-mails). Adding the function responsible for stemming or lemmatization contributed to the change in the number of occurrences of the deal word. The number increased by approximately 800. When none of the method was present in the program, word statements was the last one in the top 10 list. Once the method (either stemming or lemmatization) was defined, the word schedul(e) emerged with the significant number of occurrences (3852 for stemming and 2591 for lemmatization). Taking dataset 4 into account, the differences were less visible (Table 11). What stands out is the increased number of occurrences of the word order, which changed by almost 100 after implementing each of the two methods. With stemming, linguist appears in the top 10, pushing out the word free.

Table 9. Ten most common features for dataset 2.

Basic Methods		Stemming		Lemmatization	
Word	**Occurrences**	**Word**	**Occurrences**	**Word**	**Occurrences**
enron	6555	enron	6555	enron	6555
subject	4745	subject	4745	subject	4747
deal	2751	deal	3443	deal	3433
meter	2459	meter	2715	meter	2710
please	2230	pleas	2229	please	2230
daren	1901	thank	1945	daren	1901
thanks	1728	daren	1901	thanks	1728
corp	1644	volum	1645	corp	1644
mmbtu	1349	corp	1644	volume	1644
forwarded	1295	mmbtu	1408	mmbtu	1349

Table 10. Ten most common features for dataset 3.

Basic Methods		Stemming		Lemmatization	
Word	**Occurrences**	**Word**	**Occurrences**	**Word**	**Occurrences**
enron	7166	enron	7166	enron	7166
http	3119	deal	3879	deal	3872
deal	3073	schedul	3852	http	3108
meter	2443	http	3108	company	2839
company	2198	compani	2839	meter	2691
dbcaps	2010	meter	2697	schedule	2591
data	1996	dbcap	2010	dbcaps	2010
database	1921	data	1996	data	1996
daren	1901	databas	1908	database	1908
statements	1770	daren	1901	daren	1901

Above, significant differences were shown for only the ten most common words. Therefore, if we refer to all 200, 1500 and 3000 words, there will be even more dissimilarity in the number of word occurrences which sometimes leads to either including the word in the dictionary or not. All designed models (*k*-NN, SVM, and NB) take e-mails as input. The e-mails are represented as vectors, with the elements being the word counts, based on the content of the dictionary. Let us assume that schedul(e) is a word strongly indicating that the e-mail is not spam. For dataset 3, when the function responsible for building the dictionary does not apply stemming or lemmatization, schedul(e) is not included in the small dictionary of ten features (Table 10) and because of that it would not be taken as a valid portion of the information by the model. This could arise from the fact that the word takes many forms, such as "schedule", "schedules", "scheduling", "scheduled"—which are all counted as separate words. Using stemming or lemmatization may prevent such situations.

Table 11. Ten most common features for dataset 4.

Basic Methods		Stemming		Lemmatization	
Word	Occurrences	Word	Occurrences	Word	Occurrences
order	1190	order	1287	order	1269
report	1135	report	1213	report	1208
language	1089	mail	1107	language	1097
mail	987	languag	1097	address	996
address	959	address	1002	mail	987
e-mail	944	e-mail	960	e-mail	944
program	771	linguist	828	program	803
money	763	program	803	money	763
send	758	send	763	send	758
free	745	money	763	free	745

4.3. Spam Detection

Here, we discuss the results related to the five substages of our meta-algorithm. The substages are introduced in Table 12.

Table 12. Five substages of checking the classifiers performance.

Substage	No. of Tests	Dataset	Purpose
1	27	1	Checking the performance of the NuSVC, LinearSVC and MNB classifiers. Finding the best-performing ones for the Stage 3 testing.
2	72	1	Checking the performance of the *k*-NN classifier for various *k* values. Finding the best-performing ones for the Stage 3 testing.
3	10	2	Checking the performance of the classifiers with the specific parameters chosen in Stage 1 and 2. Finding the best-performing ones for the Stage 4 and 5 testing.
4	4	3	Checking the performance of the classifiers with specific parameters chosen in Stage 3. Recognition of the leading one.
5	4	4	Checking the performance of the classifiers with specific parameters chosen in Stage 3. Recognition of the leading one and comparison with the results obtained by Gregory Piatetsky-Shapiro and Matthew Mayo [29].

The exact results related to various substages are summarized in Appendix A given at the end of the paper. Here, we give only the main findings. Based on the Substage 1 results, the following facts may be observed:

- For all tests, the maximum accuracies were achieved by the MNB classifier.
- For each classifier, its maximum accuracy was obtained when stemming was implemented.
- The highest test average accuracy was achieved when dict = 200, with stemming.

After Substage 1, three classifiers were chosen for testing in Substage 2:

- MNB—dict = 1500, lemmatization;
- LinearSVC—dict = 200, stemming;
- NuSVC—dict = 3000, stemming.

The results based on confusion matrices are presented in Table 13. MNB classifier provides the highest probability that the e-mail classified as ham is actually a desired message (PPV = 0.887), while NuSVC performs best when predicting if the spam e-mail is in fact a spam (NPV = 0.986).

Table 13. Evaluation of the chosen classifier performance in Substage 1.

Model	Accuracy	Sensitivity	Specificity	PPV	NPV
MNB	0.923	0.969	0.877	0.887	0.966
LinearSVC	0.842	0.977	0.708	0.770	0.968
NuSVC	0.762	0.992	0.531	0.679	0.986

Substage 2 aimed to find the parameters (k and number of features in the dictionary) for which k-NN classifies the e-mails most efficiently. Because of the k-NN's computational complexity, dataset 1 (the smallest one) was chosen to conduct the experiment. The three highest accuracy values were obtained for the following parameters:

- accuracy $= 0.915$, $k = 11$, dict = 200, lemmatization;
- accuracy $= 0.912$, $k = 9$, dict = 200, no stemming or lemmatization;
- accuracy $= 0.908$, $k = 11$, dict = 200, no stemming or lemmatization.

The results of Substage 2 are interpreted with the help of graphs. Figure 3 shows the maximum accuracy obtained for k across all Substage 2 results. The maximum is obtained for $k = 11$. For values of k that are bigger than 11, the accuracy rapidly declines. This is because the greater k, the simpler the classifier. Finally, if k is too big, most of the test points will belong to the same (prevailing) class.

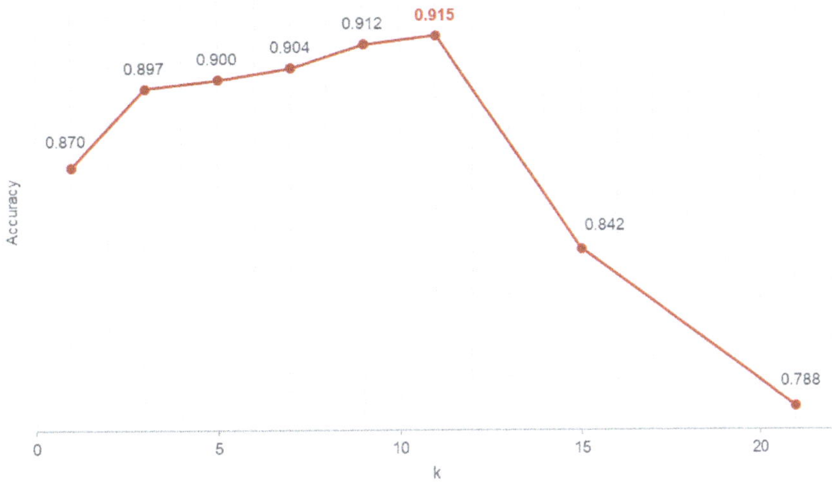

Figure 3. k-NN accuracy vs. k.

Figure 4 presents the accuracy of the average tests for each dictionary size. The higher the data dimensionality, the worse the k-NN's accuracy. The difference between k-NN when dict = 200 and dict = 1500 or dict = 3000 is significant ($\approx$0.2). To show the tendency, the power trend line was added to the graph. As we can see, the accuracy tends to change in a similar way. What is interesting, the power trend line and the exponential curve are alike. The only difference is that the arc of the first one is more symmetrical [47]. Hence, it may be concluded that in this case the accuracy experiences an exponential change.

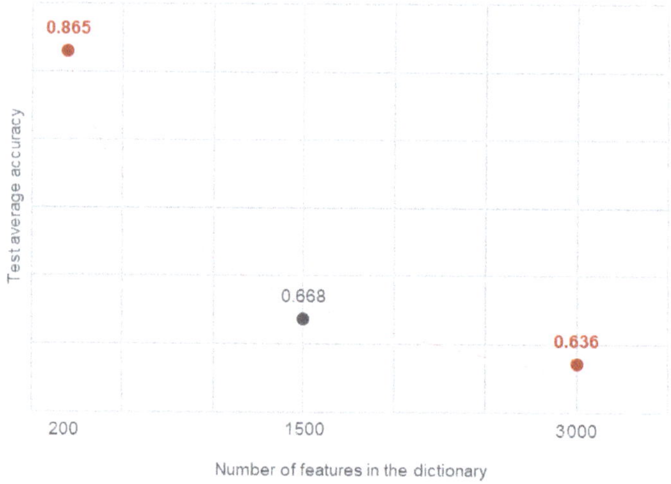

Figure 4. Accuracy of the average tests vs. dictionary length.

The three *k*-NN models with the highest accuracy were chosen to be tested in Substage 3. Table 14 presents the five indicator values. This allows performance of a more thorough evaluation.

Table 14. Evaluation of the chosen classifiers performance in Substage 2.

k	Accuracy	Sensitivity	Specificity	PPV	NPV
11 (1)	0.915	0.938	0.892	0.897	0.935
9	0.912	0.923	0.900	0.902	0.921
11 (2)	0.908	0.915	0.900	0.902	0.914

Substage 3 consisted of ten tests. The first six of them were chosen as the top results of Substage 1 and Substage 2. The other four were conducted because of their promising performance in the previous experiments. The top accuracy values were obtained for the following parameters (these four models were designated for testing in Substages 4 and 5):

- MNB (1)—accuracy = 0.914, dict = 3000, lemmatization;
- MNB (2)—accuracy = 0.909, dict = 1500, lemmatization;
- NuSVC—accuracy = 0.885, dict = 1500, stemming;
- *k*-NN $k = 11$—accuracy = 0.828, dict = 200, lemmatization.

Table 15 includes the quality metrics related to the four models that will be tested in Substages 4 and 5. When compared to MNB, NuSVC and *k*-NN have lower accuracy, sensitivity, and NPV. However, both obtained better specificity and PPV parameters. On the other hand, MNB was better at predicting the negative class.

Table 15. Evaluation of the chosen classifiers performance in Substage 3.

Model	Accuracy	Sensitivity	Specificity	PPV	NPV
MNB (1)	0.914	0.952	0.876	0.885	0.948
MNB (2)	0.909	0.950	0.868	0.878	0.945
NuSVC	0.885	0.805	0.965	0.959	0.832
k-NN $k = 11$	0.828	0.719	0.938	0.920	0.769

In Substage 4, once again, MNB models achieved the highest values of accuracy: 0.919 and 0.909. Surprisingly, k-NN with $k = 11$ performed slightly better than NuSVC. Except for NuSVC, all models obtained higher accuracy than in Substage 3.

A collection of values that facilitate assessing the performance of the classifiers in Substage 4 is presented in Table 16. A very low specificity was noted for NuSVC. The model made a considerable mistake by classifying 933 ham e-mails as spam. The number was approximately two times higher than in the case of the other classifiers.

Table 16. Evaluation of the classifiers performance in Substage 4.

Model	Accuracy	Sensitivity	Specificity	PPV	NPV
MNB (1)	0.919	0.976	0.863	0.877	0.973
MNB (2)	0.909	0.972	0.846	0.863	0.968
NuSVC	0.829	0.996	0.662	0.746	0.995
k-NN $k = 11$	0.860	0.865	0.855	0.856	0.863

Substage 5 aimed at testing the classifiers that had performed best in Substage 3, but on the dataset that was not related to Enron. The sizes of dataset 1 and the one used in this substage (dataset 4, extracted from Lingspam corpus) were the same and that is why the accuracies will be compared to those obtained in Substage 1. The training set consisted of 702 e-mails. In the test set, there were 260 messages. In both cases, when MNB classified the messages, it achieved the highest accuracy. For MNB (1), there were 3000 features in the dictionary, for MNB (2)—1500. In each case, the lemmatization was added to the program. k-NN fared the worst—much less than it achieved in Substage 1, when its accuracy was 0.915 for the same parameters. This may be a result of the source dataset content (Enron vs. Lingspam). NuSVC improved its accuracy by 0.157.

Table 17 summarizes metrics for the 4 models tested in Substage 5 and for the results obtained by G. Piatetsky-Shapiro and M. Mayo in a similar experiment on the same dataset [29]. The probability that k-NN classified a harmful message as spam is only 0.608—this is the bottom value among all results. This fact has a direct impact on the accuracy of k-NNs, which was the lowest one in this substage. Both MNB models obtained specificity and PPV equal to 1. It means there was not a single non-spam e-mail that would be misclassified as spam. Moreover, the total number of misclassified e-mails was only 10 (spam classified as ham). In Substage 5, for dataset 4, MNB classifier turned out to be nearly perfect. The results are a little better than those achieved by G. Piatetsky-Shapiro and M. Mayo [29]. This is possibly because of the more complex text-preprocessing methods that were implemented.

Table 17. Evaluation of the classifiers performance in Substage 5.

Model	Accuracy	Sensitivity	Specificity	PPV	NPV
MNB (1)	0.962	0.923	1	1	0.929
MNB (2)	0.962	0.923	1	1	0.929
NuSVC	0.915	0.862	0.969	0.966	0.875
k-NN $k = 11$	0.796	0.608	0.985	0.975	0.715
Results obtained by G. Piatetsky-Shapiro and M. Mayo [29]					
MNB	0.962	0.931	0.992	0.992	0.935

4.4. Method Validation and Discussion of Results

First, we note that text preprocessing has a significant impact on the behavior of the classifiers. There is no doubt it is always beneficial to apply the basic methods, such as conversion to lowercase (or uppercase as the effect is the same), removing stop words, digits or punctuation marks and other techniques, as described in Figure 2. Implementing advanced text-preprocessing methods (stemming or lemmatization) allows the acquisition of higher accuracy of the classification.

Second, the selected size of the dictionary (the number of features) matters. For the support vector machines and naïve Bayes classification, the results were better if the number of features was larger. On the contrary, *k*-NNs' accuracy tends to decrease rapidly for higher data dimensionality. *k*-NN obtained the highest accuracy for the smallest dictionary size. *k*-NN performs well when that data dimensionality is low. Its efficiency is also highly dependent on the *k* parameter. It might be assumed that if *k*-NN achieves the maximum accuracy for the given $k_{\max}$, the performance will experience a sharp drop for $k > k_{\max}$. Testing the support vector classification methods proved that *LinearSVC* is relatively efficient when the dataset is small. For large datasets *NuSVC* classification is more accurate.

Third, among all designed classifiers, MNB turned out to be a leader. In the relevant stages, the maximum accuracy across all results was obtained by MNB. Naïve Bayes classification is efficient in all cases but eventually returns the best outcomes when the dictionary consists of many features and the lemmatization technique is included in the application.

Fourth, the classifiers that achieved the best results when tested on the extract from the Enron Corpus, classified the e-mails even more accurately for the dataset extracted from the Lingspam corpus. This indicates that the content (words) and the structure of the data impact the model performance directly.

Fifth, the most important aspect related to validation of our work is related to the quality of the obtained results. Here, one of the most important aspect of our proposal is summarized with Table 18. It shows a signification progress in comparison with the results reported in the referenced literature (the highest values are marked in red). One can see that especially the specificity provided by our approach is attractive. It is important in the case of unbalanced datasets and applications related to anomaly detection (where spam detection is also assigned).

Table 18. Comparison of the validation results with various performance metrics.

Method	Measure	Our Result	Results Reported in the Literature
MNB	Accuracy	0.962	0.477 [7], 0.598 [24], 0.832 [23], 0.898 [11], 0.917 [14], 0.957 [10], 0.962 [29], 0.994 [5]
MNB	Sensitivity	0.923	0.496 [7], 0.897 [11], 0.931 [29]
MNB	Specificity	1.000	0.516 [7], 0.900 [11], 0.992 [29]
SVM	Accuracy	0.915	0.840 [24], 0.917 [14], 0.919 [11], 0.940 [12], 0.962 [5], 0.966 [22], 0.971 [10]
SVM	Sensitivity	0.867	0.901 [12], 0.918 [11], 0.976 [22]
SVM	Specificity	0.969	0.920 [11]
k-NN	Accuracy	0.796	0.453 [7], 0.846 [23], 0.908 [11], 0.920 [13], 0.990 [21]
k-NN	Sensitivity	0.608	0.319 [7], 0.921 [11]
k-NN	Specificity	0.985	0.478 [7], 0.887 [11]

5. Summary

The proposed multistage meta-algorithm for checking the classifiers performance, including an experimental method that involves the use of cross-validation between different datasets, allowed us to obtain reliable performance metrics in our illustrative example limited to the three important and representative classifiers. According to our results, which are consistent with other literature studies (but also typically outperform them from the viewpoint of the used metrics, especially aligned with unbalanced datasets), the multinomial naïve Bayes classifier is a method that once combined with well-thought text-preprocessing techniques as used in our meta-algorithm, may turn into the best weapon against spammers, who are becoming wiser every day. The advantage of our solution is that it can work with large datasets and give reliable results in a short time period by introducing the concept of fast recognition of the most interesting parameters. Moreover, the proposed method allows for cross-validation between different datasets in training and test phases.

Finally, the whole validation study presented in the paper based on our multistage meta-algorithm, including especially many (five) substages of cross-validation, shows that the whole method is robust. It is run on a standard desktop PC and operates within minutes to prove the results.

Author Contributions: Conceptualization, P.C.; methodology, S.R., P.C. and M.N.; software, S.R.; validation, S.R.; formal analysis, S.R., P.C. and M.N.; investigation, S.R.; resources, S.R.; data curation, S.R.; writing—original draft preparation, S.R., P.C. and M.N.; writing—review and editing, P.C. and M.N.; visualization, S.R. and M.N.; supervision, P.C.; project administration, M.N.; funding acquisition, M.N. All authors have read and agreed to the published version of the manuscript.

Funding: This research was partially supported by the National Centre for Research and Development, grant number CYBERSECIDENT/381319/II/NCBR/2018 on "The federal cyberspace threat detection and response system" (acronym DET-RES) as part of the second competition of the CyberSecIdent Research and Development Program—Cybersecurity and e-Identity and partially supported by the Polish Ministry of Science and Higher Education with the subvention funds of the Faculty of Computer Science, Electronics and Telecommunications of AGH University of Science and Technology.

Data Availability Statement: The data presented in this study are available on request from the corresponding author. The data are not publicly available due to project restrictions.

Conflicts of Interest: The authors declare no conflict of interest.

Abbreviations

Abbreviations

The following abbreviations are used in this manuscript:

ANN	Artificial Neutral Network
BN	Bayes Net(work)
BNB	Bernoulli Naïve Bayes
CNN	Convolutional Neural Network
DT	Decision Tree
EHB	Ensemble Hybrid Boosting
ET	Extra Tree
FLM	Fast Large Margin
GA	Genetic Algorithm
GB	Gradient Boosting
GLM	General Linearized Model
GNB	Gaussian Naïve Bayes (Classifier)
k-NN	*k*-Nearest Neighbours
LB	Logic Boost
LR	Linear Regression
LR	Logistic Regression
LSTM	Long Short-Term Memory
MLP	Multilayer Perceptron
MNB	Multinomial Naïve Bayes (Classifier)
NB	Naïve Bayes (Classifier)
NLTK	Natural Language Toolkit
NPV	Negative Predictive Value
PPV	Positive Predictive Value
PSO	Particle Swarm Optimization
RBF	Radial Basis Functions
RC	Random Committee
ROC	Receiver Operating Characteristic
RNN	Recurrent Neural Network
RF	Random Forest
SNR	Signal-to-Noise Ratio
SVM	Support Vector Machine
WNL	WordNet Lemmatizer

Appendix A

Here, we present the detailed results related to our numerical study, especially the ones related to the five substages.

The results obtained in Substage 1, with some additional parameters (such as test/model average and test/model maximum) are shown in Table A1.

Table A1. Accuracies obtained in Substage 1 tests.

Stem/Lem	No.			Stemming			Lemmatization			Model-Avg	Model-Max
	200	1500	3000	200	1500	3000	200	1500	3000		
NuSVC	0.750	0.757	0.758	0.758	0.758	0.762	0.754	0.735	0.738	0.752	0.762
LinearSVC	0.804	0.796	0.804	0.842	0.773	0.762	0.823	0.792	0.781	0.797	0.842
MNB	0.900	0.915	0.919	0.923	0.888	0.885	0.900	0.923	0.915	0.908	0.923
Test-avg	0.818	0.823	0.827	0.841	0.806	0.803	0.826	0.817	0.811		
Test-max	0.900	0.915	0.919	0.923	0.888	0.885	0.900	0.923	0.915		

All results of Substage 2 are presented in Table A2.

Table A2. Accuracies obtained in Substage 2 tests.

Stem/Lem	No.			Stemming			Lemmatization			Model-Avg	Model-Max
k	200	1500	3000	200	1500	3000	200	1500	3000		

The results of Substage 3 are presented in Table A3.

Table A3. Accuracies obtained in Substage 3 tests.

Test	Model	Accuracy (Substage 1/2)	Dictionary Length	Stem/Lem	Accuracy (Substage 3)
1	MNB	0.923	1500	lem	0.909
2	k-NN $k = 11$	0.915	200	lem	0.828
3	k-NN $k = 9$	0.912	200	no	0.825
4	k-NN $k = 11$	0.908	200	no	0.821
5	LinearSVC	0.842	200	stem	0.880
6	NuSVC	0.762	3000	stem	0.884
			Additional tests		
7	NuSVC	0.762	1500	stem	0.885
8	MNB	0.923	3000	lem	0.914
9	k-NN $k = 11$	0.915	1500	lem	0.795
10	LinearSVC	0.842	1500	stem	0.859

The results of Substage 4 are shown in Table A4.

Table A4. Accuracies obtained in Substage 4 tests.

Test	Model	Accuracy (Substage 3)	Dictionary Length	Stem/Lem	Accuracy (Substage 4)
1	MNB (1)	0.914	3000	lem	0.919
2	MNB (2)	0.909	1500	lem	0.909
3	NuSVC	0.885	1500	stem	0.829
4	k-NN $k = 11$	0.828	200	lem	0.860

The accuracies obtained by the classifiers in Substage 5 are presented in Table A5.

Table A5. Accuracies obtained in Substage 5 tests.

Test	Model	Accuracy (Substage 1)	Dictionary Length	Stem/Lem	Accuracy (Substage 5)
1	MNB (1)	0.915	3000	lem	0.962
2	MNB (2)	0.923	1500	lem	0.962
3	NuSVC	0.758	1500	stem	0.915
4	k-NN $k = 11$	0.915	200	lem	0.796

The confusion matrices of MNB (1) and MNB (2) are identical and presented below as Table A6.

Table A6. MNB (1) and MNB (2) confusion matrix in Substage 5.

	Ham	Spam
Ham	130	0
Spam	10	120

References

1. Bauer, E. 15 Outrageous Email Spam Statistics that Still Ring True in 2018. Available online: https://www.propellercrm.com/blog/email-spam-statistics (accessed on 6 August 2021).
2. Symantec. Internet Security Threat Report. 2019. Available online: https://www.symantec.com/content/dam/symantec/docs/reports/istr-24-2019-en.pdf (accessed on 6 August 2021).
3. Ferrara, E. The History of Digital Spam. *Commun. ACM* **2019**, *62*, 82–91. [CrossRef]
4. Dada, E.G.; Bassi, J.S.; Chiroma, H.; Adetunmbi, A.O.; Ajibuwa, O.E. Machine Learning for Email Spam Filtering: Review, Approaches and Open Research Problems. *Heliyon* **2019**, *5*, e01802. [CrossRef] [PubMed]
5. Awad, W.A.; ELseuofi, S.M. Machine Learning Methods for Spam E-Mail Classification. *Int. J. Comput. Sci. Inf. Technol.* **2011**, *3*, 173–184. [CrossRef]
6. Sharma, S.; Arora, A. Adaptive Approach for Spam Detection. *Int. J. Comput. Sci. Issues* **2013**, *10*, 23.
7. Harisinghaney, A.; Dixit, A.; Gupta, S.; Arora, A. Text and Image Based Spam Email Classification using KNN, Naïve Bayes and Reverse DBSCAN Algorithm. In Proceedings of the International Conference on Reliability Optimization and Information Technology (ICROIT), Faridabad, India, 6–8 February 2014; pp. 153–155. [CrossRef]
8. Sharma, D. Experimental Analysis of KNN with Naive Bayes, SVM and Naive Bayes Algorithms for Spam Mail Detection. *Int. J. Comput. Sci. Technol.* **2016**, *7*, 225–228.
9. Sharma, U.; Khurana, S.S. SHED: Spam Ham Email Dataset. *Int. J. Recent Innov. Trends Comput. Commun.* **2017**, *5*, 1078–1082.
10. Jawale, D.S.; Mahajan, A.G. Hybrid Spam Detection using Machine Learning. *Int. J. Adv. Res. Ideas Innov. Technol.* **2018**, *4*, 2828–2832.
11. Bassiouni, M.; Ali, M.; El-Dahshan, E.A. Ham and Spam E-Mails Classification Using Machine Learning Techniques. *J. Appl. Secur. Res.* **2018**, *13*, 315–331. [CrossRef]
12. Shajideen, N.M.; Bindu, V. Spam Filtering: A Comparison between Different Machine Learning Classifiers. In Proceedings of the Second International Conference on Electronics, Communication and Aerospace Technology (ICECA), Coimbatore, India, 29–31 March 2018; pp. 1919–1922. [CrossRef]
13. Suryawanshi, S.; Goswami, A.; Patil, P. Email Spam Detection: An Empirical Comparative Study of Different ML and Ensemble Classifiers. In Proceedings of the IEEE 9th International Conference on Advanced Computing (IACC), Tiruchirappalli, India, 13–14 December 2019; pp. 69–74. [CrossRef]
14. Shahariar, G.M.; Biswas, S.; Omar, F.; Shah, F.M.; Hassan, S.B. Spam Review Detection Using Deep Learning. In Proceedings of the IEEE 10th Annual Information Technology, Electronics and Mobile Communication Conference (IEMCON), Vancouver, BC, Canada, 17–19 October 2019; pp. 0027–0033. [CrossRef]
15. Swetha, M.S.; Sarraf, G. Spam Email and Malware Elimination Employing Various Classification Techniques. In Proceedings of the 4th International Conference on Recent Trends on Electronics, Information, Communication and Technology (RTEICT), Bangalore, India, 17–18 May 2019; pp. 140–145. [CrossRef]
16. Gaurav, D.; Tiwari, S.M.; Goyal, A.; Gandhi, N.; Abraham, A. Machine Intelligence-based Algorithms for Spam Filtering on Document Labeling. *Soft Comput.* **2020**, *24*, 9625–9638. [CrossRef]
17. Ablel-Rheem, D.M.; Ibrahim, A.O.; Kasim, S.; Almazroi, A.A.; Ismail, M.A. Hybrid Feature Selection and Ensemble Learning Method for Spam Email Classification. *Int. J. Adv. Trends Comput. Sci. Eng.* **2020**, *9*, 217–223. [CrossRef]
18. Kumar, N.; Sonowal, S. Nishant, Email Spam Detection Using Machine Learning Algorithms. In Proceedings of the Second International Conference on Inventive Research in Computing Applications (ICIRCA), Coimbatore, India, 15–17 July 2020; pp. 108–113. [CrossRef]
19. Gibson, S.; Issac, B.; Zhang, L.; Jacob, S.M. Detecting Spam Email with Machine Learning Optimized with Bio-Inspired Metaheuristic Algorithms. *IEEE Access* **2020**, *8*, 187914–187932. [CrossRef]
20. Karimovich, G.S.; Jaloldin ugli, K.S.; Salimbayevich, O.I. Analysis of Machine Learning Methods for Filtering Spam Messages in Email Services. In Proceedings of the International Conference on Information Science and Communications Technologies (ICISCT), Tashkent, Uzbekistan, 4–6 November 2020; pp. 1–4. [CrossRef]
21. Nandhini, S.; Marseline, K.S. Performance Evaluation of Machine Learning Algorithms for Email Spam Detection. In Proceedings of the International Conference on Emerging Trends in Information Technology and Engineering (ic-ETITE), Vellore, India, 24–25 February 2020; pp. 1–4. [CrossRef]
22. Saidani, N.; Adi, K.; Allili, M.S. A Semantic-Based Classification Approach for an Enhanced Spam Detection. *Comput. Secur.* **2020**, *94*, 101716. [CrossRef]
23. Hossain, F.; Uddin, M.N.; Halder, R.K. Analysis of Optimized Machine Learning and Deep Learning Techniques for Spam Detection. In Proceedings of the IEEE International IOT, Electronics and Mechatronics Conference (IEMTRONICS), Toronto, ON, Canada, 21–24 April 2021; pp. 1–7. [CrossRef]
24. Rastenis, J.; Ramanauskaitė, S.; Suzdalev, I.; Tunaitytė, K.; Janulevičius, J.; Čenys, A. Multi-Language Spam/Phishing Classification by Email Body Text: Toward Automated Security Incident Investigation. *Electronics* **2021**, *10*, 668. [CrossRef]
25. Şahin, D.Ö.; Demirci, S. Spam Filtering with KNN: Investigation of the Effect of *k* Value on Classification Performance. In Proceedings of the 2020 28th Signal Processing and Communications Applications Conference (SIU), Gaziantep, Turkey, 5–7 October 2020; pp. 1–4. (In Turkish). [CrossRef]

26. Maria; James, M.; Mruthula, M.; Bhaskaran, V.; Asha, S. Evasion Attacks On SVM Classifier. In Proceedings of the 2019 9th International Conference on Advances in Computing and Communication (ICACC), Kochi, India, 6–8 November 2019; pp. 125–129. [CrossRef]
27. Di Mauro, M.; Longo, M. Skype Traffic Detection: A Decision Theory Based Tool. In Proceedings of the 2014 International Carnahan Conference on Security Technology (ICCST), Rome, Italy, 13–16 October 2014; pp. 1–6. [CrossRef]
28. Di Mauro, M.; Longo, M. A Decision Theory Based Tool for Detection of Encrypted WebRTC Traffic. In Proceedings of the 2015 18th International Conference on Intelligence in Next Generation Networks, Paris, France, 17–19 February 2015; pp. 89–94. [CrossRef]
29. Mayo, M.; Piatetsky-Shapiro, G. Email Spam Filtering: An Implementation with Python and Scikit-Learn. 2017. Available online: https://www.kdnuggets.com/2017/03/email-spam-filtering-an-implementation-with-python-and-scikit-learn.html (accessed on 6 August 2021).
30. Radicati. Email Statistics Report, 2019–2023. Available online: https://www.radicati.com/wp/wp-content/uploads/2018/12/Email-Statistics-Report-2019-2023-Executive-Summary.pdf (accessed on 6 August 2021).
31. SpamAssasin. Available online: https://spamassassin.apache.org/old/publiccorpus/ (accessed on 6 August 2021).
32. SpamAssasin. Available online: https://spamassassin.apache.org (accessed on 6 August 2021).
33. Project Honeypot. Available online: https://www.projecthoneypot.org (accessed on 6 August 2021).
34. MailBait. Available online: https://mailbait.info (accessed on 6 August 2021).
35. Enron Email Dataset; Athens University of Economics and Business. Available online: http://www2.aueb.gr/users/ion/data/enron-spam (accessed on 6 August 2021).
36. Androutsopoulos, I.; Metsis, V.; Paliouras, G. Spam Filtering with Naive Bayes—Which Naive Bayes? In Proceedings of the CEAS Third Conference on Email and Anti-Spam 2006, CEAS 2006, Mountain View, CA, USA, 27–28 July 2006.
37. Kadhim, A. An Evaluation of Preprocessing Techniques for Text Classification. *Int. J. Comput. Sci. Inf. Secur.* **2018**, *16*, 22–32.
38. Wikipedia. Stop Words. Available online: https://en.wikipedia.org/wiki/Stopwords (accessed on 6 August 2021).
39. Jabeen, H. Stemming and Lemmatization in Python. 2018. Available online: https://www.datacamp.com/community/tutorials/stemming-lemmatization-python (accessed on 6 August 2021).
40. Trudgian, D. Spam Classification Using Nearest Neighbour Techniques. In Proceedings of the Intelligent Data Engineering and Automated Learning, IDEAL 2004, Exeter, UK, 25–27 August 2004; pp. 578–585. [CrossRef]
41. Guttag, J.V. *Introduction to Computation and Programming Using Python with Application to Understanding Data*; The MIT Press: Cambridge, MA, USA, 2017.
42. Stamp, M. *Machine Learning with Applications in Information Security*; CRC Press: Boca Raton, FL, USA, 2018.
43. Hackeling, G. *Mastering Machine Learning with Scikit Learn*, 2nd ed.; Packt Publishing: Birmingham, UK, 2017.
44. Christmann, A.; Steinwart, I. *Support Vector Machines*; Springer: New York, NY, USA, 2008.
45. Stamp, M. A Survey of Machine Learning Algorithms and Their Application in Information Security. In *Computer Communications and Networks—Guide to Vulnerability Analysis for Computer Networks and Systems*; Springer: Cham, Switzerland, 2018; pp. 33–55. [CrossRef]
46. Scikit-learn. Multinomial Naive Bayes. Available online: https://scikitlearn.org/stable/modules/naivebayes:htm (accessed on 6 August 2021).
47. Excel Trendline Types, Equations and Formulas. Available online: https://www.ablebits.com/office-addins-blog/2019/01/16/excel-trendline-types-equations-formulas (accessed on 6 August 2021).

electronics

Article

Dataset Generation for Development of Multi-Node Cyber Threat Detection Systems

Jędrzej Bieniasz * and Krzysztof Szczypiorski

Institute of Telecommunications, Faculty of Electronics and Information Technology, Warsaw University of Technology, 00-661 Warsaw, Poland; k.szczypiorski@tele.pw.edu.pl
* Correspondence: J.Bieniasz@tele.pw.edu.pl

Abstract: This paper presents a new approach to generate datasets for cyber threat research in a multi-node system. For this purpose, the proof-of-concept of such a system is implemented. The system will be used to collect unique datasets with examples of information hiding techniques. These techniques are not present in publicly available cyber threat detection datasets, while the cyber threats that use them represent an emerging cyber defense challenge worldwide. The network data were collected thanks to the development of a dedicated application that automatically generates random network configurations and runs scenarios of information hiding techniques. The generated datasets were used in the data-driven research workflow for cyber threat detection, including the generation of data representations (network flows), feature selection based on correlations, data augmentation of training datasets, and preparation of machine learning classifiers based on Random Forest and Multilayer Perceptron architectures. The presented results show the usefulness and correctness of the design process to detect information hiding techniques. The challenges and research directions to detect cyber deception methods are discussed in general in the paper.

Keywords: cybersecurity; data science; machine learning; datasets; cyber threats modeling; multi-agent systems; cyber deception

Citation: Bieniasz, J.; Szczypiorski, K. Dataset Generation for Development of Multi-Node Cyber Threat Detection Systems. *Electronics* **2021**, *10*, 2711. https://doi.org/10.3390/electronics10212711

Academic Editor: Qusay H. Mahmoud

Received: 28 September 2021
Accepted: 3 November 2021
Published: 7 November 2021

1. Introduction

In recent years, threats in cyberspace have evolved into well-organized, long-term, and resource-intensive intrusion campaigns known as Advanced Persistent Threats. As a result, there is a need to increase research into and the implementation of new cyber defense solutions, methods, operations, and procedures. Cybersecurity research activity is very broad, but it could be summarized by offering new developments and solutions for new use cases for each function of the NIST Cybersecurity Framework (CSF) [1]. Examples of a tailored solution that has been developed based on a new use case for cybersecurity would be physical unclonable functions [2]. The need for a new secure identification and authentication method was driven by the restricted requirements of cyber-physical systems. The resulting concept offers low computational cost and resource requirements, whereas Identify and Protect functions are easily provided for such systems. The same strategy for research in cyber threat detection is followed in this paper. One of the most important scientific and technological areas that are increasingly being used for cyber defense is data science and data-driven methods. The following list summarizes the five areas of work around data science in cybersecurity:

1. Modeling cyber threats to leverage across the data pipeline, from observation to flaw detection to actionable cyber threat data—for example modeling techniques: NIST Incident Response [3], Cyber Kill Chain [4], and MITRE ATT&CK [5].
2. Applying new models of cyber threats and setting up platforms to simulate them in near-production environments.
3. Elaboration of detection algorithms that are technically feasible in modern networks and systems.

4. Sharing the collected information on cyber threats and applying this information in technical solutions.
5. Accelerating the decision-making process within cybersecurity teams and departments together with the company's decision-makers.

For a more detailed overview of the challenges of data-driven cyber threats and intrusion detection, see [6]. This paper presents efforts to create an end-to-end process that combines aspects 1, 2, and 3. This is possible by extending the established approach for cyber threat detection systems [7], mainly realized by Network and Host Intrusion Detection Systems (NIDS, HIDS), to Multi-Node Cyber Threat Detection (MNCTD) systems. The cyber defense action matrix [4] shows that classical NIDS or HIDS can be used individually in four out of seven phases of the Cyber Kill Chain—weaponization, exploitation, installation, and C2 (Command and Control). MNCTD goes further and proposes the combination of the detection capabilities of all steps into a cyber threat detection system that focuses on network communications.

Any research project on such models, algorithms, and systems suffers from the availability of the right data. The recognized problem of availability of specific datasets for the particular research hypothesis and preparation of appropriate datasets for cyber threat detection is a critical challenge [8]. In the first part, this paper summarizes the current state of datasets for cyber security research available in academia and industry. It then proposes an approach to create specific datasets for hybrid cyber threat detection systems research, as such datasets are scarcely available in the public domain. Most of these available datasets focus on network attacks, such as Distributed Denial of Service (DDoS), SSH Brute Force, or botnet communication over open text protocols such as HTTP or IRC. Modern cyber threat modeling shifts thinking to identify threat phases (Cyber Kill Chain) or tactics realized through various techniques (MITRE ATT&CK) to block a threat as early as possible. Developing new solutions for cyber defense is about defining the aspects of the threat using the chosen modeling method and creating certain observable indicators that can be analyzed by detection algorithms. Such an approach could provide the desired ability to block and counter cyber threat campaigns as soon as indicators of threat are detected. Another novelty of this paper is the emphasis on the increasing importance of detecting information concealment techniques used in cyber attacks, especially in APT campaigns. One of the most important reports on the rise of stegomalware was the June 2017 McAfee report [9]. In it, steganography was identified as the emerging element used in new malware campaigns. Information hiding techniques can be used at any stage of a cyber threat campaign, but the focus is on methods that work with communication activities over networks:

- Delivery and C2 Phase designated by Cyber Kill Chain methodology.
- Defense Evasion, Exfiltration, and C2 Tactics classified by MITRE ATT&CK methodology.

This paper presents the possibility of preparing datasets with information hiding techniques to develop the concept of a Multi-Node Cyber Threat Detection platform. The created multi-agent system for collecting network packet traces was applied in the automatically generated environment of network nodes and with the random setup of malicious pairs of hosts (sender-receiver) per experimental run. Then, the collected sample datasets were used in the data science workflow for cyber threat detection. The classical pipeline of a data science experiment includes data cleaning, feature selection, under- or over-selection of rare class examples, and development of the default solution for classification problems.

The structure of the paper is as follows:

- Section 2 briefly presents the available datasets and the systematic approach to evaluating self-generated datasets. It builds the context for the need for the research in this paper:

    - Generating datasets for cyber threat detection research in the domain of information hiding techniques applied by modern malware and malicious cyber operations like APTs.
    - Establish the possibility to generate these datasets in different and randomized networking environments with a varying set of sources and destinations for the simulated cyber attacks.
- Section 3 presents the methodology used to establish the framework for end-to-end dataset generation for cyber threat detection research. It follows the context of the research established in Section 2. This section covers the concept of the system for capturing network traffic in a multi-node setup, simulation of benign and malicious network flows and scenarios for generating the final datasets, and a simple methodology for generating datasets.
- Section 4 shows examples of data science experiments enabled by the generated data. This part presents the empirical evaluation of the datasets generated by the methods introduced in Section 3.
- Section 5 concludes the paper with a summary of the results and further research directions that could be based on this paper.

Contributions of the Paper

The main contributions of the paper are:

- An approach to collect datasets for cyber threat detection research in a multi-node setup using the developed agent system. This contribution goes far beyond the state-of-the-art presented in Section 2.3. The majority of the available datasets are focused on providing indicators for simulated cyber attacks from single endpoints like central collectors, whereas this research tackles multi-node cyber data collection to follow the cyber attack path of execution.
- Application of the information hiding techniques in communication networks [10] to research cyber threats as an emerging problem in cyberspace. The paper shows how to generate network data streams using information hiding techniques. This is a key effect, as most of the state-of-the-art datasets presented in Section 2.3 include the classic types of cyber attacks only with no covert communication samples. The introduction of this paper and Section 2.4 show the increase in malware applying information hiding techniques for Command and Control channels, to exfiltrate data or to persistently maintain the presence in the compromised environments. It means that any research into cyber threat detection methods in the area of steganography used in malicious operations has never been as important.
- Development of an automated and randomized tool for setting up network configurations (nodes and links) when performing simulations of network communication scenarios. According to the state-of-the-art cyber data collection environments of the datasets presented in Section 2.3 they were mostly configured once with the chosen sources and destinations of cyber attacks. The contribution of this paper offers a solution to mitigate the biases in datasets related to the shape and topology of the environment in which they were collected.
- The execution of reference cyber threat detection experiments on the collected datasets. Most of the state-of-the-art research papers related to datasets included in Section 2.3 present the datasets and collection process. This paper contributes to the approach applied by the authors where the collected datasets were evaluated to be feasible in data-driven cyber threat detection workflows.

2. Related Work

2.1. Multi-Node Cyber Defense Solutions

The systems that could be built upon the results of this paper combine the idea of network intrusion detection systems with the concept of multi-agent systems into a multi-node cyber threat detection system. In the last 30 years, it has been investigated in different aspects related to architectures, computational aspects (for example, involving AI), effective collaboration within multi-agent platforms, and applications. One of the milestones is the paper [11], where the idea of intrusion detection using autonomous agents was proposed. Publications such as [12–14] combined are drawing a comprehensive review of the state-of-the-art in multi-agent cyber defense solutions.

Nowadays, a cyber defense based on multi-agent systems is recognized as a modern and very efficient approach with continuous emerging. Interest in such systems has been extensively revisited recently within academia, industry, law enforcement agencies, and even the military. In [15], the author developed the idea that intelligent autonomous agents will be the standard on the battlefield of the future. It means that intelligent autonomous cyber defense agents are going to become the main element of any entity involved with the battlefield, where cyberspace will become the crucial area of conflict. The paper introduced several novel ideas with summarization of the other ones into the reference architecture of any multi-agent system for cyber defense.

A current industrial application of such systems could be any Internet of Things networks or, in general, cyber-physical systems and networks. The justification behind this is that these systems are by default distributed and multi-node. Furthermore, the requirements on the lightness of the computation on the nodes implicates that only multi-agent cyber threat detection solutions would fit such environments. For example, the state-of-the-art in this field from two papers [16,17] introduce the intrusion detection system in connected vehicles (Vehicle-to-Vehicle, V2V). The system presented in [16] consists of the part that is analyzing the node of the environment—a vehicle—with the option of centralized data analytics in the cloud. The main contribution of the authors was to consider each single element of the vehicle as the valuable source of data to detect cyber threats. Next, it was proposed to combine the real time data from such different and distributed elements together for the classification algorithm based on Bayesian networks. The paper [17] investigates such multi-agent cyber threat detection within a single vehicle more deeply in terms of how to combine data from different sensors to detect intrusions. Such an approach complies with the general idea of multi-agent intrusion detection systems and it is an important example of how to apply it to solve the modern problems of security in cyberspace. As the use case of a connected vehicle will be rapidly adopted, cyber defense solutions involving multi-agent concepts crucially need to be developed.

2.2. Generation of Datasets for Cyber Threat Detection Research

The general prerequisite for any discovery problem to be addressed by data science methods is to have the right data. There are three main approaches to obtaining data for cyber threat detection:

- Collecting data from actual production networks and cyber intrusions,
- Building models of production networks and simulating network communications (malicious and benign),
- The use of mathematical, statistical, machine learning, and other algorithms to generate the data.

The first approach is highly desirable, as working on actual data should guarantee low-fault detection algorithms that are ready for actual cyber attacks. The main problem with the reliability of this approach is that few organizations could use such data for cyber threat detection research. Cyber attacks are very rare if we consider the total observation time. This means that it would take a very long time to collect enough examples to train a detection algorithm on these indicators. Another challenge with such data is the privacy concern. It is impossible to share such information, so the cybersecurity community cannot benefit from it for cyber threat detection.

The simulation approach is usually used in modern research into intrusion detection systems in industry and academia. One of the most well-known research institutes in this field is the Canadian Institute of Cybersecurity (CIC) at the University of Brunswick. The institute has published nearly 30 datasets over the past decade, while researchers have developed reference methods for generating such data. A 2018 paper [18] summarizes the current approach to generating the simulated networks and data for cyber threat detection research over them. The main outcome was the development of a parametric configuration of the network communication patterns to be simulated, called profiles. This improved the quality of the resulting datasets. Another systematic approach was presented in [19]. This paper adds the new idea of simulated datasets for cyber threat detection systems based on a novel architecture:

- Collecting sensors distributed on network nodes,
- Allowing for continuous communication and coordination between sensors,
- The use of a central processing unit to improve detection decisions,
- The automation of the network scenarios in which the data was collected,
- The use of data science methods to oversample the least representative samples of malicious data.

The details are presented in Section 3. The latter approach exploits the mathematical foundations of modeling and data analysis, in particular, to apply machine learning methods for data generation. It could help to increase the similarity of generated data with actual production or to address shortcomings of simulations (complementary between approaches). An example of applying machine learning to improve detection rates and complement the small number of malicious samples is presented in [20]. It uses adversarial machine learning methods for cyber threat detection research. Generative Adversarial Networks (GAN) are implemented to generate synthetic samples. Then, the module IDS was trained on them along with the original samples. It also fixes the problems of unbalanced or missing data on input. This approach greatly improves the performance of the IDS detection algorithm. The major challenge in applying machine learning for cyber threat detection is the explainability and transparency of such algorithms.

2.3. *Availability of Datasets for Cyber Threat Detection Research*

Historically, the first milestones in the public availability of datasets for cyber threat detection research were in 1998–1999, when the DARPA'98 and KDD'99 datasets were released. Since then, many other and different datasets have been created, but there are still not enough publicly available datasets for cybersecurity research. This section presents some examples of publicly available datasets that are generally recognized as comprehensive, well-prepared, and appropriate for cybersecurity research on cyber threat detection systems.

Canadian Institute of Cybersecurity datasets: ISCX 2012 Dataset [21] was the first participation of the Canadian Institute of Cybersecurity that provided a systematic approach for creating datasets for cyber threat detection systems research. They introduced the concept of profiles, which contain detailed descriptions of intrusions and abstract distribution models for lower-level applications, protocols, or network entities. Previously, they analyzed real-world traces to create these profiles. The dataset created included benign and malicious network traffic traces of HTTP, SMTP, SSH, IMAP, POP3, and FTP. The NSL-KDD ISCX Dataset [22] was created as a solution to the inherent problems of the original KDD'99 dataset. It still suffers from some of the problems and may not perfectly represent real-world networks. Nevertheless, it can be used as a useful benchmark dataset to help researchers compare different cyber threat detection methods. The CIC 2017 dataset [23] contains benign and the most recent widespread attacks stored as network traffic traces from actual real-world executions. Implemented attacks include Brute Force FTP, Brute Force SSH, DoS, Heartbleed, web attack, infiltration, botnet, and DDoS. Due to the nature of the prepared profiles, they can be directly applied to a variety of network protocols with different topologies to create a dataset for specific requirements. The CSE-CIC 2018 dataset [24] follows the pattern in the scaled infrastructure of 500 devices. The dataset provides the network traffic traces and system logs from each of these devices.

UNSW-NB15 Dataset [25]: The raw network packets of the UNSW-NB 15 dataset were created in the Cyber Range Lab of the Australian Center for Cyber Security (ACCS). It contains a mixture of actual normal activities and synthetic current attack behaviors from fuzzers, backdoors, DoS, exploits, generic cyber attacks, reconnaissance, shellcode, and worms. Argus, Zeek (formerly *BroIDS*), and the authors' tools were used for data collection. Class tagging was also provided. The number of records in the training set was 175,341, and the testing set was 82,332 from different types of network traffic (benign and malicious).

UGR'16 Dataset [26]: The dataset was created with real traffic and actual attacks. The network traffic was recorded by Netflow v9 collectors strategically placed in the network of one of the Internet Service Providers from Spain. It consists of two datasets split into weeks:

- CALIBRATION set was collected from March to June 2016 (four months) with accurate background traffic data.
- itemize TEST was collected from July to August 2016 with factual background and synthetically generated traffic data of various known attack types.

The main advantage of this dataset is its usefulness for evaluating cyber threat detection algorithms with a long-term perspective. The models can also take into account differentiation by day/night or working days/off days.

CAIDA Datasets: The Center for Applied Internet Data Analysis (CAIDA) collects various types of data from geographically and topologically diverse locations and makes these data available to the research community. Information was collected from active and passive measurement infrastructures that provide insights into global Internet behavior. CAIDA collects, curates, archives, and shares the datasets resulting from these measurements. It also processes and shares several derived datasets. Datasets through April 2016 are available at [27]. One of the most well-known CAIDA datasets is the DDoS 2007 dataset, which contains network traffic traces from large-scale distributed denial-of-service attacks. More recent datasets are made available on the Impact Cyber Trust Project [28] system. The Information Marketplace for Policy and Analysis of Cyber-risk Furthermore, Trust (IMPACT) project was created by the U.S. Department of Homeland Security to support the global cyber-risk research community through the coordination and development of real-world data and information sharing capabilities. The IMPACT project enables the sharing of empirical data and information among the global cybersecurity research and development (R&D) community in academia, industry, and government to accelerate solutions to cyber risk and infrastructure security. Datasets are available exclusively to researchers from the U.S. and collaborating countries.

2.4. *Malware with Information Hiding Techniques Applied*

Network steganography, as a branch of information hiding techniques, is rapidly evolving and has attracted tremendous interest from cybersecurity researchers since the paper [29]. Any network steganography technique must meet three conditions [10]:

- Modified properties of the protocol;
- Modified properties of the protocol may refer to mechanisms related to inadequacies of the communication channel, the nature of the messages exchanged, or their form;
- Communication parties trying to prevent the observer from detecting the transmission of data using information hiding techniques.

The Morto worm [30], a malware with network steganography capabilities, used records stored on Domain Name System (DNS) servers to communicate with C2 servers. This was the first actual implementation of network steganography in malware ever discovered. Over the years, DNS has proven to be one of the most popular network protocols abused for information concealment techniques. Any system from IT that has access to the Internet must use it, so port 53 is wide open and allowed by firewalls and cyber threat detection systems. The DNS protocol is characterized by open text messages that provide many opportunities to hide data in them using text steganography methods. Another protocol that has been used for network steganography in malware in recent years is the Secure Shell (SSH) protocol. It was discovered in 2013 in the Fokitor Trojan [31].

The motivation to use SSH for such operations is the same as DNS: widely used in IT systems, port 22 open and allowed. In this method, the SSH protocol connections merely carried the hidden information as a payload. The Regin malware [32], discovered in 2014, was equipped with three mechanisms to prevent network communication:

- Stealth data tunneling in ICMP protocol traffic (ping).
- Insertion of steering commands in cookies in the HTTP protocol header.
- Insertion of steering commands into specially prepared TCP protocol segments or UDP datagrams.

This is the ongoing trend of implementing different steganographic C2 channels and using them depending on the deployment conditions. Steganography, a cyber deception method, provides the ability to bypass the detection and measures of standard network security applications, such as blocking by firewalls or triggering alerts by cyber threat detection systems.

Another trend is the combination of different methods to hide information, e.g., combining multimedia steganography with hidden communication via TCP/IP protocols. The typical approach for combining multimedia steganography and network communication to form hybrid steganography is as follows:

- Use of multimedia steganography to hide the data.
- Use of standard protocols of the TCP/IP stack, especially application network traffic, to smuggle multimedia files between victims and attackers either directly or via C2 servers.

The first practical application of such an approach was a 2011 malware campaign. Duqu [33] used multimedia steganography to hide data in JPEG images and then sent them to the C2 server. This communication looks like an ordinary image file transfer, but in reality it is used to establish a covert C2 channel. A similar technique was used for the 2014 Zeus Trojan morph, Lurk [34], where images were the carriers of the hidden control commands. In the following years, the C2 channels used in modern cyber threat campaigns were considered for information hiding techniques. More recently, the techniques have evolved, spreading multimedia steganography over open social networks (OSN) and adding methods of text steganography. This introduced a new level of complexity to any forensic analysis, making it a problem similar to finding a needle in a haystack. An example of a practical application is Hammertoss APT, applied by the group APT29 [35]. They used Twitter to exchange URLs to image files that contained hidden data. Each Twitter message also contained a specially prepared hashtag needed to decode the hidden part of the image. The project attracted interest from cybersecurity researchers who were looking for models to define detection techniques, as the classical signature approach was insufficient. Interesting proofs-of-concept of steganography systems include:

- Stegobot [36]—one of the pioneering systems using OSNs as an overlay network for the technical operations.
- Instegogram [37]—a technique that uses the image feed of a given Instagram account to decode C2 messages from images. The main achievement here was using a popular internet service to smuggle malware communications.
- StegHash with SocialStegDisc [38]—The StegHash technique was used to distribute multimedia files with hidden data portions across many Internet services and accounts. The mechanism of hashtags creates an invisible chain through which the original message can be recovered. SocialStegDisc implemented the StegHash technique to address the scheme in a novel steganographic file system.

Therefore, the use of hybrid and network steganography to breach the security of computer systems, in particular, is an important area of research to identify vulnerabilities and methods to combat them. This is the critical goal of this work, to improve the security of cyberspace.

3. Generating Datasets for Cyber Threat Detection Research

3.1. Application of Multi-Node Cyber Threat Detection System

A multi-node cyber threat detection system operates in the environment of distributed network devices running open operating systems (e.g., Linux), mainly programmable routers. Each router contains the execution environment of mobile agents that are interconnected to form a platform that controls the Central Unit. For the purpose of this study, the monitoring mode of such a system is considered.

On the execution platform, it is possible to run agents with different purpose settings:

- Agents collect network traffic logs in a specific format and send this data to the Central Unit for analysis.
- Agents equipped with motion logic that follows the developed algorithm for computing anomaly metrics and cooperates in selecting additional areas of the observation network. The goal is to discover the sources of the attack.

To build a multi-agent peer-to-peer communication, the concept of the actor system [39] has been used. The main purpose of the actor system is to develop a high-level and non-blocking parallel execution model for computation. The atomic execution units, called actors, execute their assigned tasks and then share the results via the message box communication abstraction. The actor system is responsible for creating and managing the lives of the actors (agents) in various distributed environments. The scheme of the prepared platform is shown in Figure 1. It shows the main nodes of the architecture:

- The Central Unit node, which manages the actor system of the whole platform and coordinates the life cycle of the distributed agents and of itself. More details are presented in Table 1.
- A router with the actor system instance in which the single node managing agent is instanced and connected with the whole platform managed by the Central Unit. Furthermore, this agent could spawn other node agents to operate different functions. Figure 1 shows such an agent called the Interface Sniffer Agent.

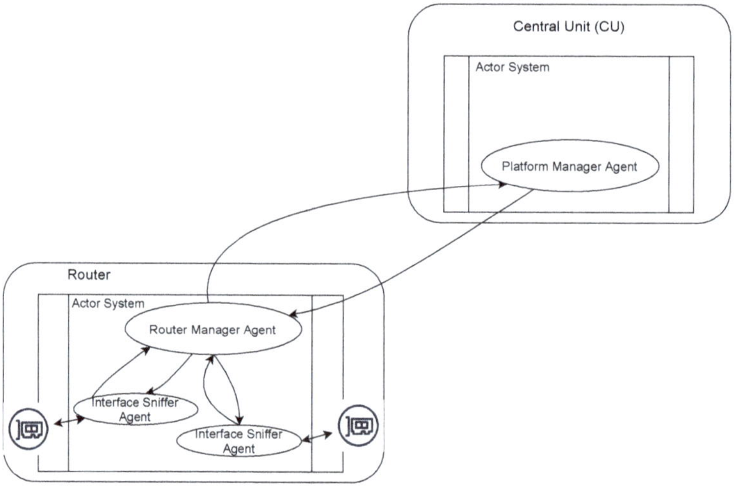

Figure 1. Scheme of monitoring platform based on actor system approach.

As the prototype of the platform is utilized in the monitor mode (sniffing and collecting network data), the main functionalities to be included within the main nodes of the system are:

- Sniffing network traffic on all interfaces of a router;
- Storing PCAP files at the nodes;

- Collecting PCAP files across nodes in the Central Unit.

The interface Sniffer Agent would provide the first and second functionality. The third is implemented by the communication scheme between the Platform Manager Agent, Router Manager Agents, and Interface Sniffer Agents. The whole platform (node agents and Central Unit) provides the other functions, such as life cycle management, PCAP file management, or controlling the operation mode of the system. Table 1 summarizes the operational aspects for each main component of the platform: Central Unit, a router, Router Controlling Agent, and Internal Sniffing Agent. It includes the functional role (*Objectives* column) realized by each of them and the communication patterns with the other components that are utilized to fulfill the role (*Communication patterns* column).

Table 1. Summary of operational aspects of multi-node Network Traffic Monitoring Platform.

Module	Objective	Communication Patterns
Central Unit	1. Hosting main control agent of entire platform 2. Managing entire platform 3. Managing joining process of routers hosting platforms 4. Managing joining process of router control agents 5. Maintaining status of remote router hosting platforms and control agents on each of them 6. Receiving message about new PCAP file available on router 7. Downloading PCAP file from selected remote router 8. Managing collection of PCAP files downloaded from remote routers	1. Initializing the connection between Central Unit and any router hosting platform 2. Requesting creation of new controlling agent on remote router platform 3. Submitting configuration settings to newly created controlling agent on remote router platform 4. Receiving notification about availability of new PCAP file 5. Retrieving PCAP file over HTTP protocol
A router	Computing and networking platform that hosts remote portion of entire agent system	Receiving requests to create new control agent
Router Controlling Agent	1. Router management 2. Joining platform agent system 3. Managing network interfaces and setting up sniffing on them 4. Providing HTTP server through which PCAP files can be downloaded from Central Unit 5. Managing status of availability of new PCAP files	1. Receiving configuration settings from Central Unit 2. Requesting creation of internal agents to sniff network interfaces 3. Controlling start and stop of network sniffing per selected interface 4. Notifying Central Unit about availability of new PCAP file
Internal Sniffing Agent	1. Creating sniffing process on bound network interface 2. Managing sniffing strategy 3. Managing end of sniffing action by passing callback to router control agent	1. Receiving request to start sniffing 2. Receiving request to stop sniffing 3. Collecting information when new PCAP file is available 4. Notifying router control agent of availability of new PCAP file

The concept of architecture realized by the presented proof-of-concept is easily expandable by embedding processing and detection algorithms together with any distributed computing strategies to be imposed within the system. Any complexity in terms of logical distribution of the processing and decision-making could be considered. However, the constraints and limitations of such an expansion for any logical workflow of the cyber threat detections and mitigations are driven by:

- The performance related to the type of the networks and its protocols.
- The data flow rates and processing performance.

- The physical bandwidth of interfaces within network nodes (routers and the other network appliances).
- The computational resources within a single network node where the cyber threat detection agent would operate.
- The multi-node cyber threat detection system management links to the performance.

Network packets need to be processed in the time imposed by the bandwidth of the interfaces within a network node. If there is an objective to detect and react to cyber threats inline, then the detection and computation architecture is required to be able to draw decisions within the time frame of the network packet processing. For 10 Gb/s networks, one packet of 300 bytes (average size in the Internet) needs to be processed in 240 nanoseconds. Otherwise, the system would process the copy of the data, so the main constraints will be limited to copy operation, transferring data to the other agents, and the size of the generated data in time (directly based on the network flow data rates).

In fact, the proof-of-concept was implemented in Python as the most efficient for fast prototyping. It was used in the monitor mode only to collect PCAPs as datasets. However, the real production multi-node cyber threat detection system should be implemented in more suitable hardware and software for technological stacking. Software programming languages for data processing within constrained environments in terms of computational resources are C, C++, or Rust. If the software processing cannot fulfill the processing requirements, then hardware solutions to accelerate the computations needs to be considered, such as ASICs or FPGAs. The Central Unit node or any other node considered in general as the "computational center" could be built upon Big Data technological stacks characterized by high scalability, efficiency, and possibility to parallelize computations. The main limitation would be related to the available hardware resources and if it is possible to implement several computational servers as the component of a production multi-node cyber threat detection system.

3.2. Network Traffic Streams Simulations

3.2.1. Malicious Network Data Streams

The network data streams within the scope of this paper must contain steganographic techniques of the various types. For this work, the implementation could be simple so that the required data can be generated for further research. The choices of such methods are:

- Method based on intentionally lost packets that can carry hidden payloads. It could be implemented in various network protocols such as SIP or RTP, with one important characteristic rule—a packet is detected as lost even if it eventually reaches the destination, it is simply discarded. No verification is performed. This fact can be directly applied to network steganography in the following way:
  - Some packets must be intentionally delayed to be detected as lost.
  - The payload of such packets could be overwritten to carry steganograms.
  - When such a packet finally arrives at its destination, it is simply discarded. If a steganographic receiver is installed, it could intercept these packets to extract the hidden payload.
- Method based on modulating the transmission time between packets to encode bits '0' and '1'. Delay-based network steganography is a type of time-based steganography. It uses modulation of the transmission times of successive packets in network traffic to encode '1' and '0' bits of hidden data. Probably any network protocol can be used for such a method. The secret between sender and receiver is to encode and decode the hidden data in the temporal relationships between the packets. The sender side must be parameterized with the type of distribution used to generate the network stream. The receiver side must also be parameterized with this distribution and with decision thresholds in the decoding module.

The dedicated applications were prepared as the element of the whole end-to-end framework for cyber threat detection research. The signalization over lost packets in the multimedia stream of packets utilizes the RTP protocol. The hidden communication over packets with the modulated time of sending uses the ICMP protocol. The prepared applications could also be executed in benign mode to generate the expected network flows of the selected protocols (RTP or ICMP).

3.2.2. Benign Network Data Streams

The approach to generating benign network traffic was developed by analyzing the typical patterns of network communications in consumer and enterprise networks LAN/WAN. Several specific applications and protocols were identified:

- Surfing the Internet and using the HTTP protocol.
- VoIP communication using SIP, RTP, UDP, HTTP, and TCP protocols.
- Video streaming using RTP and HTTP protocols.
- Data transfer using HTTP, FTP, SSH/SFTP, TCP, UDP, or email protocols.
- Using network-related protocols such as ICMP.

For this study, some publicly available applications and scripts were used to simulate such traffic. The complementary method uses publicly available network traces to replay them within a network. The applications mentioned in Section 2.4 are also used in benign mode to generate legitimate traffic without using information hiding techniques.

3.2.3. Engine of Generation of Network Topologies for Experimentation

A network emulation engine should be used for functions such as:

- Enabling rapid prototyping of new use cases.
- Enabling automatic generation of new network topologies, i.e., setting up a new dataset.

When generating a new topology, the basic features must be specified, such as:

- The number of routers and hosts,
- IP address ranges and routing,
- Whether to provide access to the Internet,
- Whether a firewall should be included,
- Placement of the Central Unit of the entire Cyber Threat Detection Agent system.

Based on these parameters, a random graph should be generated and then fed into a network emulation engine via an API or configuration file. Preparing such an automation promises to minimize any bias in the network topologies on the measured effectiveness of the newly developed cyber threat detection algorithms. This means that well-generalized cyber threat detection models should be created that can work equally efficiently in any network topology.

For this research, we developed such a tool for the automatic generation of test application scenarios, which consist of the following elements:

- The backend network engine and simulation tool—GNS3 [40];
- The text file to hold the network configuration—nodes and their types;
- the main script in Python, which
  - Interprets and validates the entered network configuration,
  - Randomly generated connections between nodes,
  - Automatically sets up the network using the GNS3 API;
- The set of scripts in Python to configure the senders and receivers of the network traffic depending on the purpose—to run benign, malicious, or mixed scenarios using the agent system presented in Section 3.1.

It should be noted that the crucial aspect of the prepared solution is the automated and randomized mechanism for:

- Generation of links between the configured set of nodes.
- Selection of senders and receivers for each network data stream profile (benign or malicious).

Such an approach allows the generation of datasets from many network scenarios and data generation configurations. It could also provide the ability to mitigate any factors associated with configuration bias across the broad spectrum of research in data-driven cyber threat detection. Data sets were collected in specific network scenarios with a small degree of variation in the sender-receiver network data configuration (benign or malicious).

3.2.4. Generation of Example Datasets

As presented in Section 3.2.3, the application to automatically generate the network configurations and network communication scenarios was implemented in this article. Figure 2 shows an example output topology of the fully working network of nodes (routers, PC hosts, firewall, Internet connection) prepared by this tool for the purpose of this article.

Among the setup presented in Figure 2, the seven different network scenarios of hidden communication between transmitters and receivers were run. The configuration for each scenario is shown in Table 2. A given scenario (Table 2, first column) consists of the setup of the sender and receiver for a hidden communication over lost packets (second column in Table 2) and the setup of the sender and receiver for a hidden communication over lost packets (third column in Table 2). The order of operations to collect the datasets is as follows:

1. Select nodes (computer hosts) that represent the pairs of a sender and a receiver of a hidden communication for both techniques in this study.
2. Run benign network communication patterns along with hidden network communication.
3. Collect PCAPs for each network node used.
4. Drag all PCAPs to the Central Unit of the platform.
5. Finish generating and collecting data.

Table 2. Setup of pairs of sender-receiver for generation of hidden communication simulations.

Scenario	Hidden Communication over Lost Packets	Hidden Communication over Time Modulation
Scenario 1	Sender: Host H1 Receiver: Host H2	Sender: Host H6 Receiver: Host H7
Scenario 2	Sender: Host H2 Receiver: Host H3	Sender: Host H1 Receiver: Host H6
Scenario 3	Sender: Host H3 Receiver: Host H4	Sender: Host H1 Receiver: Host H2
Scenario 4	Sender: Host H4 Receiver: Host H5	Sender: Host H2 Receiver: Host H3
Scenario 5	Sender: Host H1 Receiver: Host H5	Sender: Host H3 Receiver: Host H7
Scenario 6	Sender: Host H6 Receiver: Host H7	Sender: Host H4 Receiver: Host H5
Scenario 7	Sender: Host H4 Receiver: Host H7	Sender: Host H5 Receiver: Host H6

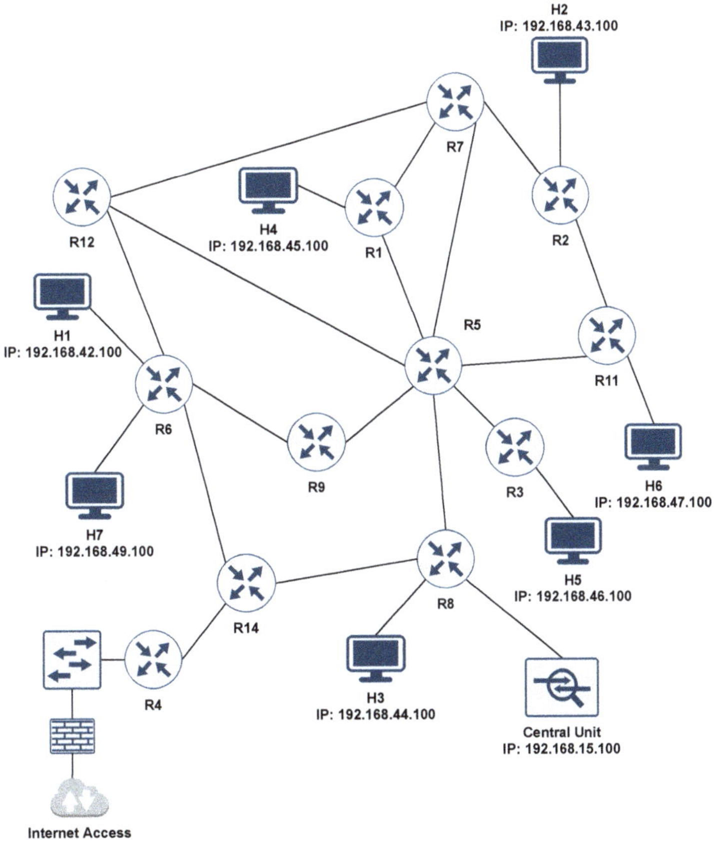

Figure 2. Network setup for simulations of hidden communication techniques.

4. Cyber Threat Detection Research Enabled

4.1. Cyber Threat Detection as Standard ML Classification Problem

In terms of machine learning, cyber threat detection is defined by the principles of the classification problem adapted to the chosen goal of detection. The two classical objectives for cyber threat detection algorithms are:

- Anomaly detection to anomalously detect observations defined as a deviation from the specified base model. The detected anomaly is examined in more detail to determine if it is a cyber threat.
- Detection of cyber threats by finding patterns of the known nature of a cyber attack. Tagged records are required.

Since the datasets generated for this work contain the detailed labels of the cyber threats, the set of experiments follows the second option to be presented. Table 3 shows the general workflow used in this work for cyber threat detection experiments on the hidden communication techniques. It presents the objectives to be achieved for each step of the workflow.

Table 3. Generic procedure to research cyber threat detection methods.

Workflow Step	Objective
Step 1	Benign and malicious network communication simulation scenarios
Step 2	Collect raw source data
Step 3	Generate network data representation from raw source data
Step 4	Data labeling
Step 5	Feature selection
Step 6	Prepare train and test datasets
Step 7	Train data augmentation to balance samples per each label
Step 8	Train, test, and evaluate a machine learning classifier

4.1.1. Network Flows Generation

The records were collected by the system presented in Section 3.1 as network data traces in PCAP format. These PCAPs were then processed into network datasets using CICFlowMeter [41].

Each network flow dataset was bidirectional and consisted of 84 metrics. Network flows were identified by the classic 5-tuple key (source IP, destination IP, source port, destination port, Layer 4 protocol code). When a flow exceeded the configured flow timeout (in seconds) or the flow was inactive (activity timeout), the status was saved and exported. In the exported flow table, the timestamp adds the 5-tuple key to distinguish the flows. In the case of this experiment, the parameters were set to:

- flow timeout—120 s
- activity timeout—30 s

The output of the application is a CSV file. Another step was the labeling. It was done manually through the script based on the coding scheme shown in Table 2. The last step in the data preparation was to combine all CSV files into a final dataset with all collected observations from the simulated scenarios. This dataset was then preprocessed in the data experimentation phase according to the state of the art in data science.

4.1.2. Training Classifiers for Cyber Threat Detection

This part shows how to use the prepared datasets to find the classifier to be used as a cyber threat detector. The first step was to analyze the metrics in the dataset. The aim was to check which metrics were more important for predicting the target class (feature selection). The process involved pairwise correlation between the metrics and the explained variable (class or label). Figure 3 shows the filtered matrix of the metrics for which the absolute value of the correlation coefficient of any metric with a `label` greater than 0.05. The most positively or negatively correlated metrics were related to time (inter-arrival time (IAT), time of activity) and volume of the network data (number of packets, number of bytes). The practical aspect of this step was to reduce the dimension of the problem. The number of metric outputs was 28 since 52 metrics were filtered and three metrics were skipped since they were not relevant to the problem (flow ID, timestamp, IP addresses).

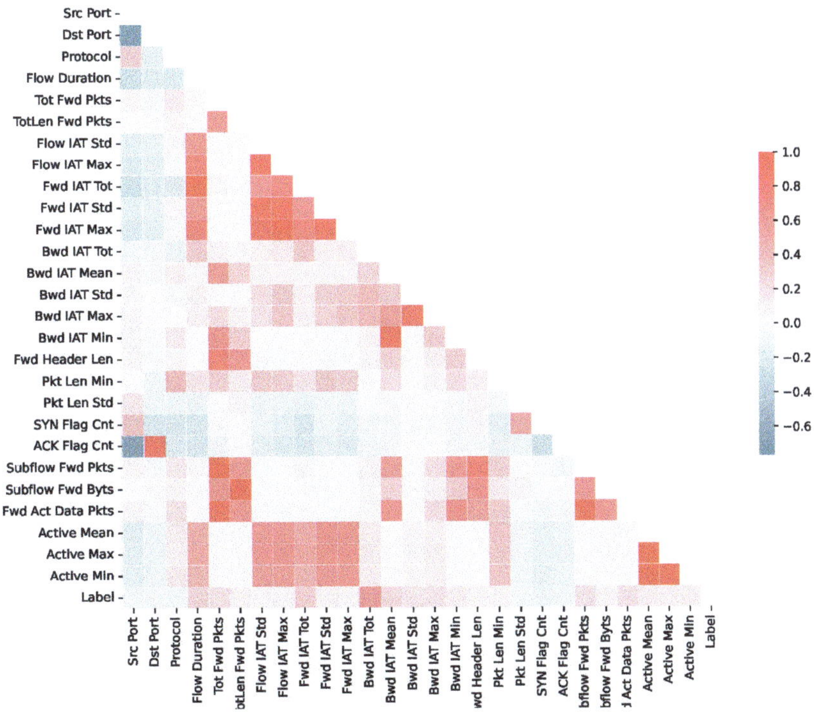

Figure 3. Pairwise correlation matrix of prepared dataset. Filtered according to level of correlation between label column and other parameters for feature selection purposes.

Subsequently, the filtered data were divided into training (80% of the datasets) and testing (20% of the datasets) datasets. The number of collected data streams distributed among the training and testing datasets after the split is shown in Table 4.

Table 4. Distribution of flows among training and test datasets split from collected dataset as described in Section 3.2.4.

Types of Flows	Flows in Training Dataset	Flows in Testing Dataset
Benign traffic (Label: 0)	92,737	23,163
Hidden communication over lost packets (Label: 1)	742	195
Hidden communication over time modulation (Label: 2)	457	127

The first problem that arises is the imbalance of the class examples in the training dataset. The imbalanced classifications poses a challenge to any predictive algorithm. In the case of imbalance in the training examples, the trained models might have poor predictive performance, especially for the minor classes. On the other hand, the minor classes are important because the context of the experiment is cyber threat detection research, where cyber attacks are sporadic compared to harmless attacks. The Synthetic Minority Oversampling Technique (SMOTE) [42] and Edited nearest neighbor (ENN) [43] were applied as data extensions to overcome the problem. SMOTE is used to increase the number of samples in the minority class by linear interpolation, and ENN is used to remove the noise from the majority samples. Table 5 contains the number of data streams before and after applying SMOTE-ENN techniques to the training dataset.

Table 5. Data augmentation results—number of flows in training dataset before and after SMOTE-ENN.

Types of Flows	Flows in Training Dataset before SMOTE-ENN	Flows in Training Dataset after SMOTE-ENN
Benign traffic (Label: 0)	92,737	92,737
Hidden communication over lost packets (Label: 1)	742	92,706
Hidden communication over time modulation (Label: 2)	457	92,643

The experimental phase was conducted to measure the possibility of predicting cyber threats by applying selected ML classifiers. Two classifiers were chosen to test the preparation of the example solution for this paper:

- Random Forest (RF) [44] was implemented based on RandomForestClassifier from the Scikit-learn [45]. The initial setup was based on the default parameters as described in [46].
- Multilayer Perceptron (MLP) classifier was implemented as the custom architecture in TensorFlow [47] using the Keras subpackage. The setup of this classifier in terms of architecture, optimizer, loss function, and evaluation metrics is presented in Table 6.

Table 6. Custom MLP classifier architecture setup per each layer (6) with selected optimizer, loss function, evaluation metric, and training procedure.

Parameter	The Custom MLP Classifier
Layer 1 (Input)	`Input`, size: 28 × 20, activation: relu
Layer 2	Dense, size: 20 × 20, activation: relu
Layer 3	BatchNormalization, size: 20 × 20, activation: relu
Layer 4	Dense, size: 20 × 150, activation: relu
Layer 5	Dense, size: 150 × 20, activation: relu
Layer 6 (Predictions)	Dense, size: 20 × 20, activation: softmax
Optimizer	Adam with the learning rate: 0.001
Loss	Categorical Cross Entropy
Evaluation metrics	accuracy
Training setup	batch size: 128, epochs: 20, validation split: 0.15

Both classifiers were trained with the SMOTE-ENN training datasets and evaluated with non-SMOTE-ENN test datasets. The metrics of accuracy, precision, recognition, and F1 score were used to evaluate the performance. The result metrics for the RF and MLP classifier are shown in Tables 7 and 8, respectively. The last rows of both tables show the overall accuracy of the respective classifier. Figures 4 and 5 show the confusion matrices of the two selected classifier models. The classification was conducted within three classes (0, 1, or 2), so the size of each confusion matrix was 3 × 3. Each cell presented the number of instances of a given true class (rows) classified into a given predicted class (columns). The diagonal of each matrix included true positives. The other cells could be classically interpreted as false positives, false negatives, and true negatives in relation to a selected class.

Table 7. Performance of RF classifier in terms of precision, recall, F1 score, and overall accuracy metrics.

Types of Flows	Precision	Recall	F1 Score
Benign traffic (Label: 0)	1.00	0.99	1.00
Lost packets attack (Label: 1)	0.49	0.92	0.64
Packet timing attack (Label: 2)	0.93	1.00	0.97
Overall accuracy	0.99		

Table 8. Performance of custom MLP classifier in terms of precision, recall, F1 score, and overall accuracy metrics.

Types of Flows	Precision	Recall	F1 Score
Benign traffic (Label: 0)	1.00	0.99	1.00
Lost packets attack (Label: 1)	0.42	0.99	0.59
Packet timing attack (Label: 2)	0.84	1.00	0.91
Overall accuracy	0.99		

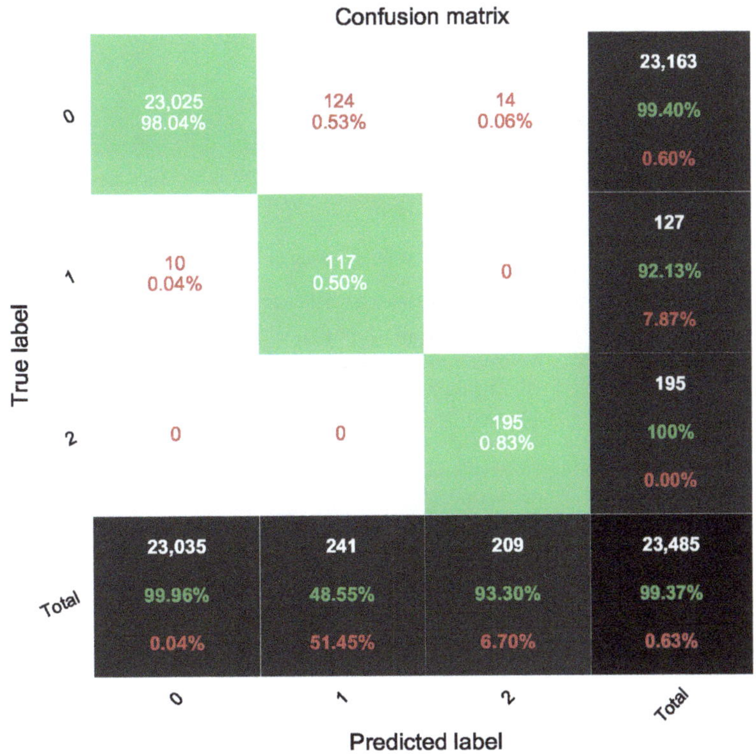

Figure 4. Confusion matrix of RF classifier.

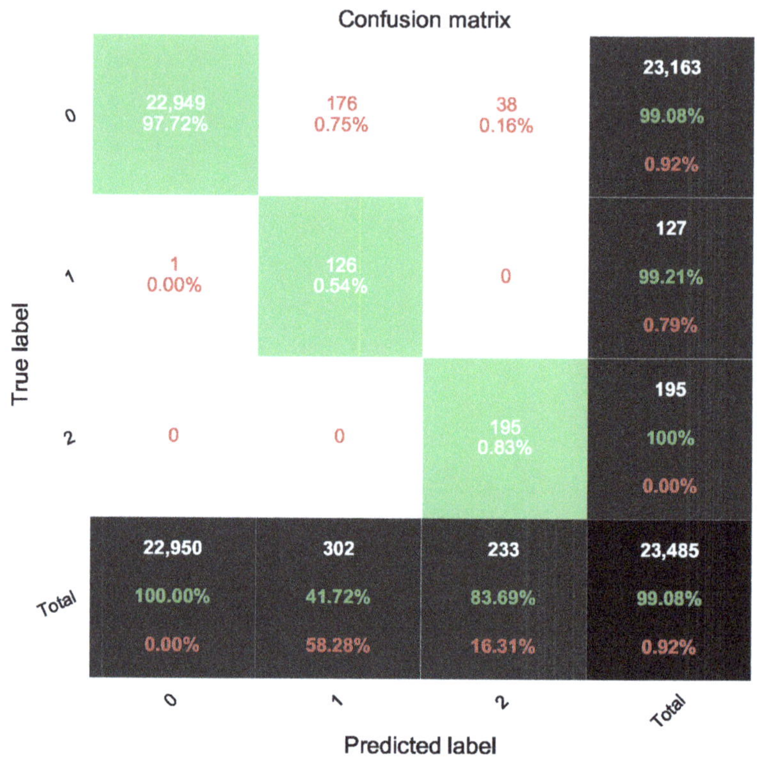

Figure 5. Confusion matrix of custom MLP classifier.

4.2. Conclusions and Future Directions

The prepared models show almost perfect results for the Label 2—Time Modulation Information Hiding Attack. The results for Label 1—Flows with Lost Packets Information Hiding Technique—were noticeably worse. The presented results should be considered as an example for the research work with the generated datasets. Table 9 supplements Table 3 with a summary of the actions performed for each step in this work.

Each step of the workflow shown in Table 9 could be considered to be different research directions, such as:

- Generating more data using the prepared application for simulations, especially for more examples of hidden communication techniques.
- Selection of a different predictive model or design of a more complex architecture combining some classifiers.
- To find a data representation that is better suited to the context of detecting information hiding techniques. The output data representation of CICFlowMeter contains several different metrics related to temporal aspects of network communication. Thus, this is the most likely answer as to why the detection of temporally modulated hidden communication had better performance.
- With other feature selection or data augmentation methods.

Table 9. Summary of realized actions per each step of generic procedure to research cyber threat detection methods.

Workflow Step	Objective	Realization in This Paper
Step 1	Benign and malicious network communication simulations	Implementation of the dedicated tool to set up networks and simulations; proof-of-concept implementation of Multi-node Cyber Threat Detection System to use in monitor and collect mode
Step 2	Collecting raw source data	Network traffic traces collected as PCAP files
Step 3	Generating network data representation from raw source data	Generation of network flows including 80 metrics from [41]
Step 4	Data labeling	Labeling based on Table 2 by adding the column Label with the respected coding to the corresponding network flows
Step 5	Feature selection	Selecting features using correlation coefficient between Label and the other features
Step 6	Preparing train and test datasets	Train/test split method from [45]
Step 7	Train data augmentation to balance samples per each label	Augmentation of the training dataset with SMOTE-ENN method
Step 8	Train, test, and evaluate a machine learning classifier	Preparing Random Forest and the custom MLP classifiers.

5. Summary

This paper presents an approach for the availability of datasets for cyber threat detection research and the application to Data Science. As a first step, the multi-agent platform for collecting network flows was implemented for this purpose. Such a platform enabled the collection of network flow data in a multi-node setup. One of the achievements was the implementation of an automated solution for generating network configurations and running application scenarios for cyber threat activities. This is a promising aspect to scale research experiments for cyber threat detection. Another future aspect to be explored is the ease of practical implementation of such solutions. The input problem of any practical cyber threat detection solution is the working environment in which the method is implemented. The standard approach is the learning phase, which assumes that the cyber detection engine must be adapted to the network environment through monitoring and learning. After reaching *readiness*, the cyber threat detection engine is partially trained with new examples of malicious activity or feedback data from human operators. The ability to scale the data generated in different network configurations would improve cyber threat detection engines for the sake of greater generality. Other practical implications could include easier implementation in working environments and reduced relearning and manual tuning efforts.

The unique value of this research was to emphasize the recognition of information concealment techniques, which are generally considered concepts within the broad domain of cyber deception. The most important prerequisite for working on cyber threat detection is the availability of the right data. All known data sets that are available focus on the publicly known types of cyber attacks. The examples of the use of cyber deception techniques are very rare or non-existent. One of the main contributions of this work was the development of a network environment integrated with the tools to collect such examples. This was achieved by proposing the implementation of a multi-agent system as a Multi-Node Cyber

Threat Detection platform utilizing the monitor mode. Based on the collected data, the reference data science workflow was evaluated by applying methods for data representation and classification of malicious network flows. The final result confirms the usefulness of the presented end-to-end approach for researching the discovery of information hiding techniques. The authors will apply it in further research and development projects. Cyber attackers are increasingly using cyber deception techniques. Moreover, advanced cyber attacks, for example, APT (Advanced Persistent Threat) campaigns, could combine more than one deception technique in two dimensions:

- Within different abstraction layers within the ISO-OSI 7-layer model, with the application layer in particular considered a significant threat.
- Per each step of a cyber attack modeled as Cyber Kill Chain. [4]

This poses a massive threat to the security of cyberspace, so more efforts need to be made in the coming years to improve cyber defense capabilities in the area of cyber deception.

Author Contributions: Conceptualization, K.S.; Investigation, J.B.; Methodology, J.B.; Software, J.B.; Supervision, K.S.; Validation, J.B. All authors have read and agreed to the published version of the manuscript.

Funding: This research was funded by The Polish National Centre for Research and Development under project No. CYBERSECIDENT/369532/I/NCBR/2017.

Data Availability Statement: The data cannot be shared due to project restrictions.

Acknowledgments: The authors wish to thank Monika Stępkowska, for her contribution to setting up the experiments.

Conflicts of Interest: The authors declare no conflict of interest.

References

1. Barrett, M. NIST Cybersecurity Framework (CSF): Framework for Improving Critical Infrastructure Cybersecurity. Version 1.1. 2018. Available online: https://nvlpubs.nist.gov/nistpubs/CSWP/NIST.CSWP.04162018.pdf (accessed on 22 October 2021).
2. Fragkos, G.; Minwalla, C.; Plusquellic, J.; Tsiropoulou, E.E. Artificially Intelligent Electronic Money. *IEEE Consum. Electron. Mag.* **2021**, *10*, 81–89. [CrossRef]
3. Cichonski, P.; Millar, T.; Grance, T.; Scarfone, K. NIST SP 800-61: Computer Security Incident Handling Guide. 2012. Available online: https://nvlpubs.nist.gov/nistpubs/specialpublications/nist.sp.800-61r2.pdf (accessed on 22 October 2021).
4. Hutchins, E.; Cloppert, M.J.; Amin, R.M. Intelligence-Driven Computer Network Defense Informed by Analysis of Adversary Campaigns and Intrusion Kill Chains. *Lead. Issues Inf. Warf. Secur. Res.* **2011**, 1, 80. Available online: https://www.lockheedmartin.com/content/dam/lockheed-martin/rms/documents/cyber/LM-White-Paper-Intel-Driven-Defense.pdf (accessed on 22 October 2021).
5. MITRE ATT&CK. Available online: https://attack.mitre.org/ (accessed on 5 September 2021).
6. Chou, D.; Jiang, M. Data-Driven Network Intrusion Detection: A Taxonomy of Challenges and Methods. *arXiv* **2020**, arXiv:2009.07352.
7. Ptacek, T.; Newsham, T. *Insertion, Evasion, and Denial of Service: Eluding Network Intrusion Detection*; Secure Networks, Inc.: Mandaluyong, Philippines, 1998. Available online: http://www.icir.org/vern/Ptacek-Newsham-Evasion-98.ps (accessed on 22 October 2021).
8. Nehinbe, J.O. A Simple Method for Improving Intrusion Detections in Corporate Networks. In *Information Security and Digital Forensics*; Weerasinghe, D., Ed.; Springer: Berlin/Heidelberg, Germany, 2010; pp. 111–122.
9. McAfee Labs Threats Report—June 2017. Available online: https://www.mcafee.com/enterprise/en-us/assets/reports/rp-quarterly-threats-jun-2017.pdf (accessed on 5 September 2021).
10. Mazurczyk, W.; Wendzel, S.; Zander, S.; Houmansadr, A.; Szczypiorski, K. Background Concepts, Definitions, and Classification. In *Information Hiding in Communication Networks: Fundamentals, Mechanisms, Applications, and Countermeasures*; IEEE-Wiley Press: New York, NY, USA, 2016; Chapter 2, pp. 39–58.
11. Balasubramaniyan, J.; Garcia-Fernandez, J.; Isacoff, D.; Spafford, E.; Zamboni, D. An architecture for intrusion detection using autonomous agents. In Proceedings of the 14th Annual Computer Security Applications Conference (Cat. No. 98EX217), Phoenix, AZ, USA, 7–11 December 1998; pp. 13–24.
12. Herrero, A.; Corchado, E. Multiagent Systems for Network Intrusion Detection: A Review. *Comput. Intell. Secur. Inf. Syst.* **2009**, *63*, 143–154.

13. Docking, M.; Uzunov, A.V.; Fiddyment, C.; Brain, R.; Hewett, S.; Blucher, L. UNISON: Towards a Middleware Architecture for Autonomous Cyber Defence. In Proceedings of the 2015 24th Australasian Software Engineering Conference, Adelaide, SA, Australia, 28 September–1 October 2015; pp. 203–212.
14. Saeed, I.A.; Selamat, A.; Rohani, M.F.; Krejcar, O.; Chaudhry, J.A. A Systematic State-of-the-Art Analysis of Multi-Agent Intrusion Detection. *IEEE Access* **2020**, *8*, 180184–180209. [CrossRef]
15. Kott, A. Intelligent Autonomous Agents are Key to Cyber Defense of the Future Army Networks. *Cyber Def. Rev.* **2018**, *3*, 57–70.
16. Pascale, F.; Adinolfi, E.A.; Coppola, S.; Santonicola, E. Cybersecurity in Automotive: An Intrusion Detection System in Connected Vehicles. *Electronics* **2021**, *10*, 1765. [CrossRef]
17. Lombardi, M.; Pascale, F.; Santaniello, D. EIDS: Embedded Intrusion Detection System using Machine Learning to Detect Attack Over the CAN-BUS. In Proceedings of the 30th European Safety and Reliability Conference and 15th Probabilistic Safety Assessment and Management Conference, Venice, Italy, 1–5 November 2020; pp. 2028–2035. Available online: https://www.rpsonline.com.sg/proceedings/esrel2020/pdf/5090.pdf (accessed on 22 October 2021).
18. Sharafaldin, I.; Lashkari, A.H.; Ghorbani, A. Toward Generating a New Intrusion Detection Dataset and Intrusion Traffic Characterization. *ICISSP* **2018**, *1*, 108–116.
19. Ring, M.; Wunderlich, S.; Grüdl, D.; Landes, D.; Hotho, A. Creation of Flow-Based Data Sets for Intrusion Detection. *J. Inf. Warf.* **2017**, *16*, 41–54.
20. Shahriar, M.H.; Haque, N.I.; Rahman, M.A.; Alonso, M. G-IDS: Generative Adversarial Networks Assisted Intrusion Detection System. In Proceedings of the 2020 IEEE 44th Annual Computers, Software, and Applications Conference (COMPSAC), Madrid, Spain, 13–17 July 2020; pp. 376–385.
21. Canadian Institute for Cybersecurity. Intrusion Detection Evaluation Dataset (ISCXIDS2012). Available online: https://www.unb.ca/cic/datasets/ids.html (accessed on 5 September 2021).
22. Canadian Institute for Cybersecurity. NSL-KDD Dataset. Available online: https://www.unb.ca/cic/datasets/nsl.html (accessed on 5 September 2021).
23. Canadian Institute for Cybersecurity. Intrusion Detection Evaluation Dataset (CICIDS2017). Available online: https://www.unb.ca/cic/datasets/ids-2017.html (accessed on 5 September 2021).
24. Canadian Institute for Cybersecurity. A Realistic Cyber Defense Dataset (CSE-CIC-IDS2018). Available online: https://www.unb.ca/cic/datasets/ids-2018.html (accessed on 5 September 2021).
25. Moustafa, N.; Slay, J. UNSW-NB15: A comprehensive data set for network intrusion detection systems (UNSW-NB15 network data set). In Proceedings of the 2015 Military Communications and Information Systems Conference (MilCIS), Canberra, ACT, Australia, 10–12 November 2015; pp. 1–6.
26. Maciá-Fernández, G.; Camacho, J.; Magán-Carrión, R.; García-Teodoro, P.; Therón, R. UGR'16: A new dataset for the evaluation of cyclostationarity-based network IDSs. *Comput. Secur.* **2018**, *73*, 411–424. [CrossRef]
27. Center for Applied Internet Data Analysis. CAIDA Datasets. Available online: https://www.caida.org/catalog/datasets/overview/ (accessed on 5 September 2021).
28. Information Marketplace For Policy and Analysis of Cyber Risk & Trust. Available online: https://www.impactcybertrust.org (accessed on 5 September 2021).
29. Szczypiorski, K. *Steganography in TCP/IP Networks—State of the Art and a Proposal of a New System—HICCUPS*; Institute of Telecommunications' Seminar, Warsaw University of Technology: Warsaw, Poland, 2003.
30. Mullaney, C. Morto Worm Sets a (DNS) Record. 2011. Available online: http://www.symantec.com/connect/blogs/morto-worm-sets-dns-record (accessed on 22 October 2021).
31. Attackers Hide Communication within Linux Backdoor. Available online: https://www.securityweek.com/attackers-hide-communication-linux-backdoor (accessed on 5 September 2021).
32. Regin: Top-Tier Espionage Tool Enables Stealthy Surveillance. 2015. Available online: https://docs.broadcom.com/doc/regin-top-tier-espionage-tool-15-en (accessed on 5 September 2021).
33. Bencsáth, B.; Pék, G.; Buttyán, L.; Félegyházi, M. Duqu: A Stuxnet-like malware found in the wild. *CrySyS Lab Tech. Rep.* **2011**, *14*, 60–141. Available online: https://www.crysys.hu/publications/files/bencsathPBF11duqu.pdf (accessed on 22 October 2021).
34. Dell Secureworks. Malware Analysis of the Lurk Downloader. Available online: https://www.secureworks.com/research/malware-analysis-of-the-lurk-downloader (accessed on 5 September 2021).
35. FireEye Threat Intelligence. HAMMERTOSS: Stealthy Tactics Define a Russian Cyber Threat Group. Available online: https://www.fireeye.com/blog/threat-research/2015/07/hammertoss_stealthy.html (accessed on 5 September 2021).
36. Nagaraja, S.; Houmansadr, A.; Piyawongwisal, P.; Singh, V.; Agarwal, P.; Borisov, N. Stegobot: A Covert Social Network Botnet. In *Information Hiding*; Filler, T., Pevný, T., Craver, S., Ker, A., Eds.; Springer: Berlin/Heidelberg, Germany, 2011; pp. 299–313.
37. Deutsch, J.; Garrie, D. Instegogram: A New Threat and Its Limits for Liability. *J. Law Cyber Warf.* **2017**, *6*, 1–7.
38. Bieniasz, J.; Szczypiorski, K. Methods for Information Hiding in Open Social Networks. *JUCS-J. Univers. Comput. Sci.* **2019**, *25*, 74–97.
39. Hewitt, C.; Bishop, P.; Steiger, R. A Universal Modular Actor Formalism for Artificial Intelligence. In Proceedings of the 3rd International Joint Conference on Artificial Intelligence (IJCAI'73), Stanford, CA, USA, 20–23 August 1973; Morgan Kaufmann Publishers Inc.: San Francisco, CA, USA, 1973; pp. 235–245.
40. GNS3 Network Simulation Tool. 2021. Available online: https://www.gns3.com (accessed on 5 September 2021).

41. Canadian Institute for Cybersecurity. CICFlowmeter—Network Traffic Bi-Flow Generator and Analyzer for Anomaly Detection. Available online: https://github.com/ahlashkari/CICFlowMeter (accessed on 5 September 2021).
42. Chawla, N.V.; Bowyer, K.W.; Hall, L.O.; Kegelmeyer, W.P. SMOTE: Synthetic Minority over-Sampling Technique. *J. Artif. Int. Res.* **2002**, *16*, 321–357. [CrossRef]
43. Beckmann, M.; Ebecken, N.F.; de Lima, B.S.P. A KNN undersampling approach for data balancing. *J. Intell. Learn. Syst. Appl.* **2015**, *7*, 104. [CrossRef]
44. Breiman, L. Random forests. *Mach. Learn.* **2001**, *45*, 5–32. [CrossRef]
45. Pedregosa, F.; Varoquaux, G.; Gramfort, A.; Michel, V.; Thirion, B.; Grisel, O.; Blondel, M.; Prettenhofer, P.; Weiss, R.; Dubourg, V.; et al. Scikit-learn: Machine Learning in Python. *J. Mach. Learn. Res.* **2011**, *12*, 2825–2830.
46. Random Forrest Classifier from Scikit-Learn Framework. 2018. Available online: https://scikit-learn.org/stable/modules/generated/sklearn.ensemble.RandomForestClassifier.html (accessed on 20 September 2021).
47. Abadi, M.; Agarwal, A.; Barham, P.; Brevdo, E.; Chen, Z.; Citro, C.; Corrado, G.S.; Davis, A.; Dean, J.; Devin, M.; et al. TensorFlow: Large-Scale Machine Learning on Heterogeneous Systems. 2015. Available online: tensorflow.org (accessed on 5 September 2021).

electronics

Article

A New Approach to the Development of Additive Fibonacci Generators Based on Prime Numbers

Volodymyr Maksymovych [1], Oleh Harasymchuk [1], Mikolaj Karpinski [2,*], Mariia Shabatura [1], Daniel Jancarczyk [2] and Krzysztof Kajstura [2]

[1] Department of Information Technology Security, Lviv Polytechnic National University, 79013 Lviv, Ukraine; volodymyr.m.maksymovych@lpnu.ua (V.M.); oleh.i.harasymchuk@lpnu.ua (O.H.); mariia.m.mandrona@lpnu.ua (M.S.)

[2] Department of Computer Science and Automatics, University of Bielsko-Biala, 43-309 Bielsko-Biala, Poland; djancarczyk@ath.bielsko.pl (D.J.); kkajstura@ath.bielsko.pl (K.K.)

* Correspondence: mkarpinski@ath.bielsko.pl

Abstract: Pseudorandom number and bit sequence generators are widely used in cybersecurity, measurement, and other technology fields. A special place among such generators is occupied by additive Fibonacci generators (AFG). By itself, such a generator is not cryptographically strong. Nevertheless, when used as a primary it can be quite resistant to cryptanalysis generators. This paper proposes a modification to AGF, the essence of which is to use prime numbers as modules of recurrent equations describing the operation of generators. This modification made it possible to ensure the constancy of the repetition period of the output pseudorandom pulse sequence in the entire range of possible values of the initial settings–keys (seed) at specific values of the module. In addition, it has proposed a new generator scheme, which consists of two generators: the first of which is based on a modified AFG and the second is based on a linear feedback shift register (LFSR). The output pulses of both generators are combined through a logic element XOR. The results of the experiment show that the specific values of modules provide a constant repetition period of the output pseudorandom pulse sequence in a whole range of possible values of the initial settings–keys (seed) and provide all the requirements of the NIST test to statistical characteristics of the sequence. Modified AFGs are designed primarily for hardware implementation, which allows them to provide high performance.

Keywords: cybersecurity; pseudorandom sequences generators; prime numbers; additive Fibonacci generator; statistical characteristics

Citation: Maksymovych, V.; Harasymchuk, O.; Karpinski, M.; Shabatura, M.; Jancarczyk, D.; Kajstura, K. A New Approach to the Development of Additive Fibonacci Generators Based on Prime Numbers. *Electronics* **2021**, *10*, 2912. https://doi.org/10.3390/electronics10232912

Academic Editor: Myung-Sup Kim

Received: 16 October 2021
Accepted: 22 November 2021
Published: 24 November 2021

1. Introduction

At the present stage of scientific development and technological progress, pseudorandom bit sequence generators have found more and more application areas. The scientific and practical importance of generating qualitative pseudorandom sequences are significant and many researchers are devoted to this area to find the best algorithms for generating pseudorandom sequences with properties that are closest to random sequences.

In particular, Professor Amalia Beatriz Orue Lopez from Isabel I University in Burgos, Spain [1–3], Professor Miguel Angel Murillo-Escoba from the Centre for Research and Higher Education in Ensenada, Baja California, Mexico [4,5] and Rafik Hamza from LAMIE Laboratory, University of Batna, Algeria [6] in their articles research the various methods of constructing pseudorandom sequence generators, in order to determine issues regarding information protection and estimation quality.

Pseudorandom sequence generators are used in various fields of science and technology. A special place among such generators is occupied by additive Fibonacci generators (AFG) [7–16].

Almost all Fibonacci generators are designed for hardware implementation are used as recurrent equation modules equal to the power of two. This greatly simplifies the hardware implementation of Fibonacci generators but narrows their functionality and does not allow for the improvement of their statistical characteristics without a significant increase in the number of bits of structural elements. The main motivation for this work was the awareness of the need to solve the hardware implementation of Fibonacci generators with an arbitrary module value. This was also facilitated by the authors' experience gained in the hardware implementation of controlled digital frequency synthesizers [17].

Creating new circuit engineering solutions for the hardware implementation of Fibonacci generators allows for the implementation of a new approach to its creation, which, unlike previous approaches, enables the design of generators with an arbitrary module of the recurrent equation, in particular with modules whose values are prime numbers. As a result, there is an opportunity to significantly improve the generator's statistical characteristics.

This work was aimed at researching the hardware implementation of an additive Fibonacci generator, based on an algorithm that uses a module of prime numbers, and to analyse their characteristics.

2. Related Works

Additive Fibonacci generators (AFG) are widely used in cybersecurity devices to generate pseudorandom sequences of bits or numbers.

By itself, such a generator is not cryptographically strong. Nevertheless, using it is fundamental to create a completely secure and resistant cryptanalysis algorithm. For example, based on these generators, the algorithms Fish, Pike and Mush are implemented [7,8].

The use of Additive Fibonacci Generators is not limited to cybersecurity systems (cryptography), they are also used for other applications.

In particular, Ref. [18] describes a method for constructing a pseudorandom number generator based on a recurrent linear sequence of Fibonacci p-numbers to generate a variable carrier frequency of pulse-width modulation (PWM) of the power converter control system to reduce acoustic noise and electromagnetic obstacle level.

The classical algorithm of AGF was formed based on the equation:

$$x_i = (x_{i-1} + x_{i-k})\ mod(m), l > k > 0 \tag{1}$$

General view:

$$x_i = (x_{i-a} + x_{i-b} + \dots + x_{i-p})\ mod(m), a > b > \dots > p > 0 \tag{2}$$

An effective hardware implementation of Equations (1) and (2) are chosen according to Equation $2 - m = 2^n$. This simplifies the hardware implementation of the generators. Compliance with the requirements for the selection of parameters l, k and $a, b, \dots, p$ in Equations (1) and (2) ensures that the repetition period of the sequence at the output of the sequence of generators will be no less than $2^n - 1$ [8].

In articles [10–19], modified additive Fibonacci generators (MAFG) were proposed, operating according to the equation:

$$x_i = (x_{i-a} + x_{i-b} + \dots + x_{i-p} + a)\ mod(2^n), \tag{3}$$

where, $a = a_0 \oplus a_1 \oplus \dots \oplus a_z$; a_i $((i = 0, 1, \dots, z), (z \le n - 1))$—values of the number x_i binary bits.

In the studies of [10,11,19], it is shown that the process of adding the number "a" causes certain "confusion"—the dependence of each bit of the number, including the youngest bit, from all its other bits, allows to significantly improve the statistical characteristics of the output signals of the MAFG. An array of initial values of numbers $x_i, x_{i-a}, x_{i-b}, \dots, x_{i-p}$, is called the cryptographic generator key and is under the condition of hardware implementation. These numbers are used as the initial values of the registers that are part of its block diagram.

Our research on AGF and MAGF [10,11,13,18,19] show a significant dependence of the statistical characteristics of the pseudorandom sequence at the output of the generator on the output parameters. In particular, they strongly depend on the value of the repetition period of the output sequence [10,13]. This means the presence of so-called "weak keys", which could be relatively easily disclosed.

This paper presents the results of research aimed at eliminating this shortcoming of AFG and MAFG. We focus on the hardware implementation of generators.

3. Case Study

3.1. The Structure Schema and the Work Principle of the New AFG

As emphasized above, the construction AGF and MAGF uses algorithms in which the modulus of recurrent Equations (1)–(3) is the power of number 2. This significantly simplifies hardware implementation.

Papers [17,18] proposed a new approach to constructing two-level frequency synthesizers using the change of the average value of the output frequency with an arbitrarily given step. These approaches can be effectively applied in the hardware implementation of our proposed generators.

Figure 1 shows a variant of one such additive Fibonacci generator [10].

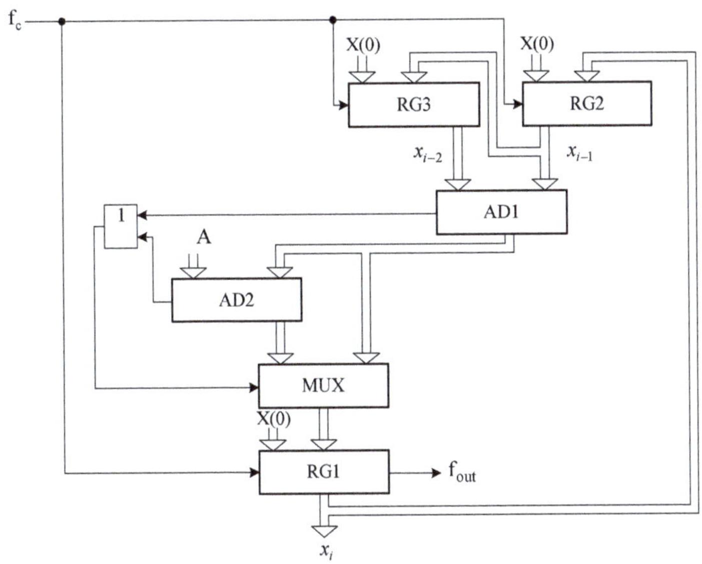

Figure 1. Structure schema of AFG.

AFG consists of registers RG1-RG3, adders AD1-AD2, multiplexer MUX and logical element OR. The generator functions according to the equation:

$$x_i = (x_{i-2} + x_{i-1})\ mod(m), \tag{4}$$

where, m—prime number; x_i, x_{i-1}, x_{i-2}—numbers in registers RG1, RG2 i RG3.

The number of binary bits n of the structural elements of the scheme (RG1-RG3, AD1, AD2) is selected based on the need to ensure the condition $2^n > m$.

Herewith, the smallest value n is selected, at which this condition is fulfilled. The number A, which is applied to one of the AD2 input groups, is determined by the equation $A = 2^n - m$. In the absence of carrying signals on the outputs of AD1 and AD2 to the RG1 information inputs through the multiplex, MUX passes a number from the output AD1, and in the presence of one of these signals the number from the output AD2. The initial number—the key (seed) $X(0)$—is written to registers RG1-RG3.

Clock pulses receive at the clock inputs of the registers RG1-RG3. The output pseudo-random bits sequence formed on one of the register's RG1 bits. The described operating mode of the generator provides a change of the numbers in registers RG1-RG3 in the range of values $0 \div m - 1$.

3.2. Research of the New AFG Characteristics

Figure 2 shows the dependences of the repetition periods of the studied pseudorandom numbers sequence generators on the value of the key $X(0)$.

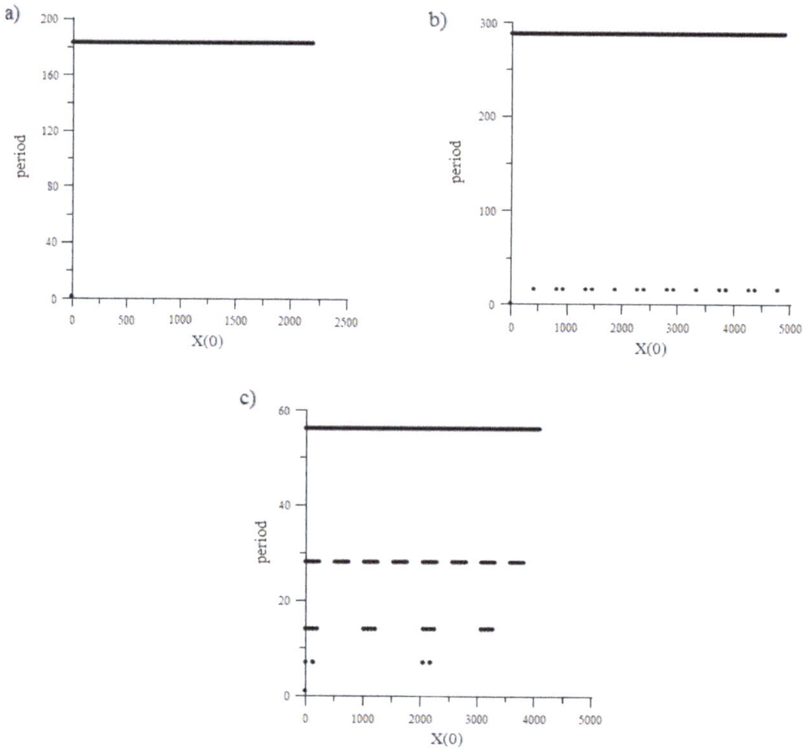

Figure 2. Dependences of AFG repetition periods on the key.

Figure 2a,b shows the dependencies for the new AFG, that function following Equation (4): $m = 13$ (Figure 2a) and $m = 17$ (Figure 2b). In Figure 2c, the corresponding dependence for the classical AFG, which operates following Equation (4) at $m = 2^4 = 16$, is given for comparison. In order to go through all possible values, the initial number is determined by the formula:

$$X(0) = (x_{i-2}(0) + m\, x_{i-1}(0) + m^2\, x_i(0), \quad (5)$$

where, $x_i(0)$, $x_{i-1}(0)$, $x_{i-2}(0)$ are the initial values of the numbers in the registers RG1-RG3, accordingly.

This article presents only some results of different AFG versions of repetition periods research. During the work, a large amount of AFG at different modulus values was analysed. This allows us to draw the following conclusions:

- A new type of AFG, in which the modulus of recurrent equations is a prime number (Figure 1), differs favourably from classical AFG, in which the modulus of recurrent equations is a power of 2, in the absence or relatively small number of "weak keys" (in which the repetition period of the pseudo-random sequence is small);
- In AFG of a new type (Figure 1), there are values of the module for which there are no "weak keys".

Table 1 presents the values of the repetition periods of the output sequence of new AFG for some m values fixed on the whole set of possible $X(0)$ values.

Table 1. The dependence of the repetition periods of the new AFG output sequence for some m values on the whole set of possible $X(0)$ values.

Prime Numbers, Max and Min Repetition Period Values								
m	2	3	5	7	11	13	17	19
period	7	13	24 4	48 6	120 10	183	288 16	180 9
m	23	29	31	37	41	43	47	53
period	506 22	871	993	1368 36	1723	231 21	2257	1404
m	59	61	67	71	73	79	83	89
period	58	930	4488 66	5113	5403	3120	2296 82	3960 44
m	97	101	103	107	109	113	127	131
period	3116	100 50	3536	2862	1485	4256	16,257	

In Table 1, for values $m = 2, 3, 13, 29, 31, 41, 47, 53, 59, 61, 71, 73, 79, 97, 103, 107, 109, 113, 127$ no "weak keys" were found in the whole $X(0)$ range values. The only fixed value of the period is indicated in the table. For other m values, a small number of "weak keys" were fixed, for which, along with the principal (predominant) reduced values of the period were indicated.

At sufficiently large m values, the procedure for finding the repetition periods of the output sequence for all possible $X(0)$ values requires a lot of machine time and, under certain conditions, is such that it is practically not implemented. Table 2 shows the values of the repetition periods for relatively large values of m prime numbers when $x_i(0) = 1$, $x_{i-1}(0) = 1$, $x_{i-2}(0) = 1$.

Table 2. Dependence of repetition periods of the new AFG output sequence for some m values, at $x_i(0) = 1$, $x_{i-1}(0) = 1$, $x_{i-2}(0) = 1$.

Prime Numbers, Repetition Period Values				
m	8191	9973	65,537 (Fermat number)	2,147,483,647 (Marsenn number)
period	22,366,291	99,46,728	1,431,699,455	$>10^{10}$

The tendencies revealed at small values of the module m (Table 1) allow us to state with a high probability that, at relatively large values, the number of "weak keys" will be small or absent.

According to this property, the proposed Fibonacci generator, in which the modules of the recurrent equation are prime numbers, differs favourably from the known Fibonacci generators, in which the value of the modulus is equal to the power of two. For comparison, Table 3 shows some research results of the repetition periods of the output pseudo-random sequence of the classical additive Fibonacci generator, which functions according to Equation (4), at $n = 2^m$. The results are obtained by imitation modelling.

Table 3. Dependence of repetition periods of the output sequence of classical AGF for some values of m on the whole set of values $X(0)$.

m Values, Max and Min Repetition Period Values								
m	2	4	8	16	32	64	128	256
period	7	14 7	28 7	56 7	112 7	224 7	448 7	896 7

Thus, in contrast to the proposed device, in the known device, at $2 > m$, there are different values of the repetition periods, including those that have critically small values. This indicates the presence of "weak keys". In addition, the maximum values of the repetition periods are usually smaller than the relative values of module m in the proposed device. These trends are also observed for arbitrary and much larger values of the modulus m.

Research of the statistical characteristics of the output pseudorandom bit sequences of new AFGs were carried out with NIST tests package [20]. If the proportion fell outside of this interval (0.98–1.0), then this was evidence that the data were non-random. Testing was carried out at different values. As a result, the sequence was entirely non-random, so requirements of statistical security were not accepted. For example, Figure 3 presents a statistical portrait of the output sequence at $m = 2{,}147{,}483{,}647$.

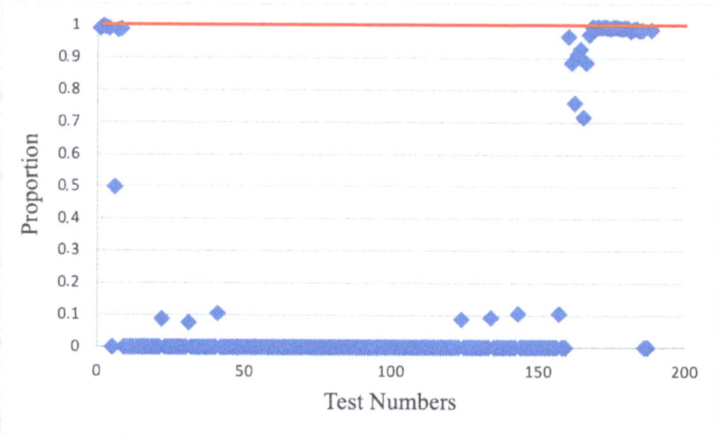

Figure 3. Statistical portrait of AFG at $m = 2{,}147{,}483{,}647$.

As can be seen from Figure 3, the sequence of the investigated generator does not meet the requirements of randomness as most of the tests were valued at 0 and did not fall within the specified interval.

Thus, new AFGs, built using prime numbers as modules of recurrent equations, provide the absence or the small number of "weak keys". At the same time, they do not accord to the criteria of statistical security; however, they can be used in conjunction with other pseudorandom bit sequence generators (PRBSGs). In this case, their useful property can be used to ensure the constancy of the repetition period of the output sequence for all possible values of the initial settings, and for many values of the module m.

3.3. Structure Scheme and Operation Principle of the Combined PRBSG

The structure scheme of the combined PRBSG is given in Figure 4.

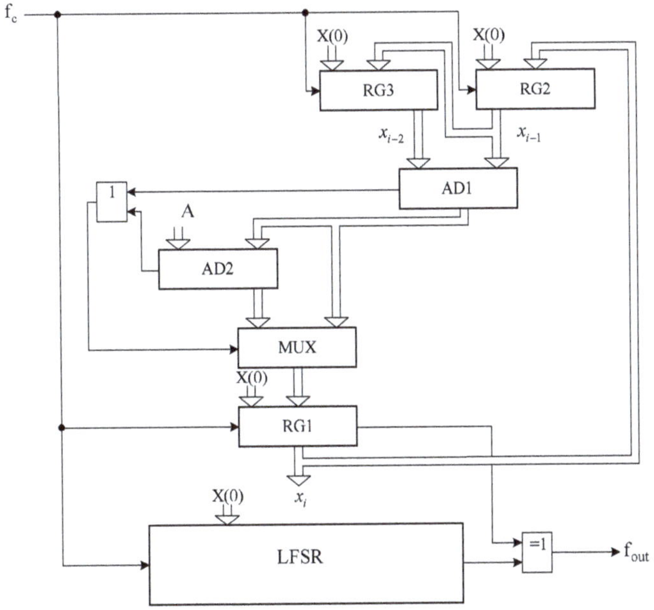

Figure 4. Structure scheme of the combined PRBSG.

The combined PRBSG consists of two generators: a generator based on a new AFG (Figure 1) and a generator based on the shift register with linear feedbacks LFSR. The output pulses of both generators are combined through a logic element XOR. The choice of type and LFSR bit number depends on the need to provide the specified characteristics of the output bit sequence. Instead of LFSR, other types of PRBSGs be used, which requires additional research.

In this work, the combined PRBSG used LFSR work according to the forming equation $F(x) = 1 + 18x + 31x$. The matrix $T1$ and the power of the matrix $r = 10$ [8] are used.

Figure 5 shows a statistical portrait of the LFSR, from which it follows that the output pseudo-random sequence does not pass only two tests from the NIST set.

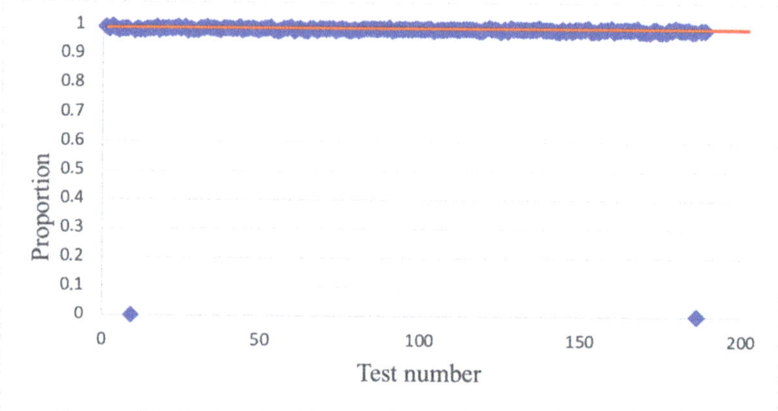

Figure 5. Statistical portrait of the LFSR ($F(x) = 1 + 18x + 31x$, matrix $T1$, power of the matrix $r = 10$).

Figure 6 shows the result of testing combined PRBSG (Figure 4) with such parameters: new AFG $m = 2{,}147{,}483{,}647$, LFSR $F(x) = 1 + 18x + 31x$, matrix $T1$ and the power of the matrix $r = 10$. The output sequence passes all tests from the NIST set.

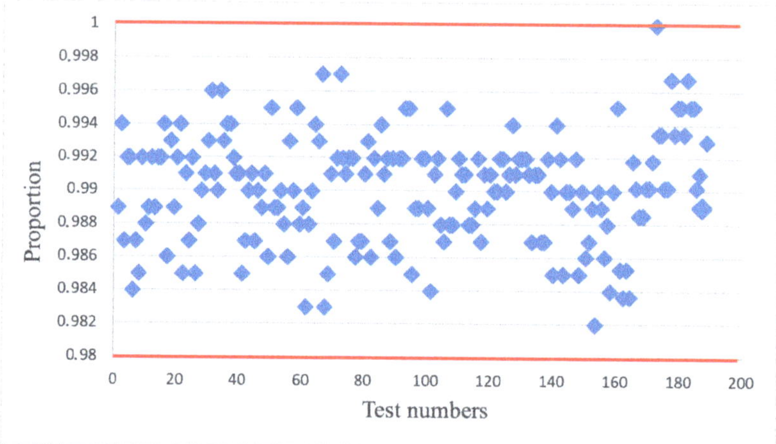

Figure 6. Statistical portrait of the combined PRBSG.

Thus, as can be seen from Figure 6, the results of all tests are within the allowable range. This suggests that the combined PRBSG generator provides the formation of the output pseudo-random sequence with high statistical characteristics.

4. Conclusions

New AFG (Figure 1), built using prime numbers as modules of recurrent equations at specific values of modules, provide a constant repetition period of the output pseudo-random pulse sequence in the whole range of possible values of the initial settings keys (seed).

According to this property, the proposed Fibonacci generator differs favourably from the known Fibonacci generators, in which the value of the modulus is equal to the power of two. In contrast to the proposed device, in the known device, at $m > 2$, there are different values of the repetition periods, including those with critically small values. This indicates the presence of "weak keys". In addition, the maximum values of the repetition periods in the known device are usually smaller than the relative values of the module m of the proposed device.

When two pseudorandom pulse sequences combine through a logical element XOR, the period of the combined sequence is not less than the repetition period of each of them.

Combined PRBSG (Figure 4), under specific requirements for their construction, can provide the specified statistical characteristics and the absence of "weak keys" in the whole range of possible values of the initial settings–keys (seed).

The results obtained and presented in the article show that the proposed generators can be effectively used in cyber security, particularly as components of cryptographic information protection or noise generators for information security, or as noise-like code sequences of modern communication systems.

Perspective for further research is the development and analysis of other types of combined PRBSG with the possibility of their hardware implementation, as well as expanding the scope of such generators.

Author Contributions: Conceptualization, V.M., O.H., M.S.; Methodology, D.J., K.K., M.K.; Validation, D.J., K.K., V.M.; Formal Analysis, M.K., O.H., M.S.; Investigation, V.M., O.H., M.S., D.J., K.K., M.K.; Data Curation, V.M., O.H., M.S., M.K.; Writing—Original Draft Preparation, D.J., K.K., O.H., M.S.; Writing—Review and Editing, V.M., O.H., M.S, D.J., K.K., M.K. All authors have read and agreed to the published version of the manuscript.

Funding: This research received no external funding.

Institutional Review Board Statement: Not applicable.

Informed Consent Statement: Not applicable.

Conflicts of Interest: The authors declare no conflict of interest.

References

1. Cardell, S.D.; Requena, V.; Fuster-Sabater, A.; Orue, A.B. Randomness Analysis for the Generalized Self-Shrinking Sequences. *Symmetry* **2019**, *11*, 1460. [CrossRef]
2. Blanco, A.; Orúe, A.B.; López, A.; Martín, A. On-the-Fly Testing an Implementation of Arrow Lightweight PRNG Using a LabVIEW Framework. In *Advances in Intelligent Systems and Computing*; Springer: Cham, Switzerland, 2019; pp. 175–184.
3. Orúe, A.B.; Encinas, L.H.; Fernández, V.; Montoya, F. A Review of Cryptographically Secure PRNGs in Constrained Devices for the IoT. In *Advances in Intelligent Systems and Computing*; Springer: Cham, Switzerland, 2017; pp. 672–682.
4. Murillo-Escobar, M.A.; Cruz-Hernández, C.; Cardoza-Avendaño, L.; Méndez-Ramírez, R. A novel pseudorandom number generator based on pseudorandomly enhanced logistic map. *Nonlinear Dyn.* **2017**, *87*, 407–425. [CrossRef]
5. Meranza-Castillón, M.O.; Murillo-Escobar, M.A.; López-Gutiérrez, R.M.; Cruz-Hernández, C. Pseudorandom number generator based on enhanced Hénon map and its implementation. *J. AEU-Int. J. Electron. Commun.* **2019**, *107*, 239–251. [CrossRef]
6. Hamza, R. A novel pseudo random sequence generator for image-cryptographic applications. *J. Info. Secur. Appl.* **2017**, *35*, 119–127. [CrossRef]
7. Ivanov, M.A.; Chugunkov, I.V. *Theory, Application and Evaluation of the Quality of Pseudorandom Consequences Generators*; KUDITS-OBRAZ: Moskow, Russia, 2003; p. 240.
8. Schneier, B. *Applied Cryptography: Protocols, Algorithms, and Source Code in C*; John Wiley & Sons: Hoboken, NJ, USA, 2007; p. 675.
9. Orue, A.B.; Montoya, F.; Encinas, L.H. Trifork, a New Pseudorandom Number Generator Based on Lagged Fibonacci Maps. *J. Comput. Sci. Eng.* **2010**, *2*, 46–51.
10. Maksymovych, V.; Harasymchuk, O.; Mandrona, M. Additive Fibonacci Generators Using Prime Numbers. In Proceedings of the VIIth International Scientific and Technical Conference "Information protection and Information Systems Security", Lviv, Ukraine, 30–31 May 2019; pp. 66–68.
11. Mandrona, M.; Maksymovych, V.; Harasymchuk, O.; Kostiv, Y. Generator of pseudorandom bit sequence with increased cryptographic immunity. *Metall. Min. Ind.* **2014**, *6*, 24–28.
12. Aluru, S. Lagged Fibonacci Random Number Generators for Distributed Memory Parallel Computers. *J. Parallel Distrib. Computing* **1997**, *45*, 1–12. [CrossRef]
13. Mandrona, M.; Maksymovych, V. Investigation of the statistical characteristics of the modified Fibonacci generators. *J. Autom. Inf. Sci.* **2014**, *46*, 48–53. [CrossRef]
14. Baldoni, S.; Battisti, F.; Carli, M.; Pascucci, F. On the Use of Fibonacci Sequences for Detecting Injection Attacks in Cyber Physical Systems. *IEEE Access* **2021**, *9*, 41787–41798. [CrossRef]

15. Agarwal, A.; Agarwal, S.; Singh, B.K. Algorithm for data encryption & decryption using Fibonacci primes. *J. Math. Control. Sci. Appl.* **2020**, *6*, 63–71.
16. Yacoab, M.; Sha, M.; Mustaq Ahmed, M. Secured Data Aggregation Using Fibonacci Numbers and Unicode Symbols for Wsn. *Int. J. Comput. Eng. Technol.* **2019**, *10*, 218–225. [CrossRef]
17. Wang, J.; Przystupa, K.; Maksymovych, V.; Stakhiv, R.; Kochan, O. Computer Modelling of Two-level Digital Frequency Synthesizer with Poisson Probability Distribution of Output Pulses. *Meas. Sci. Rev.* **2020**, *20*, 65–72. [CrossRef]
18. Maksymovych, V.; Mandrona, M.; Garasimchuk, O.; Kostiv, Y. A study of the characteristics of the Fibonacci modified additive generator with a delay. *J. Autom. Inf. Sci.* **2016**, *48*, 76–82. [CrossRef]
19. Maksymovych, V.; Mandrona, M.; Harasymchuk, O. Dosimetric Detector Hardware Simulation Model Based on Modified Additive Fibonacci Generator. *Adv. Intell. Syst. Comput.* **2020**, *938*, 162–171.
20. NIST SP 800-22 version 1a. *A Statistical Test Suite for Random and Pseudorandom Number Generators for Cryptographic Applications*; NIST: Gaithersburg, MD, USA, 2010; p. 131.

electronics

MDPI

Article

BrainShield: A Hybrid Machine Learning-Based Malware Detection Model for Android Devices

Corentin Rodrigo [1], Samuel Pierre [1], Ronald Beaubrun [2] and Franjieh El Khoury [1,*]

1 Mobile Computing and Networking Research Laboratory (LARIM), Department of Computer and Software Engineering, Polytechnique Montreal, Montreal, QC H3T 1J4, Canada; corentin.rodrigo@polymtl.ca (C.R.); samuel.pierre@polymtl.ca (S.P.)
2 Department of Computer Science and Software Engineering, Laval University, Pavillon Adrien-Pouliot, Quebec, QC G1V 0A6, Canada; Ronald.Beaubrun@ift.ulaval.ca
* Correspondence: Franjieh.el-khoury@polymtl.ca

Abstract: Android has become the leading operating system for mobile devices, and the most targeted one by malware. Therefore, many analysis methods have been proposed for detecting Android malware. However, few of them use proper datasets for evaluation. In this paper, we propose BrainShield, a hybrid malware detection model trained on the Omnidroid dataset to reduce attacks on Android devices. The latter is the most diversified dataset in terms of the number of different features, and contains the largest number of samples, 22,000 samples, for model evaluation in the Android malware detection field. BrainShield's implementation is based on a client/server architecture and consists of three fully connected neural networks: (1) the first is used for static analysis and reaches an accuracy of 92.9% trained on 840 static features; (2) the second is a dynamic neural network that reaches an accuracy of 81.1% trained on 3722 dynamic features; and (3) the third neural network proposed is hybrid, reaching an accuracy of 91.1% trained on 7081 static and dynamic features. Simulation results show that BrainShield is able to improve the accuracy and the precision of well-known malware detection methods.

Keywords: android device; BrainShield; hybrid model; machine learning; malware detection; Omnidroid

Citation: Rodrigo, C.; Pierre, S.; Beaubrun, R.; El Khoury, F. BrainShield: A Hybrid Machine Learning-Based Malware Detection Model for Android Devices. *Electronics* **2021**, *10*, 2948. https://doi.org/10.3390/electronics10232948

Academic Editor: Krzysztof Szczypiorski

Received: 30 September 2021
Accepted: 23 November 2021
Published: 26 November 2021

1. Introduction

Due to their popularity, mobile devices are becoming more and more part of our daily life. However, these devices, which handle both private and confidential data, are vulnerable to attacks by malicious people, and these are known as cyberattacks. Some of the most well-known recent examples of cyberattacks include the distributed denial of service (DDoS) attack by Mirai Botnet [1], and the massive data hijacking carried out by the WannaCry ransomware [2]. Therefore, this situation makes malware detection techniques worth investigating and improving. Malware can be any type of software that serves illegal purposes, such as spoofing or extortion [3]. This is often the case with adware that sends a lot of ads. In this paper, we generally focus on mobile app malware, and therefore, we use packages to allow us to install them.

Malware can be found in Google Play, which is the official market for Android apps. Indeed, since 2015, the number of malware has increased rapidly, which has encouraged many researchers to develop a number of malware detection methods, such as antivirus available on the Google Play Store, static method, dynamic method and hybrid method, as detailed in Section 3. However, these methods still have some limitations in terms of performance to detect the malware on the newly installed applications on Android devices, as presented in Section 3.5.

Therefore, this paper presents an extension of our previously published research work in [4] regarding malware detection on Android devices.

In [4], the proposed model is based on client/server architecture to reduce the heavy computation of data on the mobile device and perform the processing remotely on the server for prediction of the newly installed applications. We focused on the static analysis method for malware detection using the random forest regression algorithm ranging from −100 (benign) to 100 (malware) to manage the uncertainty predictions. We obtained good prediction results in terms of performance with good correlation coefficients, minimum computation time, and the smallest number of errors for malware detection.

Consequently, in this paper, we propose BrainShield, a hybrid malware detection model trained on the Omnidroid dataset [5] to reduce the attacks on Android devices, by improving the accuracy and the precision of well-known malware detection methods. More specifically, our main goal is to determine whether new samples provided to our classification model are malware or not, based on the rules previously established by the learning algorithm.

The main contributions of this paper are as follows:

1. Describe the architecture of the proposed Model called BrainShield, which is based on (1) a hybrid machine learning malware detection model for Android devices; (2) the fully connected neural networks (i.e., dense layers) composed of three layers (i.e., input layer, hidden layer and output layer) adopting one vector Tensorflow algorithm; and (3) a binary class classification that provides, as output, a probability value between 0 (i.e., benign apps) and 1 (i.e., malware apps);
2. Implement the proposed model to perform the prediction;
3. Provide the methodology that brings the detection results;
4. Train the Model with Omnidroid [5], which is the most known and diversified dataset. This dataset contains 22,000 samples and about 32,000 features (i.e., 26,000 static features and 6000 dynamic features);
5. Use the machine learning techniques [6], such as dropout and feature selection, to increase the accuracy of the proposed neural networks.

The rest of the paper is organized as follows. Section 2 details the technical background. Section 3 presents an overview of the existing malware detection methods for Android devices and their limitations. All the components of BrainShield's architecture are detailed in Section 4. In Section 5, the implementation of the BrainShield prototype and the methodology that brings the detection results are described. In Section 6, the results obtained in terms of accuracy, recall, precision, area under curve (AUC) and F1 score are illustrated, and a discussion of these results is presented. Finally, Section 6 concludes the paper by emphasizing our contribution and future work.

2. Background

In this section, we introduce a set of definitions related to Android apps, machine learning, and features used by BrainShield to detect malware apps.

Malware detection is a classification problem [7], which consists of determining the class of an app. The different classes presented in this paper are of two types: (1) malware app; and (2) benign app. Malware is an Android package kit (APK), also known as an Android app, used to serve illegal purposes, such as espionage or extortion. An app is benign if it is legitimate and harmless.

Machine learning [8] is a discipline that consists of many different methods and objectives. We use: (1) the fully connected neural networks (i.e., dense layers) with one vector Tensorflow algorithm; the Dropout regularization on the hidden layer for reducing overfitting and improving the generalization error of deep neural networks; (2) the Sigmoid activation function on the output layer to give a probabilistic distribution between 0 and 1; and (3) the optimizer ADAM to optimize the error.

The common point of these machine learning methods is to provide them with many features, labeled for supervised learning or not for unsupervised learning, which serve as input to the learning algorithm. The quantity of data and a balance of data are very important to build a precise classification model that we adopt in our proposed model.

Labeling is the act of considering an app as a malware app (i.e., value = 1) or as a benign app (i.e., value = 0). Therefore, we use the binary class classification method that gives, as output, a probability value between 0 (i.e., benign apps) and 1 (i.e., malware apps).

Features are needed in the case of supervised learning. They represent an app as faithfully as possible. Static features are those obtained using static tools, while dynamic features are those obtained using dynamic tools [9].

Evaluation metrics [10] are quantifiable measures, which determine if the detection model efficiently differentiates malware from benign apps. Among these metrics, let us quote the ones used for the evaluation of the performance of our proposed model. The accuracy represents the proportion of correct predictions. The precision is the proportion of correct positive predictions. A detection model producing no false positive has an accuracy of 1. The recall is the proportion of actual positive results that have been correctly identified. In addition, the recall is called the true positive rate (TPR). A detection model producing no false negative has a recall of 1. The F1 score is the harmonic mean of the precision and the recall. Therefore, this score considers both false positive and false negative. The area under the receiver operating characteristic (AUROC) curve measures the two-dimensional area underneath the receiver operating characteristic (ROC) curve. It gives an aggregate measure of performance across all classification thresholds.

3. Related Work

In this section, we present a literature review based on four categories of malware detection methods for mobile devices using the Android operating system: (1) company solutions; (2) static method; (3) dynamic method; and (4) hybrid method. At the end of this section, we present the limitations of the existing methods.

3.1. Company Solutions

In this section, we present a non-exhaustive list of the most popular Android apps, known as antivirus, available on the Google Play Store. This list provides solutions proposed by companies that have additional features to detect malicious apps. Table 1 illustrates a comparison of these Android apps, including the descriptive information for each app, according to Google Play Store in autumn 2019, as well as the prices offered by each app publisher.

Table 1. List of antivirus software on Google Play Store.

Play Store Name	Publisher	Free App	Price	Number of Downloads
Security Master	Cheetah Mobile	Yes	20 $/year	500,000,000+
AVG Antivirus	AVG Mobile	Yes	16 $/year	100,000,000+
Avast Antivirus	Avast Software	Yes	13 $/year	100,000,000+
Kaspersky Mobile	Kaspersky Lab	Yes	20 $/year	50,000,000+
Security Center	Hyper speed	Yes	No	50,000,000+
Mobile Security	ESET	No	7 $/year	10,000,000+
McAfee	McAfee LLC	No	41 $	10,000,000+
Malware Bytes	Malwarebytes	No	14 $/year	10,000,000+
Norton Antivirus	Norton Mobile	Yes	20 $/year	10,000,000+
Sophos Mobile	Sophos Limited	Yes	No	1,000,000+

The detection methods used by Android apps and presented in Table 1 are not known. This opacity does not allow us to develop our own detection method, but guides us to study more existing detection methods on the market. In addition, most of these Android apps provide additional functionalities besides malware detection, such as network scanner, virtual private network (VPN) service, AppLock, and permissions scanner. Typically, these features are accessible through a monthly or annual paid subscription.

Even Google Inc. cannot be certain of the 100% detection rate. Although Google Inc. made huge strides in 2019, its Google Bouncer in 2012 detection system was bypassable.

Indeed, the official announcement of its existence in February 2012 [11] caused a boom in the field of research. Several researchers have studied Google Bouncer to find out more. On 4 June 2012, Jon Oberheide and Charlie Miller [12] presented interesting results. They were able to explore the system using a command system to search for attributes of the Bouncer environment, such as the version of the kernel running, the contents of the file system, or information on some of the devices emulated by the Bouncer environment. Against all these new and increasingly virulent threats, Google Inc. has revised its policy and established Google Play Protect [13], which is the integrated malware protection platform for most Android devices. The Google Play Protect is supported by machine learning techniques to analyze more than 50 billion apps per day. Despite those advancements, malware is still found in the Google Play Store [14].

3.2. Static Method

The static analysis method does not require running the app on a device. It focuses on the app code rather than on its actual behavior when it is executed, since the app code is supposed to be faithful to the app functionality.

Fournier et al. [4] proposed a static detection method based on 151 Android system permissions trained with Waikato environment for knowledge analysis (WEKA). The model is based on training a set of 10,000 apps, consisting of 5000 benign apps and 5000 malware. Malware is from the Drebin dataset [15] dated from 2010 to 2012. The benign apps come from the top 500 in each category of the Google Play Store. The inconvenience is that no security check was offered to verify that such apps were non-malware. In the same vein, the accuracy announced on the test set is 94.62%.

IntelliAV [16] is an on-device malware detection system, which uses static analysis coupled with machine learning. The app is available on Google Play Store. Based on a training and validation set of 19,722 apps, including 9664 malware ones, the authors obtained a TPR of 92.5% and a false positive rate (FPR) of 4.2% on the validation set, with 1000 attributes generated by the training process. Moreover, the authors evaluated their model on a set of 2898 benign apps and 2311 malware from VirusTotal dated from February 2017. The accuracy is 71.96%.

MaMaDroid [17] detects malware from a behavioral perspective, modeled as a sequence of abstract API calls. It is based on a static analysis system that collects API calls made by an app, and then builds a model from the sequences obtained from the call graph in the form of Markov chains. This ensures that the model is more resilient to API changes, and that the feature set is manageable in size. MaMaDroid has been tested using a dataset of 8500 benign apps, and 35,500 malware collected over a six-year period, with F-measure reaching 99%.

DroidSieve [18] adopts a combination of features, which is suggested by authors as crucial for the robust detection of simple and obfuscated malware. Thus, syntactic features (e.g., API calls and system permissions) are integrated into such a detection method. These features have been used to build a classifier that is robust for both old and new malware, which tend to be increasingly obfuscated. To enrich all the syntactic functionalities, new features based on explicit intentions, meta-information and Dalvik Virtual Machine (DEX) files have been added. The authors created a ranking system of the most relevant features for detecting malware, where Android permissions and intents come first. The system achieves an accuracy of 99.82% with zero false positives.

FlowDroid [19] is a tool that performs taint analysis on the app code, which enables the discovery of connections where the device's International Mobile Equipment Identity (IMEI) is sent to a third party, using the network. It achieves 93% as recall, and 86% as precision.

Maldozer [20] is based on the classification of raw sequences of calls to API methods, using deep learning techniques. Maldozer can be used as a malware detection system on servers, on mobile devices, and even on Internet of Things (IoT) devices. It achieves an

F1-Score of 96–99% and a false positive rate of 0.06%. The datasets used were from the Malgenome project (2010–2011).

AndroGuard [21] is a Python library that extracts various information from code, resources or the AndroidManifest.xml file from Android. It is used for static feature extraction.

3.3. Dynamic Method

The dynamic analysis method requires running the app code on a device. Dynamic analysis is used in the literature, since techniques, such as encryption, obfuscation of code, dynamic loading of code or reflection, can be implemented to avoid detection by the static analysis method. A significant number of searches attempt to work around this problem by monitoring the actions of the app in an emulator or on a real device.

TaintDroid [22] introduces and prototypes a taint tracking method, which is widely used. The authors had to manually explore the apps, which greatly limits the number of apps that can be analyzed. Indeed, only 30 random apps have been selected.

AppsPlayground [23] takes the concept of taint tracking and develops an intelligent method of input generation and app path for dynamic analysis, which makes the detection automatic, and where the tests are performed on emulator. On the other hand, like TaintDroid, it requires a modification of the Android operating system to track data via taint tracking. AppsPlayground was evaluated with 3968 apps from the Google Play store.

Chen et al. [24] proposed the detection of systems based on data mining by ransomware for automatic detection. The actual behavior of the apps is controlled and generated in the call flow graph API (Application Programming Interface) as a set of functionalities.

Emulator vs. real phone [25] offers a detailed study of the differences between the execution environments. This study is recommended to perform the detection on a real device.

DroidBox [26] allows monitoring a wide range of events, such as file access, network traffic or DEX files loaded dynamically at runtime. DroidBox uses API 16, which covers 99.6% of smartphones according to Android. It is used for feature extraction in the context of dynamic analysis.

3.4. Hybrid Method

We define the hybrid analysis method as a method that combines static and dynamic analysis methods.

MADAM [27] is a hybrid framework using machine learning to detect malware. It classifies them based on suspicious behavior observed at different levels of Android: kernel, application, user, and package. MADAM requires administrator privileges on the phone used, since it works at the kernel level. Thus, the authors specify that their solution is not intended for the general public, but seeks to prove the strength of such an approach (i.e., multi-level, dynamic, and on the device). The 2018 version offers real-world experiments on 2800 malware of 125 different families from three datasets.

SAMADroid [28] uses machine learning to detect malware. It works on both local hosts (i.e., on-devices) to perform dynamic analysis, and remote hosts, to obtain static analysis and prediction. The SAMADroid client app is developed for Android devices. The dataset for neural network training is Drebin (2010–2012) [15], which contains old malware. However, SAMADroid claims to achieve an accuracy of 99.07%.

AndroPyTool adopted by Martin et al. [29] presents two tools that are of great importance for our own detection method: (1) the AndroPyTool framework; (2) and the Omnidroid dataset. AndroPyTool is developed in Python, and the code is hosted on GitHub. It can perform a complete extraction of static and dynamic features. It integrates the most used Android malware analysis tools (i.e., FlowDroid [19], DroidBox [26], AndroGuard [21] and Strace [30]) to perform a source code inspection, and to retrieve information on behavior when the sample is run in a controlled environment.

3.5. Limitations of the Existing Methods

Static, dynamic or hybrid approaches have the following shortcomings:

1. Little or no diversified features [4,16,20]
2. Evaluation of the model based on a poor dataset in terms of sample quantities [4];
3. An evaluation of the model based on a dataset containing old apps [4,28];
4. Obsolete methods due to new Android versions [31].

For dynamic analysis, (1) manual intervention may be required [22,32] to guarantee full exploration of the app; and (2) the app could determine if the runtime environment is an emulation. In this case, the malicious code would not be triggered, which would prevent its detection [23].

In addition to the previous shortcomings, hybrid approaches may have the following drawbacks: (1) the average performance; and (2) the device must be necessarily rooted.

Finally, all the methods presented above have high accuracy only if they are associated with: (1) many apps in the dataset for training and evaluation; and (2) recent malware. Indeed, any method that claims to achieve an accuracy of around 99%, while using old databases for evaluating the model, is considered to be obsolete.

4. BrainShield

In this section, we present the proposed hybrid malware detection model called BrainShield. BrainShield consists of three components: (1) dataset; (2) neural networks; and (3) client/server architecture.

4.1. Dataset

Omnidroid has obtained, thanks to AndroPyTool [29], a hybrid malware detection tool. The Omnidroid dataset [5] is selected for its static and dynamic features. To the best of our knowledge, it is the most diversified existing dataset in terms of types of features (i.e., static or dynamic). The static features are permissions, receivers, services, activities, Application Programming Interface (API) calls, API packages, opcodes, system commands and FlowDroid, whereas the dynamic features are opennet, sendnet, fdaccess, dataleaks, recvnet, cryptousage, and dexclass. Moreover, this dataset is balanced in terms of a number of copies, sample dates, and features. In addition, the number of apps in this dataset is significant, since there are 22,000 samples, as well as the number of features, is substantial: 25,999 static features and 5932 dynamic features. Omnidroid's samples date from 2012 to 2018, which covers a long period.

Moreover, we define a test set, which allows us to avoid over-fitting on the validation set. The test set is used at the very end of learning, and only once, to verify that the model can adapt to new samples. We have chosen to distribute our dataset as follows:

1. 70% (15,400) for the training set;
2. 15% (3300) for the validation set;
3. 15% (3300) for the test set.

4.2. Neural Networks

The use of neural networks is preferred, since it offers advantages of adaptation to new samples, which cannot be overlooked, unlike traditional detection systems operating with security rules. In particular, we chose to use fully connected neural networks (i.e., dense layers) with one vector Tensorflow algorithm. We build our model with three layers: (1) one input layer; (2) one hidden layer; and (3) one output layer. We use: (1) the Relu activation function on the input layer, which, as input, the features (i.e., static and dynamic) collected from the Omidroid dataset [5]; (2) the Dropout regularization on the hidden layer for reducing overfitting and improving the generalization error of deep neural networks; (3) the Sigmoid activation function on the output layer to give a probabilistic distribution between 0 and 1; and (4) the optimizer ADAM to optimize the error. During the training, we aim to minimize the loss function [33]. In our case, it is the binary cross entropy, which measures the performance of a classification model whose output is a probability value between 0 (i.e., benign apps) and 1 (i.e., malware apps). The binary cross entropy is denoted in Equation (1) [33]: This is example 1 of an Equation:

$$\text{Binary cross entropy} = -(y\log(p) + (1 - y) \times \log(1 - p)), \quad (1)$$

where p is the predicted probability observation and y is the binary indicator (0 or 1). To carry out good training on a dataset, it is necessary to adjust certain parameters, called hyperparameters, such as (1) the number of iterations (i.e., 200) as presented in Section 6; the dropout rate (i.e., 0.3), the learning rate (i.e., 0.002) and the activation function (i.e., Relu) proposed by Keras [34]; and (3) the number of neurons as input, as discussed in Section 5.2.1.

4.3. Architectural Design

A client/server architecture is chosen, in order to perform more flexibility with solutions and to use the feature extraction tools. Static, dynamic and hybrid proposed methods can be executed separately. Moreover, dynamic methods need to run the apps on a device, thus it makes sense to run the app in a specific environment. This device can be a real phone or a virtual machine.

BrainShield's architecture consists of two parts, as shown in Figure 1: (1) client; and (2) server.

1. The client is the Android device on which the apps must be analyzed;
2. The server is the place on which malware is detected, and it is developed with Python. Feature extraction and prediction are done on the server.

After feature extraction, each app corresponds to a feature vector. This vector is the input of the neural networks. The architecture is common to the static method, dynamic method, and hybrid method. The differences between these three methods rely on the static and dynamic feature extractions, as well as on the neural networks. Indeed, AndroGuard [21] is used for static features, while DroidBox [26] is used to extract dynamic features. The hybrid prediction is based on both static and dynamic features.

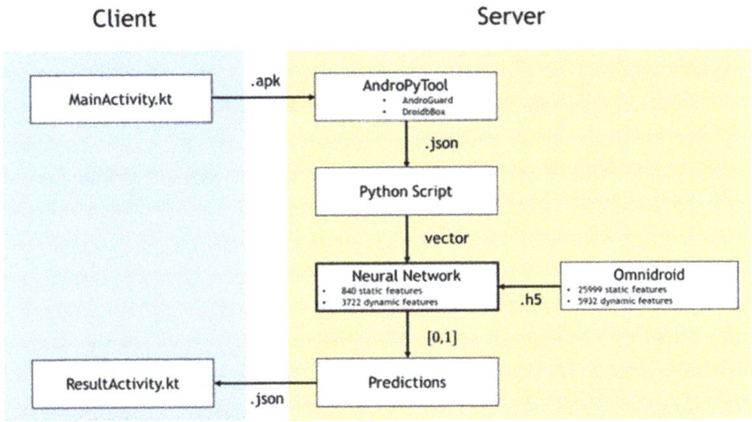

Figure 1. Architecture of BrainShield.

Chen et al. [24] proposed the detection of systems based on data mining by ransomware for automatic detection. The actual behavior of the apps is controlled and generated in the call flow graph API as a set of functionalities.

Emulator vs. real phone [25] offers a detailed study of the differences between the execution environments. This study is recommended to perform the detection on a real device.

DroidBox [26] allows monitoring a wide range of events, such as file access, network traffic, or DEX files loaded dynamically at runtime. DroidBox uses API 16, which covers

99.6% of smartphones according to Android. It is used for feature extraction in the context of dynamic analysis.

5. Implementation

In this section, we present the architecture of BrainShield's prototype and the steps of the implementation to perform the prediction. Then, we provide the methodology that brings the detection results.

5.1. BrainShield's Prototype Implementation

The architecture of BrainShield's prototype, as shown in Figure 1, is divided into two parts: (1) the client; (2) and the server.

1. The client is the Android app written in Kotlin 1.3.70 (https://kotlinlang.org/ (accessed on 20 November 2021)) that runs on the Huawei P20 Lite (ANE-LX3) (Huawei, Shenzen, China) phone;
2. The server is a Python 3.7.6 app (Python, Wilmington, DE, USA) developed with Flask 1.1 that runs on an Amazon server. This is an Ubuntu Server 18.04 LTS (HVM), SSD Volume, type t2.xlarge server with 4 vCPUs, x86_64 architecture, 16384 MiB of RAM and 16 GB of SSD memory.

In this architecture, we present the nine steps to perform the prediction, as shown in Figure 2: (1) labelling by assigning a class as benign or malicious to each app; (2) training the fully connected neural networks; (3) acquisition of APKs to be able to predict unknown apps; (4) client sends the analysis request to the server; (5) server returns the missing APKS to the client; (6) client sends APKs missing on the server; (7) feature extraction on the server; (8) prediction provided by the neural network for each app; and (9) sending prediction to the client.

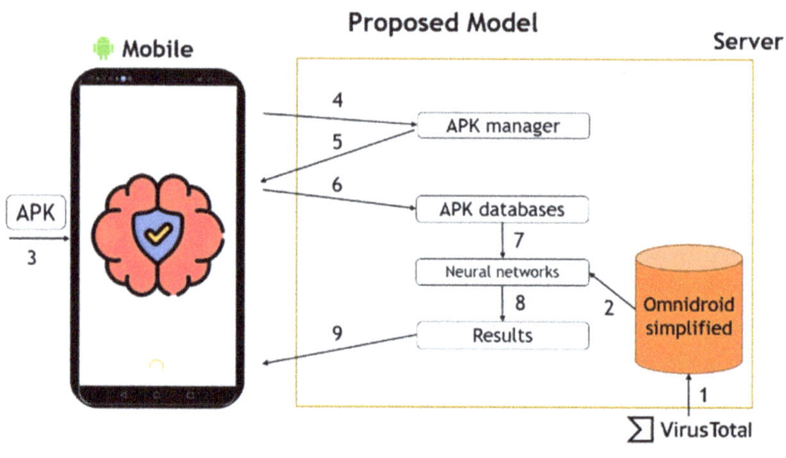

Figure 2. Steps of BrainShield.

Figure 3 depicts how we train the neural networks. We first load the database and then randomly shuffle it to have different sets of malicious and benign applications between each different training. After choosing the features, setting some parameters, creating our neural network and splitting the database into three groups, the neural network training can take place. The number of iterations can be varied, and then we finish by evaluating the neural network on the test set.

```
Training Algorithm
Loading Omnidroid dataset into memory

Random shuffling of the applications
Selection of the required features
Declaration of hyperparameters

Creation of the neural networks

Splitting the database into three groups

for 200 epochs do
        Training on the training set
        Validation on the validation set
end
Evaluation of the test set;
```

Figure 3. Pseudocode for training the neural network.

5.2. Methodology

We present hereafter the methodology, which consists of four parts: (1) hyperparameter tuning; (2) feature selection; (3) impact of feature selection; and (4) relabeling.

5.2.1. Hyperparameter Tuning

This section describes how to set the training hyperparameters values of our proposed model BrainShield based on the neural networks to detect malware.

As initial settings, the dropout rate (i.e., 0.3), the learning rate (i.e., 0.002) and the activation function (i.e., Relu) are by default proposed by Keras [34]. Moreover, the number of 50 iterations and the number of 1119 neurons as input are chosen as those large enough to have viable results and to complete hundreds of training in a suitable time. The final values of hyperparameters are illustrated in Table 2.

Table 2. Final values of hyperparameters.

Epoch Number	Static	Dynamic	Hybrid
Epoch number	200	250	200
Learning rate	0.002	0.002	0.002
Dropout rate	0.5	0.5	0.5
Batch size	512	512	512
Number of features	840	2800	Static: 840; Dynamic: 2800
Number of neurons as input	280	933	Static: 280; Dynamic: 933
Loss function	Binary crossentropy	Binary crossentropy	Binary crossentropy
Activation function	Relu	Relu	Relu
Function of prediction	Sigmoid	Sigmoid	Sigmoid

In order to obtain the best value for the epoch number (i.e., number of iterations), we consider 7 values for the epoch number (i.e., 50, 100, 150, 200, 250, 300, 400), and we compare the different results of training using the evaluation metrics (i.e., the accuracy, the recall, the precision, the AUC, and the F1 score). The results are obtained from statistical averages over 10 training sessions and are illustrated in Figure 4. Such results show that the evaluation metrics are being improved for up to 250 iterations. Then, for higher values of epoch number, no improvement in the evaluation metrics is observed (e.g., 70 training sessions with a total duration of 157 min were performed).

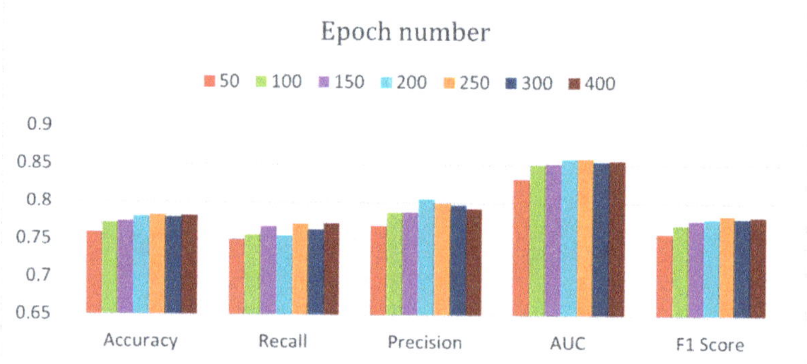

Figure 4. Epoch number comparison.

Hereafter, we justify the choice of (1) the dropout rate; (2) the learning rate; (3) the number of neurons, and (4) the activation functions.

Dropout Rate

In the same vein, we vary the value of the dropout rate from 0 (no dropout at all) to 0.9 (we forget 90% between each epoch) to obtain its best value. In Figure 5, we observe that a dropout rate of 0.3 makes it possible to obtain the best accuracy, as well as the best F1 score.

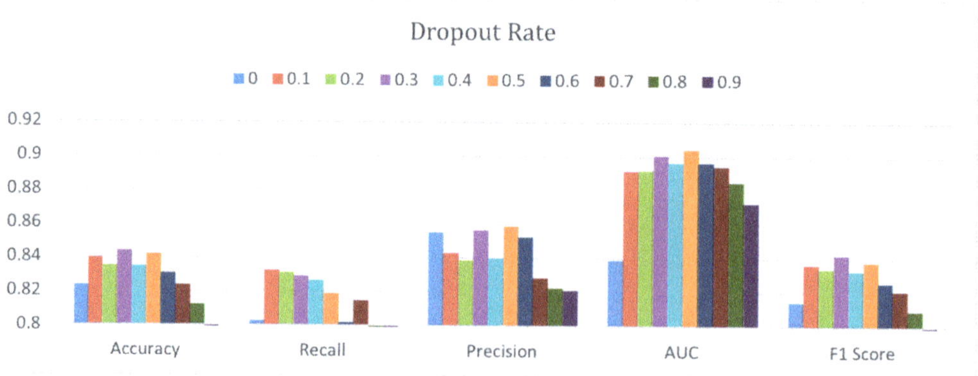

Figure 5. Dropout rate comparison.

Learning Rate

A too high learning rate may result in exceeding the minimum value of the loss function, while a too low learning rate may lead to an unnecessary too long learning process [35]. In order to obtain the appropriate learning rate value, we vary the learning rate from 0.00002 to 0.2. Figure 6 illustrates that neural networks with the default value of 0.002, as well as those with the value of 0.0002, enable us to obtain the best results, in terms of F1 Score. In this context, we choose to keep the default value of 0.002, as proposed by Keras. Indeed, both recall and AUC are being improved for detecting the false negatives (malware not detected), which constitutes the basis of malware detection.

Figure 6. Learning rate comparison.

Number of Neurons

In order to obtain the best value for the number of neurons as input, we vary the number of neurons from 10 to 4359 neurons. To attain these limits, we started by choosing a number of neurons equal to the number of features (i.e., 3359). Then, we increased and decreased this number with a pace of 250. Moreover, when the number of neurons is less than 100, we tightened the pace. In Figure 7, we observe that increasing the number of neurons above the number of features does not improve the results, in terms of accuracy, recall, precision, AUC, as well as F1 score. In addition, we can notice that the results are roughly the same from 3359 neurons to 350 neurons. Beyond this threshold, the results deteriorate. In light of such results, we estimate that the minimum number of neurons as input must be equal to 10% of the number of features to keep the same results.

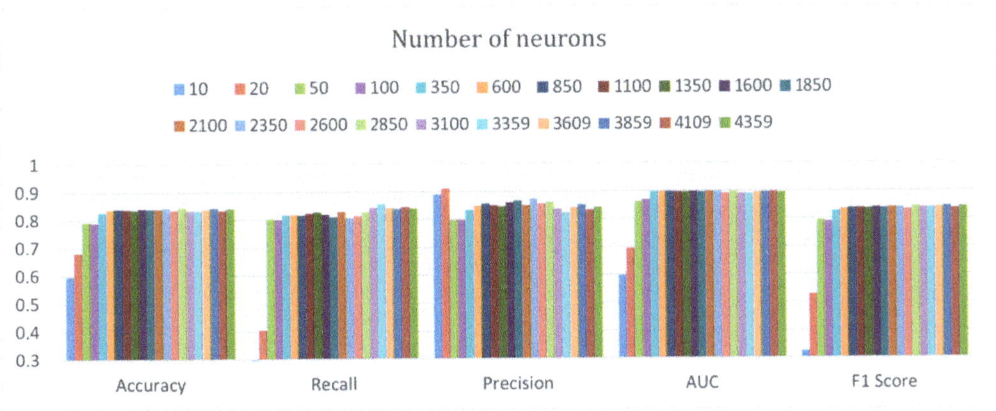

Figure 7. Number of neurons comparison.

Activation Function

In order to choose the activation function allowing us to obtain the best results, we choose to compare all the activation functions offered by Keras [34]. Depending on the dataset, each activation function has its advantages and disadvantages. In Figure 8, we note that all the activation functions barely give the same results, except the softmax and the linear activation functions.

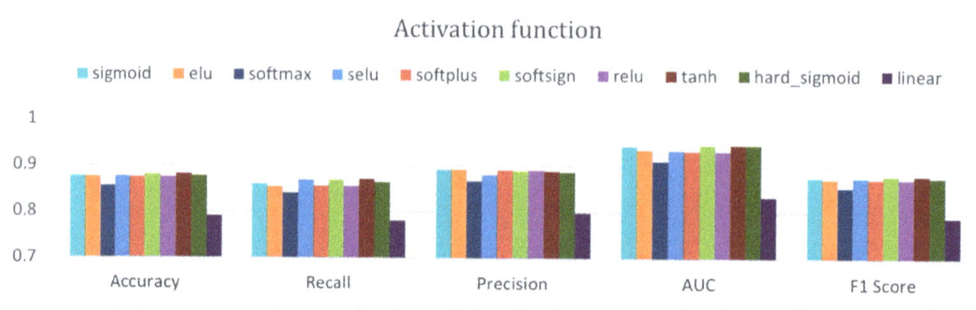

Figure 8. Activation functions comparison.

5.2.2. Feature Selection

Omnidroid initially consists of 25,999 static features. In Table 3, we present three distributions of different features. The initial distribution is Omnidroid, to which no filter is applied. The other two distributions come from the results of the selection method that we proposed.

Table 3. Static features repartition.

	Initial		Step 1		Step 2	
Features	Nb	%	Nb	%	Nb	%
Receivers	6415	24.7	216	6.4	171	8.7
Activities	6089	23.4	27	0.8	1	0.1
Permissions	5501	21.2	710	21.1	82	4.2
Services	4365	16.8	82	2.4	3	0.2
API calls	2129	8.2	1914	57.0	1369	69.4
FlowDroid	961	3.7	95	2.8	83	4.2
Opcodes	224	0.9	224	6.7	221	11.2
API Packages	212	0.8	0	0.0	0	0.0
Commands	103	0.4	91	2.7	43	2.2
Total	25,999	100	3359	100	1973	100

Step 1 consists of removing the empty features, as well as eliminating the features for which the sum of features of an app is equal to 1. In other words, only one feature over 25,999 is equal to 1. As a result, the number of features gets reduced from 25,999 to 3359. The first step proposed is therefore relevant.

Step 2 consists of removing the features whose sum does not exceed 220 for permissions, opcodes, API calls, system commands, and activities; and which does not exceed 22 for services, receivers, API packages and for FlowDroid, thus going from 3359 to 1973. The objective is to reduce the size of the dataset to allow faster training, as well as greater simplicity when loading the dataset into the RAM. We noticed a reduction of 96.8% in loading time. Although the results for the recall and precision assessment metrics are different in Figure 9, the F1 score shows that the results are very similar for the dataset of 25,999 and 3359 features, and we lose 0.1% of F1 score for that of 1973 features. Therefore, we confirm that the empty columns do not allow the neural network to improve the detection results. These features, although not useful for learning, slow down the learning time and considerably increase the allocated resources.

We now try a selection of static features, using Pearson's correlation method [36]. It allows us to select the features with the highest correlation between the features and the malware or benign apps. In Figure 10, we observe that the selection of features with Pearson's correlation enables us to improve the results obtained from the model with 3359 features. Indeed, the neural networks of 1680 and 840 features enable us to obtain an F1 score of 86.44%, compared to 85.3% in Figure 9.

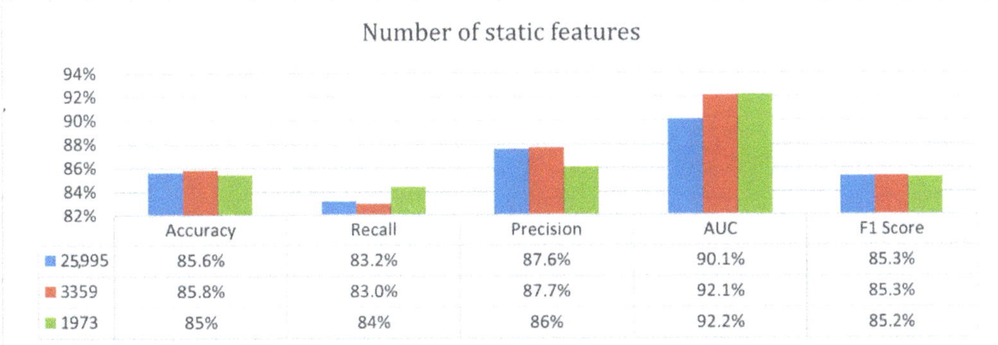

Figure 9. Static Omnidroid dataset vs. static Omnidroid simplified.

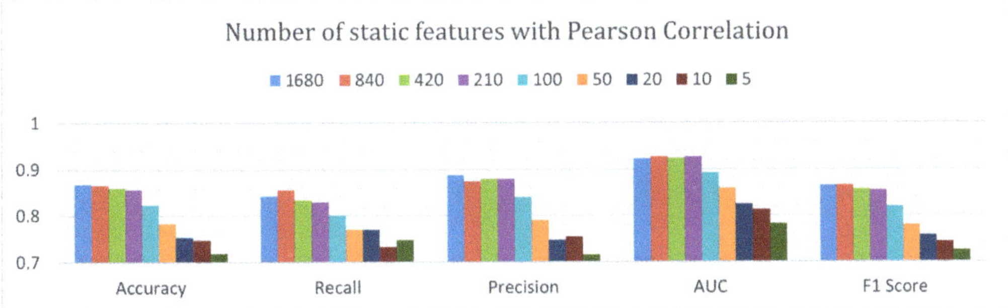

Figure 10. Feature selection with Pearson Correlation.

In the same vein, the approach that we propose for the selection of static features is carried out for the selection of dynamic features. According to this approach, we remove the features whose columns are empty or equal to 1, which reduces the number of features from 5.932 to 3722, as illustrated in Table 4.

Table 4. Dynamic features repartition.

	Initial		Step 1		Step 2	
Features	Nb	%	Nb	%	Nb	%
Opennet	1483	25.0	862	23.2	0	0.0
Sendnet	1186	20.0	718	19.3	0	0.0
Fdaccess	937	15.8	630	16.9	0	0.0
Dataleaks	867	14.6	509	13.7	134	43.2
Recvnet	837	14.1	540	14.5	0	0.0
Cryptousage	488	8.2	369	9.9	156	50.3
Dexclass	134	2.3	94	2.5	20	6.5
Total	5932	100	3722	100	310	100

In Figure 11, we note that 2210 dynamic features of Omnidroid are empty. To obtain the dataset of the 310 most diverse features, we remove all the features whose sum was less than or equal to 20. In Figure 11, we note an improvement in the results for 3722 features, which is not the case for 310 features, as presented in Table 3. In this context, we have chosen to keep 3722 dynamic features.

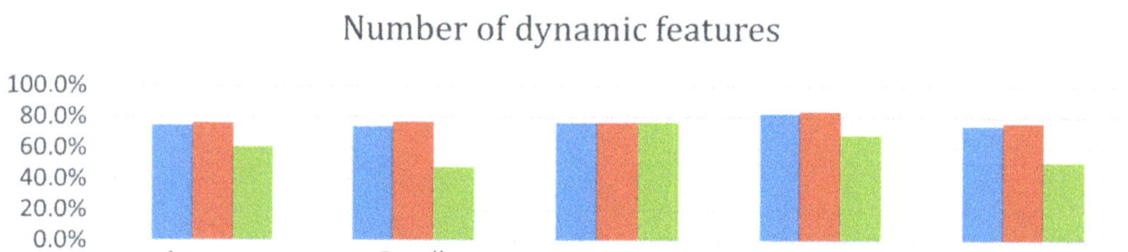

	Accuracy	Recall	Precision	AUC	F1 Score
5932	74.6%	74.0%	76.3%	82.5%	74.6%
3722	76.2%	77.1%	76.7%	83.7%	76.7%
310	61%	48%	77%	68.6%	51.1%

Figure 11. Dynamic Omnidroid dataset vs. dynamic Omnidroid simplified.

5.2.3. Impact of Feature Selection

Figure 12 illustrates the impact of the number of features on the accuracy, showing the differences when training and validating a model with 3359 and 840 static features, respectively.

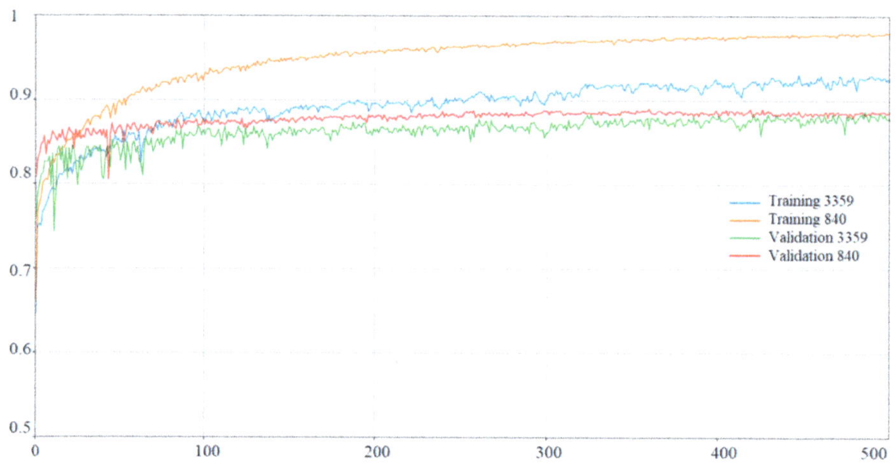

Figure 12. Impact of feature selection on accuracy.

In particular, we observe that the reduction in the number of features makes it possible to increase the accuracy on the training set and the validation set. Indeed, the training curve and the validation curve of the model with 840 features are above the training curve and the validation curve of the model with 3359 features, respectively.

5.2.4. Relabeling

The Omnidroid dataset was labeled by VirusTotal with a threshold ε equal to 1. This threshold ε represents the number of antiviruses that detects an app as malicious. Thus, a threshold ε set to 1 means that if only one of the sixty antiviruses on VirusTotal detects an app as malicious, then this app will be labeled as malicious in Omnidroid. Therefore, we track the number of apps detected by antivirus according to the number of antiviruses.

The results in Figure 13 are valid as of September 2019. We used the VirusTotal service [37] thanks to an academic API, to obtain a report for each app identified by its hash. Therefore, such results date from little more than a year and a half after the initial results obtained by [29]. Thus, we notice that a year and half later, only 9024 apps are detected as benign, with a threshold ε equal to 1, which corresponds to 0 antivirus on the abscissa axis. In addition, we note that 1807 apps are classified as malware, even though only one antivirus has detected them as malicious. A threshold ε equal to 2 would have been enough to classify them as benign.

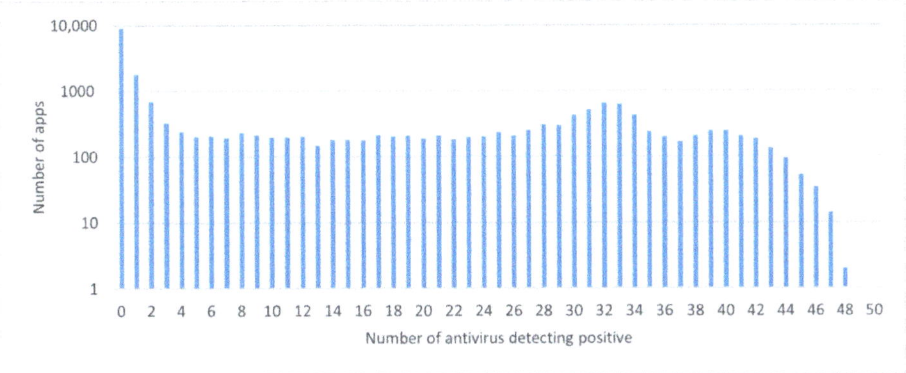

Figure 13. Scans on VirusTotal of 22,000 apps.

In Figures 14 and 15, the representations are based on 22,000 app reports from September 2019, which are collected using a python script and a VirusTotal academic key. We have relabeled Omnidroid for static trainings by setting the threshold ε from 1 to 4 for 2018′s reports, and from 1 to 10 for 2019′s reports.

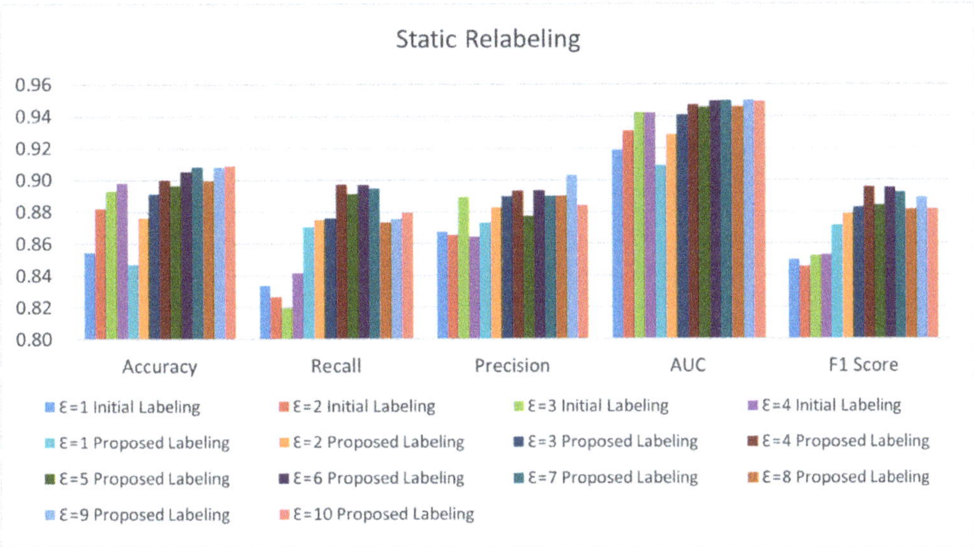

Figure 14. Static relabeling comparison.

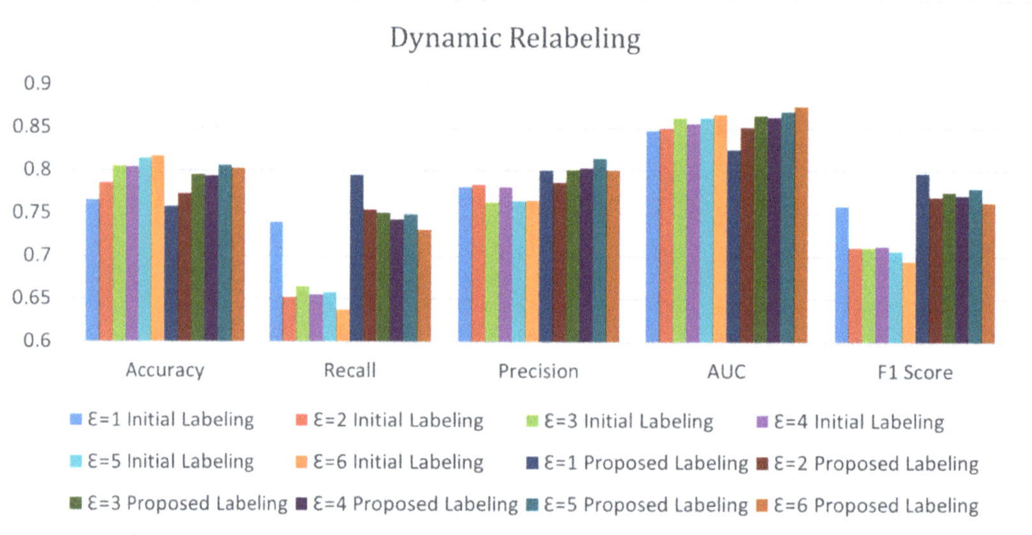

Figure 15. Dynamic relabeling comparison.

In Figure 14, we note an improvement in the results following the relabeling. We deduce that there have been new scans of antivirus among VirusTotal for a year and a half, and that this has an impact on the detection of malware.

Moreover, we have relabeled Omnidroid for dynamic trainings in Figure 15, but this time by varying the threshold ε from 1 to 6 for the reports. As with static relabeling, we notice an improvement in the results following dynamic relabeling. In particular, the recall is the metric that is improved the most (i.e., up to 10%). Relabeling has made it possible to detect more malware that were previously undetected by Omnidroid labeling.

6. Results

Following the relabeling, we set the number of iterations to the value that enables us to obtain the best scores for each type of neural network. Table 2 shows the final values of the hyperparameters.

The static feature selection was achieved with a manual method consisting in removing the empty features from 25,999 to 3359 features, as well as with Pearson's Correlation, from 3359 to 840 features. Table 5 illustrates the results of the static neural network on the test set.

Table 5. Static results.

	Accuracy	Precision	Number of Features
Martin et al. [29]	89.3%	89.3%	25,999
Proposed model	92.9%	92.1%	840
Gain	3.6%	2.8%	96.8%

Table 6 presents the results of the dynamic neural network on the test set.

Table 6. Dynamic results.

	Accuracy	Precision	Number of Features
Martin et al. [29]	78.6%	78.6%	5932
Proposed model	81.1%	83.4%	3722
Gain	2.5%	4.8%	37.3%

Table 7 shows the results of the hybrid neural network on the test set.

Table 7. Hybrid results.

	Accuracy	Precision	Number of Features
Martin et al. [29]	89.7%	89.7%	Static: 25,999; Dynamic: 5932
Proposed model	91.1%	91%	Static: 3359; Dynamic: 3722
Gain	1.4%	1.3%	77.8%

7. Discussion

We have chosen the Android operating system, since it is the most widely used operating system in the world, with more than 80% of the mobile market [38]. Moreover, the results presented in Figure 16 are valid on the date of the samples in the dataset. For a reminder, the collection of malware is registered in Omnidroid dates from 2012 to 2018. We have observed the relabeling impact clearly with more recent reports from VirusTotal on the evaluation metrics. Accordingly, we can wonder what will happen to the accuracy and the precision of neural networks in several years.

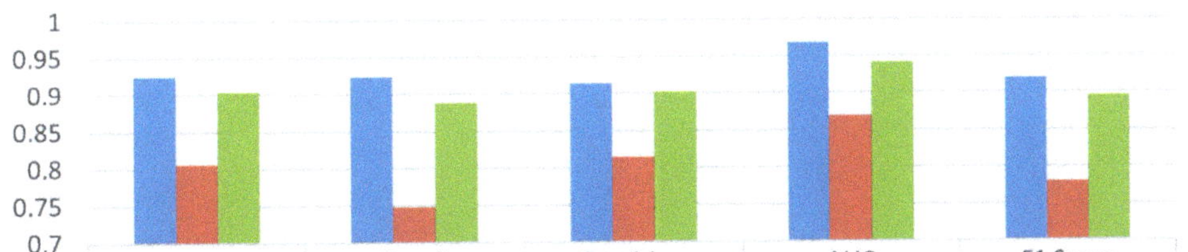

	Accuracy	Recall	Precision	AUC	F1 Score
Static	0.9252	0.9238	0.9141	0.9677	0.9189
Dynamic	0.8058	0.7483	0.8144	0.8689	0.7794
Hybrid	0.904	0.8881	0.902	0.9409	0.8949

Figure 16. Detection results comparison.

Moreover, the apps may have evolved too much to be run on an API 16 emulator, which is that of DroidBox [26], the tool for extracting dynamic features. API 16 (summer 2012) enables to run of relatively old apps for the period of summer 2020. However, it is possible that app developers consider this API too old and set the API minimum to 21. Thus, it would be impossible to run their apps on the emulator. However, this can be an advantage, since we can analyze old apps. In addition, the extraction of dynamic features can be difficult in the context of apps requiring identification (Facebook, Instagram, WhatsApp, Messenger type apps, etc.). In fact, the automatic feature extraction tool would be blocked at the displaying of identification, and would not be able to explore all the app's functionalities. As a reminder, the emulator has a feature called Monkey that enables us to randomly click on the screen to simulate user clicks. The extracted features would be either nonexistent or in too little representative quantity to be able to make a prediction. This is an intrinsic limitation of dynamic analysis.

8. Conclusions

We have proposed static, dynamic and hybrid methods for detecting malware targeting Android mobile devices. Our three methods are based on fully connected neural networks trained by the Tensorflow/Keras libraries. The static network, reaching an accuracy of 92.9% and a precision of 91.1%, is trained on 840 static features. The dynamic neural network, reaching an accuracy of 81.1% and a precision of 83.4%, is trained on

3722 dynamic features. The hybrid neural network, reaching an accuracy of 91.1% and a precision of 91.0%, is trained on 7081 features (i.e., 3359 statics and 3722 dynamics). Feature selection techniques are used, such as Pearson correlation and a manual method. In addition, we have presented that 22,636 static features and 2210 dynamic features of the Omnidroid dataset are empty for a total of 24,846 out of 31,931 (i.e., 77.81%).

As future work, this research could be generalized to other operating systems, such as iOS, which represents about 20% of the mobile market [38]. At that point, new tools for extracting static and dynamic features should be developed, in order to build a new dataset that we would be labeled by using VirusTotal. In addition, all results related to the learning techniques, the evaluation metrics, as well as the hyperparameter configuration, could be reused for training the neural networks. For further research, it would be necessary to update the dataset with the most recent labelling techniques, and to develop an automation tool for updating neural networks automatically.

Author Contributions: Conceptualization, C.R., F.E.K., R.B. and S.P.; methodology, C.R., F.E.K., R.B. and S.P.; software, C.R. and F.E.K.; validation, C.R., F.E.K., R.B. and S.P.; formal analysis, C.R. and F.E.K.; data curation, C.R. and F.E.K.; writing—original draft preparation, F.E.K.; writing—review and editing, F.E.K.; visualization, F.E.K.; supervision, S.P.; project administration, S.P. All authors have read and agreed to the published version of the manuscript.

Funding: This research was funded by Natural Sciences and Engineering Research Council of Canada (NSERC), Prompt Quebec, Flex Groups and Q-Links.

Data Availability Statement: Publicly available dataset for Omnidroid [5] is analyzed in this study. These data can be found here: (http://aida.etsisi.upm.es/download/omnidroid-dataset-v1 (accessed on 12 September 2021)).

Acknowledgments: We would like to express our gratitude to Flex Groups teams for their technical support.

Conflicts of Interest: The authors declare no conflict of interest. The funders had no role in the design of the study; in the collection, analyses, or interpretation of data; in the writing of the manuscript, or in the decision to publish the results.

References

1. Woolf, N. DDoS Attack that Disrupted Internet was Largest of its Kind in History, Experts Say. Available online: https://www.theguardian.com/technology/2016/oct/26/ddos-attack-dyn-mirai-botnet (accessed on 7 September 2021).
2. Fruhlinger, J. What is WannaCry Ransomware, How Does it Infect, and Who was Responsible? Available online: https://www.csoonline.com/article/3227906/what-is-wannacry-ransomware-how-does-it-infect-and-who-was-responsible.html (accessed on 7 September 2021).
3. MalwareBytes. Malware. Available online: https://fr.malwarebytes.com/malware/ (accessed on 10 September 2021).
4. Fournier, A.; El Khoury, F.; Pierre, S. A Client/Server Malware Detection Model Based on Machine Learning for Android Devices. *IoT* **2021**, *2*, 355–374. [CrossRef]
5. AI + DA. OmniDroid Dataset Individual Features V2. Available online: http://aida.etsisi.upm.es/download/omnidroid-dataset-v1/ (accessed on 12 September 2021).
6. Machine Learning Glossary. Available online: https://developers.google.com/machine-learning/glossary (accessed on 10 September 2021).
7. Machine Learning Crash Course. Available online: https://developers.google.com/machine-learning/crash-course/classification/ (accessed on 10 September 2021).
8. Witten, I.H.; Frank, E. Data mining: Practical machine learning tools and techniques with Java implementations. *ACM Sigmod. Record.* **2002**, *31*, 76–77. [CrossRef]
9. Tounsi, W. *Cybervigilance et Confiance Numérique: La Cyber Sécurité à l'ère du Cloud et des Objets Connectés*; ISTE Group: London, UK, 2019; pp. 1–238.
10. Microsoft. Évaluation des Performances d'un Modèle Dans Azure Machine Learning Studio (Classique). 2021. Available online: https://github.com/MicrosoftDocs/azure-docs.fr-fr/blob/master/articles/machine-learning/classic/evaluate-model-performance.md (accessed on 19 September 2021).
11. Lockheimer, H. Android and Security. Available online: https://googlemobile.blogspot.com/2012/02/android-and-security.html (accessed on 11 September 2021).
12. Oberheide, J. Dissecting Android's Bouncer. Available online: https://duo.com/blog/dissecting-androids-bouncer (accessed on 11 September 2021).

13. Android. Google Play Protect. Available online: https://www.android.com/intl/en_ca/play-protect/ (accessed on 11 September 2021).
14. Then, E. BeiTaAd Adware Discovered in 238 Google Play Store Apps. Available online: https://www.slashgear.com/beitaad-adware-discovered-in-238-google-play-store-apps-05579281/ (accessed on 12 September 2021).
15. Arp, D.; Spreitzenbarth, M.; Hübner, M.; Gascon, H.; Rieck, K.; Siemens, C. Drebin: Effective and explainable detection of android malware in your pocket. In Proceedings of the NdSS Symposium, San Diego, CA, USA, 23–28 February 2014; pp. 23–26. Available online: https://www.ndss-symposium.org/wp-content/uploads/2017/09/11_3_1.pdf (accessed on 14 September 2021).
16. Ahmadi, M.; Sotgiu, A.; Giacinto, G. IntelliAV: Building an Effective On-Device Android Malware Detector. 2018. Available online: http://arxiv.org/abs/1802.01185 (accessed on 17 September 2021).
17. Onwuzurike, L.; Mariconti, E.; Andriotis, P.; De Cristofaro, E.; Ross, G.; Stringhini, G. Mamadroid: Detecting android malware by building markov chains of behavioral models. *ACM Trans. Priv. Secur.* **2019**, *22*, 1–34. [CrossRef]
18. Suarez-Tangil, G.; Dash, S.K.; Ahmadi, M.; Kinder, J.; Giacinto, G.; Cavallaro, L. Droidsieve: Fast and accurate classification of obfuscated android malware. In Proceedings of the Seventh ACM on Conference on Data and Application Security and Privacy (CODASPY '17), Scottsdale, AZ, USA, 22–24 March 2017; pp. 309–320. [CrossRef]
19. Arzt, S.; Rasthofer, S.; Fritz, C.; Bodden, E.; Bartel, A.; Klein, J.; Le Traon, Y.; Octeau, D.; McDaniel, P. Flowdroid: Precise context, flow, field, object-sensitive and lifecycle-aware taint analysis for android apps. *ACM SIGPLAN Not.* **2014**, *49*, 259–269. [CrossRef]
20. Karbab, E.B.; Debbabi, M.; Derhab, A.; Mouheb, D. MalDozer: Automatic framework for android malware detection using deep learning. In Proceedings of the Fifth Annual DFRWS Europe, Florence, Italy, 21–23 March 2018; pp. S48–S59. [CrossRef]
21. Kali Tutorials, Androguard: Reverse Engineering, Malware and Goodware Analysis of Android Applications. 2019. Available online: https://kalilinuxtutorials.com/androguard/ (accessed on 12 September 2021).
22. Enck, W.; Gilbert, P.; Han, S.; Tendulkar, V.; Chun, B.-G.; Cox, L.P.; Jung, J.; McDaniel, P.; Sheth, A.N. TaintDroid: An information-flow tracking system for realtime privacy monitoring on smartphones. *ACM Trans. Comput. Syst. (TOCS)* **2014**, *32*, 1–29. [CrossRef]
23. Rastogi, V.; Chen, Y.; Enck, W. AppsPlayground: Automatic security analysis of smartphone applications. In Proceedings of the Third ACM Conference on Data and Application Security and Privacy, San Antonio, TX, USA, 18–20 February 2013; pp. 209–220. [CrossRef]
24. Chen, Z.-G.; Kang, H.-S.; Yin, S.-N.; Kim, S.-R. Automatic ransomware detection and analysis based on dynamic API calls flow graph. In Proceedings of the International Conference on Research in Adaptive and Convergent Systems (RACS'17), Krakow, Poland, 20–23 September 2017; pp. 196–201. [CrossRef]
25. Alzaylaee, M.K.; Yerima, S.Y.; Sezer, S. Emulator vs. real phone: Android malware detection using machine learning. In Proceedings of the 3rd ACM on International Workshop on Security and Privacy Analytics (IWSPA '17), Scottsdale, AZ, USA, 24 March 2017; pp. 65–72. [CrossRef]
26. The Honeynet Project. Droidbox: An Android Application Sandbox for Dynamic Analysis. Available online: https://www.honeynet.org/projects/active/droidbox/ (accessed on 12 September 2021).
27. Saracino, A.; Sgandurra, D.; Dini, G.; Martinelli, F. MADAM: Effective and Efficient Behavior-based Android Malware Detection and Prevention. *IEEE Trans. Dependable Secur. Comput.* **2018**, *15*, 83–97. [CrossRef]
28. Arshad, S.; Shah, M.A.; Wahid, A.; Mehmood, A.; Song, H.; Yu, H. SAMADroid: A Novel 3-Level Hybrid Malware Detection Model for Android Operating System. *IEEE Access* **2018**, *6*, 4321–4339. [CrossRef]
29. Martín, A.; Lara-Cabrera, R.; Camacho, D. Android malware detection through hybrid features fusion and ensemble classifiers: The AndroPyTool framework and the OmniDroid dataset. *Sci. Inf. Fusion* **2019**, *52*, 128–142. [CrossRef]
30. Utilisation de Strace. Available online: https://source.android.com/devices/tech/debug/strace (accessed on 12 September 2021).
31. Enck, W.; Ongtang, M.; McDaniel, P. On lightweight mobile phone application certification. In Proceedings of the 16th ACM conference on Computer and communications security, Chicago, IL, USA, 9–13 November 2009; pp. 235–245. [CrossRef]
32. Yan, L.K.; Yin, H. DroidScope: Seamlessly reconstructing the OS and Dalvik semantic views for dynamic Android malware analysis. In Proceedings of the 21st ACM USENIX Conference on Security Symposium, Security'12, Bellevue, WA, USA, 8–10 August 2012; p. 29.
33. Docs. Loss Functions. Available online: https://ml-cheatsheet.readthedocs.io/en/latest/loss_functions.html (accessed on 12 September 2021).
34. Keldenich, T. Fonction D'activation, Comment ça Marche? 2021. Available online: https://inside-machinelearning.com/fonction-dactivation-comment-ca-marche-une-explication-simple/ (accessed on 12 September 2021).
35. Machine Learning Crash Course. Reducing Loss: Learning Rate. 2020. Available online: https://developers.google.com/machine-learning/crash-course/reducing-loss/learning-rate (accessed on 12 September 2021).
36. Chatterjee, S. Good Data and Machine Learning. 2017. Available online: https://towardsdatascience.com/data-correlation-can-make-or-break-your-machine-learning-project-82ee11039cc9 (accessed on 12 September 2021).
37. VirusTotal. Available online: https://support.virustotal.com (accessed on 12 September 2021).
38. MMA. Android S'approche des 80% de Parts de Marché. 2019. Available online: https://www.mobilemarketing.fr/android-sapproche-des-80-de-parts-de-marche/ (accessed on 12 September 2021).

electronics

MDPI

Article

Detection of Image Steganography Using Deep Learning and Ensemble Classifiers

Mikołaj Płachta *, Marek Krzemień, Krzysztof Szczypiorski and Artur Janicki *

Faculty of Electronics and Information Technology, Warsaw University of Technology, Nowowiejska 15/19, 00-665 Warsaw, Poland; marek.krzemien.stud@pw.edu.pl (M.K.); krzysztof.szczypiorski@pw.edu.pl (K.S.)
* Correspondence: mikolaj.plachta.dokt@pw.edu.pl (M.P.); artur.janicki@pw.edu.pl (A.J.)

Abstract: In this article, the problem of detecting JPEG images, which have been steganographically manipulated, is discussed. The performance of employing various shallow and deep learning algorithms in image steganography detection is analyzed. The data, images from the BOSS database, were used with information hidden using three popular steganographic algorithms: JPEG universal wavelet relative distortion (J-Uniward), nsF5, and uniform embedding revisited distortion (UERD) at two density levels. Various feature spaces were verified, with the discrete cosine transform residuals (DCTR) and the Gabor filter residuals (GFR) yielding best results. Almost perfect detection was achieved for the nsF5 algorithm at 0.4 bpnzac density (99.9% accuracy), while the detection of J-Uniward at 0.1 bpnzac density turned out to be hardly possible (max. 56.3% accuracy). The ensemble classifiers turned out to be an encouraging alternative to deep learning-based detection methods.

Keywords: steganography; machine learning; image processing; BOSS database; ensemble classifier; deep learning; steganalysis; stegomalware

Citation: Płachta, M.; Krzemień, M.; Szczypiorski, K.; Janicki, A. Detection of Image Steganography Using Deep Learning and Ensemble Classifiers. *Electronics* **2022**, *11*, 1565. https://doi.org/10.3390/electronics11101565

Academic Editor: Stefanos Kollias

Received: 14 April 2022
Accepted: 11 May 2022
Published: 13 May 2022

1. Introduction

Steganography is a method of hiding classified information in non-secret material. In other words, we can hide a secret message in data that we publicly send or deliver, hiding the very existence of a secret communication. Steganographic methods pose a significant threat to users, as they can be used to spread malicious software, or can be used by such malware (so-called stegomalware [1]), for example, for C&C communications or to leak sensitive data.

An important share of steganographic methods use multimedial data, including images, as a carrier. These methods are often referred to as digital media steganography and image steganography, respectively. An example is a method used by the Vawtrak/Neverquest malware [2], the idea of which was to hide URL addresses within favicon images. Another example would be the Invoke-PSImage [3] tool, where developers hid PowerShell scripts in image pixels using a commonly used least-significant bit (LSB) approach. Yet another variance may be hiding information in the structure of GIF files [4], which is quite innovative due to the binary complexity of the GIF structure.

It is observed that a growing number of malware infections take advantage of some kinds of hidden transmission, including that based on image steganography. Since malware infections pose a significant threat to the security of users worldwide, finding efficient, reliable, and fast methods of detecting hidden content becomes very important. Therefore, numerous initiatives and projects have been recently initiated to increase malware and stegomalware resilience–one of them is the Secure Intelligent Methods for Advanced RecoGnition of malware and stegomalware (SIMARGL) project [5], realized within the EU Horizon 2020 framework.

The experiments presented in this article are part of this initiative. The aim of our research was to find the most effective automatic methods for detecting digital steganography in JPEG images. JPEG-compressed images are usually stored in files with extensions:

.jpeg, .jpg, .jpe, .jif, .jfif, and .jfi. JPEG compression is commonly used for image storage and transfer; according to [6], 74.3% of web pages contain JPEG images. Therefore, these images can also be easily used for malicious purposes. In this study, we researched various machine learning (ML) methods of creating predictive models able to discover steganographically hidden content that can be potentially used by malware. Such a detection method can be integrated with antimalware software or any other system performing file scanning for security purposes (e.g., a messaging system).

The great advantage of our research is that we experimented both with shallow ML algorithms and with deep learning methods. As for shallow algorithms, we focused on ensemble classifiers, which have been recently shown to yield good results in detection tasks. When dealing with deep learning methods, we concentrated on a lightweight approach, which did not involve computationally-intensive convolutional layers in a neural network architecture. However, for the sake of simplicity, we did not research the impact of hidden content on detection accuracy—all experiments were conducted with random hidden messages.

Our article is structured as follows: first, in Section 2, we briefly review the state of the art in the area of hiding data in digital images and its detection. Next, in Section 3, we describe the experimental environment, including the test scenarios and the evaluation metrics used. The results are described in Section 4. The article concludes with discussion of the results in Section 5 and a summary in Section 6.

2. Related Work

This article focuses on JPEG images as carriers of steganographically embedded data. The popularity of this file format has resulted in a number of data-hiding methods being proposed, as well as various detection methods. In this section, we briefly review the basics of JPEG-based image steganography, including the most commonly used algorithms. Next, we proceed to the detection methods.

2.1. *JPEG-Based Image Steganography*

While multiple steganographic algorithms operate in the spatial domain, there are some that introduce changes on the level of discrete cosine transform (DCT) coefficients stored in JPEG files. Moreover, certain algorithms are designed to minimize the probability of detection through the use of content-adaptiveness: they embed data predominately in less predictable regions, where changes are more difficult to identify. Such modifications are the most challenging to detect; this is why we selected them for our study. Following other studies, e.g., [7], we chose nsF5 [8], JPEG universal wavelet relative distortion (J-Uniward) [9], and uniform embedding revisited distortion (UERD) [10]. They are briefly characterized in the following subsections.

2.1.1. nsF5

The nsF5 [8] algorithm embeds data by modifying the least significant bits of AC ("alternating current", having at least one non-zero frequency) DCT coefficients of JPEG cover objects. Data is hidden using syndrome coding. Assuming that the sender has a p-bit message $m \in \{0,1\}^p$ to embed using n AC DCT values with their least significant bits $x \in \{0,1\}^n$ while only k coefficients $x_i, i \in I$ are non-zero, only some bits $x_i, i \in I$ are modified, thus receiving $y \in \{0,1\}^n$. This vector needs to satisfy:

$$Dy = m,$$

where D is a binary $p \times n$ matrix that is shared between the sending and receiving party. The embedding party needs to find the solution for the aforementioned equation that does not require modifying the bits of zero-valued coefficients ($x_i = y_i, i \notin I$). The solution needs to minimize the Hamming weight between the modified and unmodified least-significant-bit vectors ($x - y$). Using this coding method allows the sender to introduce fewer changes than there are bits to embed, thus decreasing the impact of embedding on the carrier object. While the example provided shows how syndrome coding works, usually a more sophisticated coding scheme, syndrome trellis coding (STC) [11], using a parity-check matrix in place of D, is applied. The y vector represents a path through a trellis built based on the parity-check matrix.

2.1.2. J-Uniward

J-Uniward [9] is a method for modeling steganographic distortion caused by data embedding. It aims to provide a function that determines which regions of the cover object are less predictable and harder to model. Changes introduced during steganographic data embedding in those areas are harder to detect than if they were introduced uniformly across the carrier. Through computation of relative changes of values based on directional filter bank decomposition this method is able to detect smooth edges that are easy to model. By detecting these predictable and unpredictable areas, this method provides a way of determining where embedding changes would be least noticeable. This method is paired with a coding scheme, such as syndrome trellis coding (STC), to create a content-adaptive data-hiding algorithm.

2.1.3. UERD

UERD [10] is a steganographic embedding scheme that aims to minimize the probability of steganographically encoded information's presence being detected, by minimizing the embedding's impact on the statistical parameters of the cover information. It achieves this by analyzing the parameters of DCT coefficients of given modes, as well as whole DCT blocks and their neighbors. Through this, the method can determine whether the region can be considered "noisy" and whether embedding will impact statistical features such as histograms of the file. "Wet" regions are those where statistical parameters are predictable and where embedding would cause noticeable changes. The scheme does not exclude values such as the DC mode coefficients or zero DCT coefficients from being used when embedding, as their statistical profiles can make them suitable from the security perspective. UERD attempts to uniformly spread the relative changes of statistics resulting from embedding. UERD employs syndrome trellis coding (STC) to hide message bits in the desired values.

Figure 1 shows a sample clean image, the same image with random data hidden using the UERD algorithm at 0.4 bpnzac (bits per non-zero AC DCT coefficient) rate, and an image which is the difference between them. As can be observed, despite there being almost 5% hidden data in the image (b) no artifacts can be perceived. What is more, it is hardly possible to observe any difference between the clean and steganographically-modified image, even if they are displayed next to one another. It is only the differential image (c) that proves the manipulation. The same refers to nsF5, J-Uniward, and other modern algorithms realizing image steganography—their manipulations are often imperceptible and difficult to detect, considering that the original image is rarely available.

2.2. *Detection Methods*

In recent years, several methods of detecting image steganography have been researched. They usually involve the extraction of some sort of parameters out of analyzed images, followed by applying a classification algorithm. They are usually based on an ML approach, employing either shallow or deep learning algorithms. Therefore, in this subsection, we first describe the features most frequently used with steganalytic algorithms, and then briefly describe typical examples of shallow and deep learning-based detection algorithms.

Figure 1. (a) Clean image, (b) image with data hidden using UERD algorithm, and (c) differential image between them, scaled 100 times. Density of steganographic data: 0.4 bpnzac, which means here 2638 B of hidden data in each 53 kB image file. Clean image source: unsplash.com (accessed on 8 April 2022).

2.2.1. Feature Extraction

In the literature, several feature spaces for image steganalysis have been researched. One of them is based on discrete cosine transform residuals (DCTR) [12], the main purpose of which is to analyze the data resulting from obtaining the DCT value for a given image. First, in this method, a random 8 × 8 pixel filter is created that will be applied to the entire analyzed set. Then, a histogram is created after applying the convolution function with the previously mentioned filter, iterating through each fragment of the analyzed image. In [13], an example of using DCTR parameters in connection with a multi-level filter is proposed. A different variation of this approach is a method based on gabor filter residuals (GFR) [14]. It works in a very similar way to DCTR, but instead of a random 8 × 8 filter, Gabor filters are used. Article [15] describes a successful application of GFR features in JPEG steganography detection. Another approach to parameterization is using the phase aware projection model (PHARM) [16]. In this approach, various linear and non-linear filters are used, while the histogram is constructed from the projection of values for each residual image fragment.

2.2.2. Shallow Machine Learning Classifiers

A number of shallow classification methods have been proposed for JPEG steganalysis. These include the use of algorithms such as support vector machines (SVM) [17–19] or logistic regression [20]. A method often appearing in recent publications is an ensemble classifier built using the Fisher linear discriminant (FLD) as the base learner [21]. In certain cases [7], parameter extractors coupled with this ensemble classifier outperformed more recent deep learning-based systems. As such, this algorithm has become a point of reference when looking into the performance of shallow ML methods in detecting steganography. The rationale driving attempts to increase its detection accuracy is the fact that data is split randomly into subsets used to train each base learner. Thus, it may be possible that certain base learners are assigned less varied datasets. Their detection accuracy may suffer from poor generalization capabilities. Simple ensemble vote-combining methods such as the one used by default do not take such effects into consideration.

2.2.3. Deep Learning Methods

In recent years, neural networks have often been reported as being used for detecting steganographically hidden data in digital images. As input data, extracted image parameters based on decompressed DCT values such as DCTR, GFR, or PHARM have been used. Proprietary variants of convolutional networks such as XuNet [22], ResNet [23], DenseNet [24], or AleksNet [25] are most often used for this purpose. The common feature of these networks is combining the convolution-batch normalization-dense structures, i.e., the convolutional function, the normalization layer, and the basal layer of neurons with the appropriate activation function. Functions such as sigmoid [26], TLU [27] (threshold linear unit), and Gaussian [28] are used, but the most common are rectified linear unit (ReLU) [29] or TanH [22].

3. Materials and Methods

In our experiments, we compared how shallow and deep learning methods cope with detecting hidden data in JPEG images. We tested a variety of deep and shallow ML-based classifiers and various feature spaces. Initially, we used raw DCT coefficients as input for the tested methods. As it did not produce satisfactory results, we extracted various parameters from the images. We performed experiments in DCTR, GFR, and PHARM feature spaces. We taught our models features extracted from pairs of images: without and with steganographically hidden data. Details of the data and the classifiers used are presented in the next subsections.

3.1. Datasets Used

We used the "Break Our Steganograhic System" (BOSS) image collection [30], which contains 10,000 black and white photos (with no hidden data). The photos were converted into JPEG with a quality factor of 75. Then, we generated three other sets of images, hiding random data with a density of either 0.4 or 0.1 bpnzac, using three different steganographic algorithms: J-Uniward, nsF5, and UERD. We used their code published at [31]. All experiments, including generation of the steganographic files, were run on a virtual machine with 64 GB RAM and 8 vCPU cores of Intel Xeon Gold 5220 processor, running on a DELL PowerEdge R740 server. Each dataset was divided in parallel into training and test subsets, in the ratio of 90:10.

3.2. Configuration of Ensemble Classifier

The base component of the shallow classifier is the ensemble classifier based on the FLD model [21]. A diagram presenting the way the ensemble classifier operates is shown in Figure 2. The set of feature vectors created by extracting DCTR, GFR, or PHARM characteristics from pictures is used to generate smaller subsets through a random selection of samples from the original set (a process called bootstrapping). These subsets are then used to train individual base learners independently from each other to diversify their classification logic. Throughout the training process, the size of the subset and the population of the ensemble (the number of base learners) is adjusted to minimize the out-of-bag error of the system. These subsets are then used to train individual base learners. Upon testing, each base learner reaches its decision independently of others and the results from the whole "population" are aggregated to produce a single decision.

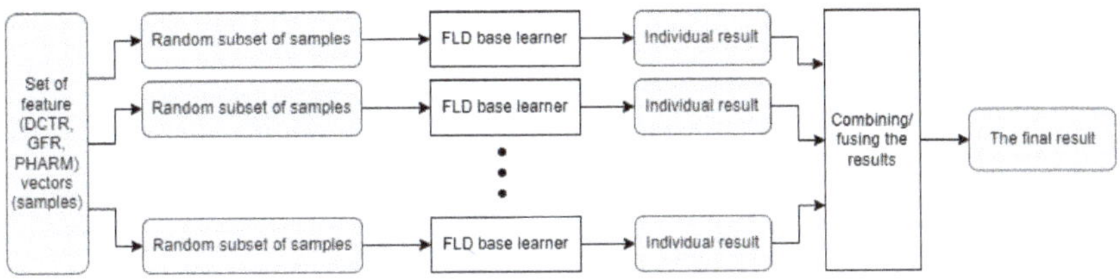

Figure 2. A diagram showing the structure of the ensemble classifier.

In our work, we focused on maximizing the detection ability of this classifier through the use of various methods to combine the votes of base learners. While the individual votes in the original ensemble were fused by simply choosing the more popular classification decision, we decided to explore the potential gain of employing machine learning for this. We trained the original ensemble classifier and then used it to obtain the decisions of all base learners for both the training and testing sets. The resulting data formed new feature vectors, which were used for further analysis with different ways of combining the votes of individual base learners. We performed this analysis using primarily methods implemented in the scikit-learn library [32]. As such, the original ensemble became a dimension-reducing layer.

3.3. Deep Learning Environment

The neural network environment was based on the Keras [33] and Tensorflow [34] library due to the simplicity of the model definition. The network architecture was mainly based on the Dense-BatchNormalization structure, but not using the convolution part as described in the available literature. We also tested various activation functions for the dense layer, such as sigmoid, softsign, TanH, and softmax, but the best results were obtained for the ReLU function. We used two optimizers: adaptive moment estimation (Adam) [35] and stochastic gradient descent (SGD) [36], which gave different results depending on the type of input parameters. The last parameter that significantly influenced the model learning efficiency was the learning rate. We found that lowering it gave very promising results without changing the network architecture and the optimizer. One of the network configurations used is displayed in Figure 3.

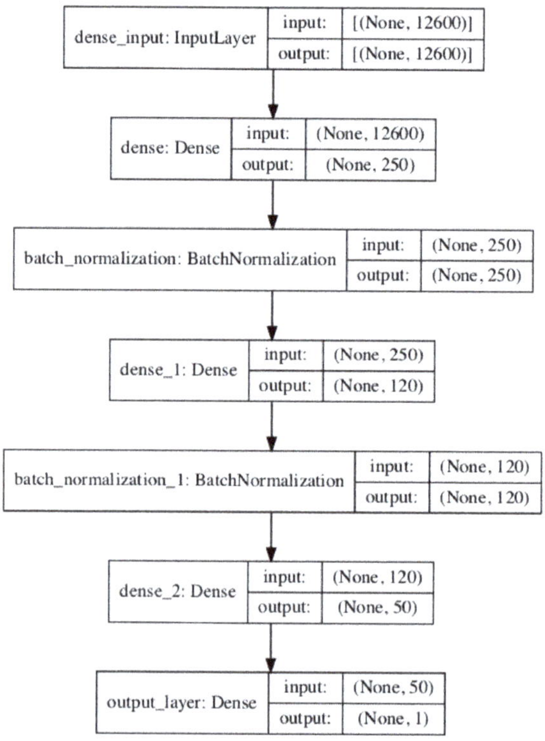

Figure 3. Example of 3 Dense-BatchNormalization neural networks used for detecting data hidden by a JPEG-based steganographic method.

3.4. Testing Scenarios

For the shallow ML-based algorithms, we decided to focus on the ensemble classifier, which has been reported in related studies as one of the most promising. During our experiments, we verified a number of ML-based methods used to combine the set of votes coming from all base learners to return the final classifier decision. These include: linear regression, logistic regression, linear discriminant analysis (LDA), and k nearest neighbors (k-NN). Moreover, the majority voting scheme (i.e., choosing the most popular classification decision, which is the original ensemble vote fusion method), as well as unquantized majority voting (i.e., classification based on the sum of non-quantized decisions of the whole ensemble) was included for comparison.

As for deep learning methods, two network architectures were selected for experiments:

- Three dense layers with the ReLU activation function, with 250 neurons in the first, 120 in the second, and 50 in the third, used in four reference models;
- Two dense layers also with the ReLU function, having 500 neurons in the first layer and 250 in the second, used in the last (fifth) reference model.

We decided not to use any convolutional layers due to their high computational requirements. However, we used additional normalization layers (BatchNormalization) between the dense layers. Half of the three-layer dense models used the Adam optimizer and half used SGD models, while the two-layer dense model used only the Adam optimizer. In the case of learning rate for the Adam optimizer, the values $1 \times e^{-4}$ or $1 \times e^{-5}$ were used, while for SGD, $1 \times e^{-3}$ or $1 \times e^{-4}$ were used. The version of the SGD optimizer with learning rate $1 \times e^{-3}$ or $1 \times e^{-4}$ and the $1 \times e^{-4}$ version of the Adam optimizer were omitted here, because they yielded much worse results compared to the version with three dense layers. In total, five different neural network configurations were tested for steganography detection.

3.5. Evaluation Metrics

To evaluate the models created, we employed commonly used metrics. The first is accuracy, which indicates what percentage of the entire set of classified data is the correct classification. The second metric is precision, which determines what proportion of the results indicated by the classifier as belonging to a given class actually belongs to it. Another metric is recall, which determines what part of the classification results of a given class is detected by the model. The fourth metric analyzed is the F1-score, which is the harmonic mean of precision and recall. It reaches 1.0 when both components give maximum results. The last metric we used to test the effectiveness of the model is the area under the ROC curve (AUC). We will also present the ROC curves themselves, as they visually present the effectiveness of the detection model.

In our results, we focus on evaluating the accuracy for each model combination, while for the best parameters we also provide the values of the other metrics. Since the testset is ideally balanced, the accuracy score is not biased and reflects well the detection ability of a given classifier.

4. Results

Tables 1 and 2 show the results of steganography detection obtained by shallow and deep methods, respectively. We display the accuracy values achieved for various steganographic algorithms and various hidden data densities, accompanied by average accuracies for each classifier/parameter combination.

Table 1. Accuracy of image steganography detection (in percentages) for various classifiers and ensemble configurations. The best values in each column are shown in bold.

Classifier	Parameters	J-Uniward		nsF5		UERD		Avg.
		0.1	0.4	0.1	0.4	0.1	0.4	
Majority voting	DCTR	50.9	84.9	78.7	**99.9**	66.1	95.3	79.3
	GFR	54.9	89.4	70.4	99.2	65.5	**95.9**	79.2
	PHARM	53.9	84.9	70.5	98.7	64.5	94.6	77.9
Unquant. majority voting	DCTR	53.3	85.0	79.4	**99.9**	66.3	95.3	79.9
	GFR	55.4	89.5	70.3	99.2	65.9	95.7	79.3
	PHARM	54.1	50.8	50.7	51.7	50.0	50.0	51.2
Linear regression	DCTR	54.2	85.6	**79.7**	**99.9**	66.1	95.2	**80.1**
	GFR	**56.3**	**89.8**	70.2	99.2	66.2	95.7	79.6
	PHARM	54.5	85.7	70.1	98.8	64.0	94.0	77.9
Logistic regression	DCTR	54.3	85.4	79.4	99.7	66.2	94.9	79.9
	GFR	**56.3**	89.5	70.1	99.1	65.9	95.5	79.4
	PHARM	54.6	62.0	–	98.5	64.5	94.7	70.7
LDA	DCTR	54.2	85.6	**79.7**	**99.9**	66.1	95.2	**80.1**
	GFR	**56.3**	89.7	70.2	99.1	66.2	95.7	79.5
	PHARM	54.4	85.7	70.1	98.8	64.0	94.0	77.8
k-NN	DCTR	53.8	85.0	78.9	**99.9**	**66.8**	95.2	79.9
	GFR	56.1	**89.8**	70.2	99.3	66.1	**95.9**	79.6
	PHARM	54.8	–	–	93.9	63.5	94.8	67.8

Table 2. Accuracy of image steganography detection (in percentages) for various architectures of neural networks and optimizers. The best values in each column are shown in bold.

Network Arch.	Optimizer	Parameters	J-Uniward		nsF5		UERD		Avg.
			0.1	0.4	0.1	0.4	0.1	0.4	
250 × BN × 120 × BN × 50 (3 layers)	Adam $1e^{-4}$	DCTR	–	83.1	**76.3**	98.8	**66.5**	**94.5**	**78.3**
		GFR	–	86.5	68.3	95.5	63.4	92.9	76.1
		PHARM	–	74.7	62.3	95.9	51.4	88.5	70.5
	Adam $1e^{-5}$	DCTR	–	83.0	74.2	**99.7**	64.7	93.1	77.5
		GFR	–	**88.4**	68.0	98.2	62.6	92.5	76.6
		PHARM	–	76.1	66.1	93.4	55.5	89.4	71.8
	SGD $1e^{-3}$	DCTR	–	77.0	73.8	99.6	62.8	91.4	75.8
		GFR	–	78.6	68.8	97.5	58.5	91.9	74.2
		PHARM	–	58.4	51.4	59.8	50.6	61.5	55.3
	SGD $1e^{-4}$	DCTR	–	73.3	52.1	99.1	51.6	91.9	69.7
		GFR	–	82.2	58.6	97.6	52.1	91.9	72.1
		PHARM	–	60.9	–	69.7	50.6	68.9	58.4
500 × BN × 250 (2 layers)	Adam $1e^{-5}$	DCTR	–	80.8	73.5	99.6	61.9	93.5	76.6
		GFR	**53.6**	86.4	67.6	97.4	64.2	91.9	76.9
		PHARM	–	75.0	54.1	94.2	54.0	87.9	69.2

On average, the use of ML for ensemble vote combination allowed for higher detection accuracy when using the systems based on DCTR or GFR features, despite marginally worse performance in certain cases (such as GFR features extracted from nsF5-modified files at 0.1 bpnzac). The PHARM-features-based classifiers sometimes yielded results worse than when using the default, majority-based scheme, or failed to converge altogether. There was no combination of type of parameters used (DCTR, GFR, PHARM) and method of fusing base-learner votes into the final decision that outperformed the others in all testing scenarios. The configuration that, on average, achieved the best results for the steganographic algorithms tested turned out to be the linear regression classifier fed with DCTR features. While using linear discriminant analysis (LDA) to fuse votes coming from a system operating on DCTR parameters achieved equal averaged accuracy, linear regression is considered in further sections due to slightly better performance with GFR and PHARM features.

As for the deep learning algorithms (Table 2), the lowest accuracy was obtained for the set based on J-Uniward. Better results in terms of accuracy were obtained for the sets based on UERD, and the best were achieved for nsF5. When analyzing the tested configurations, the worst results are those based on the SGD optimizer, while the configurations based on Adam performed better at higher learning rates. Comparing the configuration based on three layers and two layers, the results are rather similar for the Adam optimizer.

Looking at the various feature spaces, it can be seen that the least accurate results were always obtained for PHARM. On the other hand, the results obtained for the DCTR and GFR parameters for all combinations were much better and rather similar, which means that most probably they can be used interchangeably in JPEG steganalytic tools.

These observations are further confirmed in Figure 4. The PHARM parameters always yielded the worst results. The GFR features usually gave slightly better results for the higher embedding rate (0.4 bpnzac), while for the lower embedding rate (0.1 bpnzac) it was the DCTR feature space that turned out to be slightly better for most of the tested classifiers, both shallow and deep learning-based.

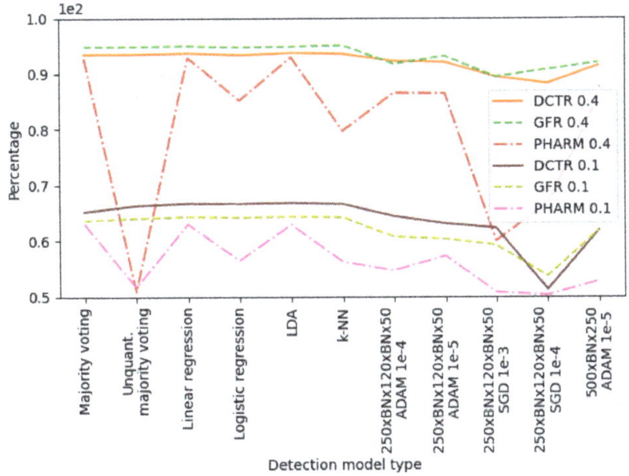

Figure 4. Accuracy achieved for various feature vectors against classifiers or network architectures.

After conducting the research, we selected the best configurations for specific types of sets, differentiating for shallow and deep learning methods, and calculated the remaining metrics. Their outcomes are visualized in Figure 5, while the details are shown in Tables 3 and 4. Based on Figure 6, one can notice that the differences between the main evaluation metrics for the best shallow and deep methods for density 0.4 bpnzac are only minor. A somewhat higher difference can be observed for all the tested steganographic algorithms applied at the lower embedding rate: 0.1 bpnzac. Here, the ensemble (shallow) classifier usually turned out to be slightly better.

Table 3. Results of image steganography detection (in percentages) for the best shallow method (linear regression).

Metric	J-Uniward		nsF5		UERD		Avg.
	0.1	0.4	0.1	0.4	0.1	0.4	
Accuracy	54.2	85.6	79.7	99.9	66.1	95.2	80.1
Precision	54.2	86.4	80.2	99.9	67.9	96.1	80.8
Recall	54.1	84.4	78.9	99.8	61.1	94.1	78.7
F1-score	54.1	85.4	79.5	99.9	64.3	95.1	79.7
AUC	54.9	91.8	87.7	99.9	72.4	98.8	84.3

Table 4. Results of image steganography detection (in percentages) for the best deep learning method ($250 \times BN \times 120 \times BN \times 50$ with Adam $1 \times e^{-4}$ based on DCTR parameters).

Metric	J-Uniward		nsF5		UERD		Avg.
	0.1	0.4	0.1	0.4	0.1	0.4	
Accuracy	50.2	83.1	76.3	98.8	66.5	94.5	78.2
Precision	50.2	80.2	74.5	98.5	63.6	94.6	76.9
Recall	46.1	87.7	80.7	99.0	78.7	94.3	81.1
F1-score	48.1	83.8	77.3	98.8	70.1	94.4	78.8
AUC	50.3	91.6	84.5	99.8	72.8	98.7	83.0

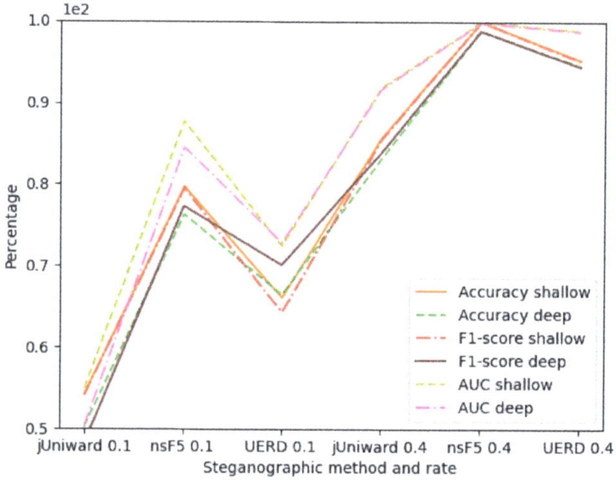

Figure 5. Visualization of evaluation results for the best shallow and deep steganalytic algorithms.

These observations are confirmed by the scores shown in Tables 3 and 4. The highest difference is for the J-Uniward 0.1 set, where the difference is about 4% relative, while for other sets we usually observe about 1–2% relative advantage in favor of the ensemble classifier, which means that these differences are only minor.

In total, the parameters of detecting data hidden using nsF5 at 0.4 embedding rate are close to 100%, regardless of the method. In contrast, the metrics for detection of data hidden with J-Uniward at 0.1 bpnzac are very poor. For the ensemble classifier with linear regression, all metrics are around 54%, while for the best neural network for most of the results are at the chance level. In general, the detection of all the tested JPEG-based steganographic methods working at the embedding rate of 0.4 bpnzac can be conducted with accuracy, with F1-score and AUC scores above 85%. The detection of hidden content embedded at a low rate of 0.1 bpnzac is problematic both for shallow and deep methods.

In the best case, the detection accuracy reached 85% for the easiest, nsF5 algorithm, while it was significantly lower for UERD and J-Uniward.

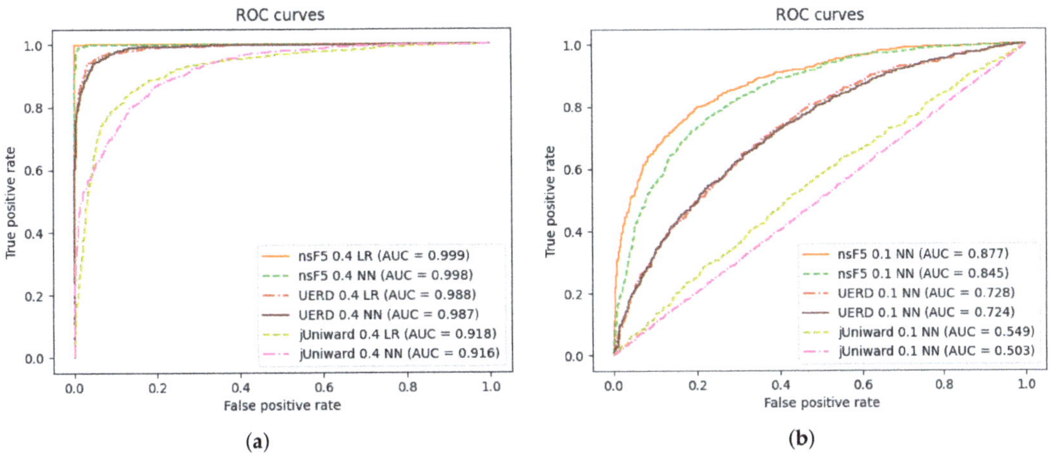

Figure 6. Comparison of ROC curves for the best neural network and the best ensemble classifier for data hidden (**a**) with density 0.4 bpnzac and (**b**) with density 0.1 bpnzac.

5. Discussion

The results obtained were compared to the results of similar studies. In article [37], the research was also carried out for the BOSS dataset using an alternative version of J-Uniward—an SUniward algorithm. The model was based on the Convolution-BatchNormalization Dense neural network scheme. The authors obtained AUC at the level of 97.9% for density 0.4 bpnzac. It was a similar result to our best model, which is much less computationally complex, due to the lack of a convolutional layer.

Articles [13,14] also conducted research on the BOSS dataset using the DCTR parameters and the Gabor filters on the J-Uniward algorithm, while using different decision models. They obtained a detection error, calculated based on false-alarm and missed-detection probabilities, as proposed in [37], of around 0.04–0.05 for DCTR, and the out-of-bag error for the Gabor filter was around 0.39, both assessed for density 0.4. Unfortunately, these results are difficult to compare with ours due to the different metrics used and the testing methodology.

During experimentation with various types of neural network layers, we noticed that adding a normalization layer significantly improved the effectiveness of a model. For example, for the nsF5 method, it improved the results by about 15–20% relative, while for the J-Uniward 0.4 case it made it possible to build a reasonable model. Without this layer, the network tended to classify all images into one class. It indicates that normalization layers in the Convolution-Dense-BatchNormalization model are indispensable, in contrast to convolution layers, the lack of which can be compensated for by, for example, choosing a different feature space.

The results achieved in our study for the three-layer and two-layer network configurations were quite similar for the Adam optimizer. This may indicate that enlarging the architecture of a neural network is pointless, as it can only have a negative impact on the computational efficiency of the neural model.

It is noteworthy that the BOSS dataset used in this study is comprised exclusively of grayscale images. Thus, only the luma channel was present in each JPEG file. However, the steganographic algorithms tested (nsF5, J-Uniward, and UERD) typically introduce changes only to DCT coefficient values of the luma channel. As such, non-grayscale (colored) images can easily be analyzed in the same way as the files from the BOSS dataset, with the chrominance channels being ignored in the detection process.

6. Conclusions and Future Work

In this article, we analyzed the effectiveness of detecting hidden content in JPEG images using either shallow, ensemble classifiers, or deep learning methods. We found that performance depended heavily on the steganographic method used and on the density of the embedded hidden data. While detecting the presence of content hidden with the nsF5 algorithm at the density 0.4 bpnzac is almost perfect, the detection of data hidden using J-Uniward at 0.1 bpnzac is hardly possible, regardless of the analysis method used.

One of the aims of our study was to find the best feature space for image steganalysis. DCTR and GFR parameters yielded the best results, while the feature space built on the PHARM parameters returned worse scores. Therefore, we recommend extracting either DCTR or GFR features when scanning JPEG files for security purposes, e.g., by antimalware software.

We also found that the performance of the best deep learning algorithm (with the network architecture: 250 × BN × 120 × BN × 50 and the Adam $1 \times e^{-4}$ optimizer) was either similar or slightly inferior to that of the best ensemble classifier built on linear regression. Therefore, we claim that carefully selected ensemble classifiers could be a promising alternative to deep learning methods in the field of image steganalysis.

Future work could concentrate on searching for effective detection methods for rates of embedding hidden data lower than 0.4 bpnzac, bearing in mind malware or advanced persistent threats (APTs) exchanging lower amounts of data. Researchers should especially focus on steganalysis of algorithms such as J-Uniward, which turned out to be particularly difficult to detect. It would be also interesting to see an application of elaborated algorithms, e.g., in an intrusion detection system (IDS). A study on the impact of characteristics of the hidden data (random, text, script) on the detectability of a JPEG-based steganographic method would also be beneficial.

Author Contributions: Conceptualization, M.P., K.S. and A.J.; methodology, M.P., M.K. and A.J.; software, M.P. and M.K.; validation, M.P. and M.K.; investigation, M.P., M.K., K.S. and A.J.; data curation, M.P.; writing—original draft preparation, M.P., M.K. and A.J.; writing—review and editing, M.P., M.K., K.S. and A.J.; visualization, M.P.; supervision, A.J. and K.S.; project administration, A.J.; funding acquisition, A.J. and K.S. All authors have read and agreed to the published version of the manuscript.

Funding: The study has been supported by the SIMARGL Project—Secure Intelligent Methods for Advanced Recognition of malware and stegomalware, with the support of the European Commission and the Horizon 2020 Program, under Grant Agreement No. 833042.

Data Availability Statement: Image data are freely available at https://www.kaggle.com/datasets/h2020simargl/jpeg-stegochecker-dataset (accessed on 10 May 2022).

Conflicts of Interest: The authors declare no conflict of interest.

Acknowledgments: The authors wish to thank the creators of the BOSS dataset: Tomáš Pevný, Tomáš Filler and Patrick Bas, for creating and publishing their data.

References

1. Caviglione, L.; Choraś, M.; Corona, I.; Janicki, A.; Mazurczyk, W.; Pawlicki, M.; Wasielewska, K. Tight Arms Race: Overview of Current Malware Threats and Trends in Their Detection. *IEEE Access* **2021**, *9*, 5371–5396. [CrossRef]
2. Cabaj, K.; Caviglione, L.; Mazurczyk, W.; Wendzel, S.; Woodward, A.; Zander, S. The New Threats of Information Hiding: The Road Ahead. *IT Prof.* **2018**, *20*, 31–39. [CrossRef]
3. Encodes a PowerShell Script in the Pixels of a PNG File and Generates a Oneliner to Execute. Available online: https://github.com/peewpw/Invoke-PSImage (accessed on 18 January 2022).
4. Puchalski, D.; Caviglione, L.; Kozik, R.; Marzecki, A.; Krawczyk, S.; Choraś, M. Stegomalware Detection through Structural Analysis of Media Files. In Proceedings of the 15th International Conference on Availability, Reliability and Security, ARES'20, New York, NY, USA, 25–28 August 2020. [CrossRef]
5. Secure Intelligent Methods for Advanced RecoGnition of Malware and Stegomalware (SIMARGL) Project. Available online: http://simargl.eu/ (accessed on 18 January 2022).
6. W3Techs—Web Technology Surveys. Available online: https://w3techs.com/technologies/overview/image_format (accessed on 6 May 2022).

7. Yang, Z.; Wang, K.; Ma, S.; Huang, Y.; Kang, X.; Zhao, X. IStego100K: Large-scale Image Steganalysis Dataset. In Proceedings of the International Workshop on Digital Watermarking, Chengdu, China, 2–4 November 2019; Springer: Berlin/Heidelberg, Germany, 2019.
8. Fridrich, J.; Pevný, T.; Kodovský, J. Statistically undetectable JPEG steganography: Dead ends, challenges, and opportunities. In Proceedings of the the 9th ACM Multimedia & Security Workshop, Dallas, TX, USA, 20–21 September 2007; Association for Computing Machinery: New York, NY, USA, 2007; pp. 3–14.
9. Holub, V.; Fridrich, J.; Denemark, T. Universal distortion function for steganography in an arbitrary domain. *EURASIP J. Multimed. Inf. Secur.* **2014**, *2014*, 1. [CrossRef]
10. Guo, L.; Ni, J.; Su, W.; Tang, C.; Shi, Y.Q. Using Statistical Image Model for JPEG Steganography: Uniform Embedding Revisited. *IEEE Trans. Inf. Forensics Secur.* **2015**, *10*, 2669–2680. [CrossRef]
11. Filler, T.; Judas, J.; Fridrich, J. Minimizing Embedding Impact in Steganography using Trellis-Coded Quantization. In Proceedings of the Media Forensics and Security II, San Jose, CA, USA, 17–21 January 2010; Memon, N.D., Dittmann, J., Alattar, A.M., Delp, E.J., III, Eds.; International Society for Optics and Photonics, SPIE: Bellingham, WA, USA, 2010; pp. 38–51.
12. Holub, V.; Fridrich, J. Low-complexity features for JPEG steganalysis using undecimated DCT. *IEEE Trans. Inf. Forensics Secur.* **2015**, *10*, 219–228. [CrossRef]
13. Wang, C.; Feng, G. Calibration-based features for JPEG steganalysis using multi-level filter. In Proceedings of the 2015 IEEE International Conference on Signal Processing, Communications and Computing (ICSPCC), Ningbo, China, 19–22 September 2015; pp. 1–4. [CrossRef]
14. Song, X.; Liu, F.; Yang, C.; Luo, X.; Zhang, Y. Steganalysis of adaptive JPEG steganography using 2D Gabor filters. In Proceedings of the 3rd ACM Workshop on Information Hiding and Multimedia Security, IH&MMSec'15, Portland, OR, USA, 17–19 June 2015; Association for Computing Machinery: New York, NY, USA, 2015; pp. 15–23.
15. Xia, C.; Guan, Q.; Zhao, X.; Xu, Z.; Ma, Y. Improving GFR Steganalysis Features by Using Gabor Symmetry and Weighted Histograms. In Proceedings of the 5th ACM Workshop on Information Hiding and Multimedia Security, IH&MMSec'17, Philadelphia, PE, USA, 20–22 June 2017; Association for Computing Machinery: New York, NY, USA, 2017; pp. 55–66. [CrossRef]
16. Holub, V.; Fridrich, J. Phase-aware projection model for steganalysis of JPEG images. In *Media Watermarking, Security, and Forensics 2015*; Alattar, A.M., Memon, N.D., Heitzenrater, C.D., Eds.; International Society for Optics and Photonics, SPIE: Bellingham, WA, USA, 2015; pp. 259–269.
17. Fridrich, J.; Kodovský, J.; Holub, V.; Goljan, M. Breaking HUGO—The Process Discovery. In *Information Hiding*; Lecture Notes in Computer Science; Springer: Berlin/Heidelberg, Germany, 2011.
18. Bhat, V.H.; Krishna, S.; Shenoy, P.D.; Venugopal, K.R.; Patnaik, L.M. HUBFIRE—A multi-class SVM based JPEG steganalysis using HBCL statistics and Fr Index. In Proceedings of the 2010 International Conference on Security and Cryptography (SECRYPT), Athens, Greece, 26-28 July 2010; IEEE: Piscataway, NJ, USA, 2010; pp. 1–6.
19. Shankar, D.D.; Azhakath, A.S. Minor blind feature based Steganalysis for calibrated JPEG images with cross validation and classification using SVM and SVM-PSO. *Multimed. Tools Appl.* **2020**, *80*, 4073–4092. [CrossRef]
20. Lubenko, I.; Ker, A.D. Steganalysis using logistic regression. In *Media Watermarking, Security, and Forensics III*; SPIE: Bellingham, WA, USA, 2011; Volume 7880, p. 78800K.
21. Kodovsky, J.; Fridrich, K.; Holub, V. Ensemble Classifiers for Steganalysis of Digital Media. *IEEE Trans. Inf. Forensics Secur.* **2012**, *7*, 432–444. [CrossRef]
22. Xu, G.; Wu, H.Z.; Shi, Y.Q. Structural Design of Convolutional Neural Networks for Steganalysis. *IEEE Signal Process. Lett.* **2016**, *23*, 708–712. [CrossRef]
23. He, K.; Zhang, X.; Ren, S.; Sun, J. Deep Residual Learning for Image Recongnition. In Proceedings of the IEEE Conference on Computer Vision and Pattern Recognition (CVPR), Las Vegas, NV, USA, 27–30 June 2016.
24. Huang, G.; Liu, Z.; van der Maaten, L.; Weinberger, K.Q. Densely connected convolutional networks. In Proceedings of the IEEE Conference on Computer Vision and Pattern Recognition (CVPR), Honolulu, HI, USA, 21–26 July 2017.
25. Mohamed, N.; Rabie, T.; Kamel, I.; Alnajjar, K. Detecting Secret Messages in Images Using Neural Networks. In Proceedings of the 2021 IEEE International IOT, Electronics and Mechatronics Conference (IEMTRONICS), Toronto, ON, Canada, 21–24 April 2021 ; pp. 1–6.
26. Tan, S.; Li, B. Stacked convolutional auto-encoders for steganalysis of digital images. In Proceedings of the Signal and Information Processing Association Annual Summit and Conference (APSIPA), 2014 Asia-Pacific, Siem Reap, Cambodia, 9–12 December 2014; pp. 1–4.
27. Yang, J.; Shi, Y.; Wong, E.K.; Kang, X. JPEG steganalysis based on DenseNet. *arXiv* **2018**, arXiv:1711.09335
28. Qian, Y.; Dong, J.; Wang, W.; Tan, T. Deep learning for steganalysis via convolutional neural networks. In Proceedings of the Media Watermarking, Security, and Forensics 2015, San Francisco, CA, USA, 8–12 February 2015; Alattar, A.M., Memon, N.D., Heitzenrater, C.D., Eds.; International Society for Optics and Photonics, SPIE: Bellingham, WA, USA, 2015; Volume 9409, pp. 171–180.
29. Pibre, L.; Pasquet, J.; Ienco, D.; Chaumont, M. Deep learning is a good steganalysis tool when embedding key is reused for different images, even if there is a cover sourcemismatch. *Electron. Imaging* **2016**, *2016*, 1–11. [CrossRef]
30. Break Our Steganographic System Base Webpage (BossBase). Available online: http://agents.fel.cvut.cz/boss/ (accessed on 18 January 2022).

31. Digital Data Embedding Laboratory. Steganographic Algorithms. Available online: http://dde.binghamton.edu/download/stego_algorithms/ (accessed on 18 January 2022).
32. Pedregosa, F.; Varoquaux, G.; Gramfort, A.; Michel, V.; Thirion, B.; Grisel, O.; Blondel, M.; Prettenhofer, P.; Weiss, R.; Dubourg, V.; et al. Scikit-learn: Machine Learning in Python. *J. Mach. Learn. Res.* **2011**, *12*, 2825–2830.
33. Ketkar, N. Introduction to Keras. In *Deep Learning with Python: A Hands-on Introduction*; Apress: Berkeley, CA, USA, 2017; pp. 97–111. [CrossRef]
34. Pang, B.; Nijkamp, E.; Wu, Y.N. Deep Learning With TensorFlow: A Review. *J. Educ. Behav. Stat.* **2020**, *45*, 227–248. [CrossRef]
35. Kingma, D.P.; Ba, J. Adam: A Method for Stochastic Optimization. *arXiv* **2014**, arXiv:1412.6980.
36. Ketkar, N. Stochastic Gradient Descent. In *Deep Learning with Python: A Hands-on Introduction*; Apress: Berkeley, CA, USA, 2017; pp. 113–132. [CrossRef]
37. Wang, H.; Pan, X.; Fan, L.; Zhao, S. Steganalysis of convolutional neural network based on neural architecture search. *Multimed. Syst.* **2021**, *27*, 379–387. [CrossRef]

electronics

MDPI

Article

Comparison of Hash Functions for Network Traffic Acquisition Using a Hardware-Accelerated Probe

Mateusz Korona, Paweł Szumełda, Mariusz Rawski and Artur Janicki *

Faculty of Electronics and Information Technology, Warsaw University of Technology, Nowowiejska 15/19, 00-665 Warsaw, Poland; m.korona@tele.pw.edu.pl (M.K.); pawel.szumelda@gmail.com (P.S.); mariusz.rawski@pw.edu.pl (M.R.)

* Correspondence: artur.janicki@pw.edu.pl

Abstract: In this article we address the problem of efficient and secure monitoring of computer network traffic. We proposed, implemented, and tested a hardware-accelerated implementation of a network probe, using the DE5-Net FPGA development platform. We showed that even when using a cryptographic SHA-3 hash function, the probe uses less than 17% of the available FPGA resources, offering a throughput of over 20 Gbit/s. We have also researched the problem of choosing an optimal hash function to be used in a network probe for addressing network flows in a flow cache. In our work we compared five 32-bit hash functions, including two cryptographic ones: SHA-1 and SHA-3. We ran a series of experiments with various hash functions, using traffic replayed from the CICIDS 2017 dataset. We showed that SHA-1 and SHA-3 provide flow distributions as uniform as the ones offered by the modified Vermont hash function proposed in 2008 (i.e., with low means and standard deviations of the bucket occupation), yet assuring higher security against potential attacks on a network probe.

Keywords: traffic analysis; network probe; hash function; SHA-3; FPGA

Citation: Korona, M.; Szumełda, P.; Rawski, M.; Janicki, A. Comparison of Hash Functions for Network Traffic Acquisition Using a Hardware-Accelerated Probe. *Electronics* **2022**, *11*, 1688. https://doi.org/10.3390/electronics11111688

Academic Editor: Taeshik Shon

Received: 29 April 2022
Accepted: 23 May 2022
Published: 25 May 2022

1. Introduction

At present, society is witnessing an unparalleled pace of technological development and global expansion of the Internet. An increasing number of ventures rely on network connectivity, both in the public sector and in business. Entities connected to the Internet range from those used for leisure purposes to elements of critical infrastructure, such as industrial process control or transportation management systems. In the background, a new technology paradigm known as Internet of Things (IoT) is evolving, which consists of objects that collect, process, and exchange data via diverse networks, often operating without direct human supervision [1]. This automation is one of the reasons why people have been already surrounded by massive numbers of IoT devices; it is estimated that about 75 million IoT devices will be connected to the network by 2025 [2].

In parallel, computer networks enable criminal activities named cybercrimes [3]. Constantly, new cybercrime types are being developed [4]. Some methods were previously associated only with mafia and now are a threat in the virtual world. This includes extortion using distributed denial of service (DDoS) attacks or ransomware—software that encrypts user data for ransom. According to the NETSCOUT Threat Intelligence Report [5], 9.7 million DDoS attacks were encountered in 2021. As Cybersecurity Ventures estimates [6], global cybercrime costs will grow yearly by 15%, reaching 10.5 trillion US dollars annually by 2025. Even though general awareness of various cybersecurity threats is increasing, as is the overall level of safety, constant effort to improve countermeasures is required. The growing number of targets, new attack vectors, and the fact that malware constantly evolves do not make this an easy task. It is estimated that over 450,000 new malicious programs and potentially unwanted applications (PUA) are registered every day [7].

In response to numerous network threats, various cybersecurity methods have been proposed. The first safeguards of a network are firewalls and intrusion detection/prevention systems (ID/PS), whose task is to analyze incoming traffic and intercept packets when a malicious signature is detected. Collecting IP traffic information for network monitoring is a common practice of network operators and researchers. To build a coarse-grained understanding of network traffic, the concept of network flows is used. It records traffic statistics in the form of flow records. Each record contains important information about a flow, such as its source and destination Internet Protocol (IP) addresses, start and end timestamps, types of service, and application ports, along with the volume of packets or bytes, etc. IP packets are assigned into flows based on their characteristics, such as source or destination address, protocol type carried, and protocol port numbers (for TCP and UDP) that can be referred to as *flow keys*. As a result of the analysis procedure, which often incorporates the most cutting-edge approaches, including machine learning [8–10], disallowed flows can be eliminated.

Flow-based network monitoring is today the most widespread technology, and NetFlow [11–13] is a widely used tool in network measurement and analysis. It is now gradually evolving into one of the most important means of ensuring network cybersecurity.

Performance of NetFlow monitoring tools has been identified as a crucial factor in network security allowing for the application of immediate countermeasures. It has been widely addressed, including the possibility for its hardware acceleration [14–18]. However, it is important to note that also the monitoring device itself can be a target of a specialized cyberattack [19], especially when the assailant has appropriate knowledge and is willing to spend their resources and time for initial reconnaissance. *Crossfire* [20] is an example of such a sophisticated attack (in comparison to the brute-force DDoS attack), tailored to a targeted enterprise, that can isolate a target area by flooding carefully selected network links.

NetFlow-like tools face great challenges when both the speed and complexity of the network traffic increase. To keep up with the multigigabit speed of network traffic, especially on high-bandwidth backbone links, *NetFlow probes* incorporate advanced techniques to efficiently store and manipulate flow records [21]. A fast local memory inside the probe, known as *flow cache*, is used to store the active flows. The flow cache is organized in a data structure called a *flow table*, which consists of a list of flow records, one for each active flow.

To efficiently process incoming packets and access the database gathered based on the flow key of the current packet often requires the use of sophisticated data structures, which vastly reduces computational complexity. Hash-based data structures are commonly proposed for this purpose as a solution allowing high-speed packet processing. Such data structures are usually coupled with a hashing function that maps a flow key to a flow cache location. Unfortunately, applying a perfect hashing function that maps each flow key to a distinct flow cache location is not possible in practice. Thus, it is crucial to select a hashing function that maps a small number of flow keys on to the same flow cache location, so-called *hash buckets*. If the number of collisions is sufficiently small, then hash tables work quite well and give $O(1)$ search times. To ensure optimal utilization of the hash table and reduce the vulnerability of a NetFlow probe to cyberattacks, the hash function needs to be carefully chosen. If it is not, malicious traffic may be able to create collisions that degenerate the hash table to linked lists with worst-case lookup times of $O(n)$ and greatly reduce the performance of the flow cache modules.

In [19], the authors evaluated the resilience of hash functions used in the software-based NetFlow probes nProbe and Vermont. Theoretical analysis and real attacks proposed by the authors show how easily flow monitors can be overloaded if the hash algorithm has not been carefully chosen. The paper also presents a hash function that seems to offer protection against hash collision attacks and computes fast enough to be deployed in high-speed flow meters.

The obvious countermeasure against hash collision-based attacks (hash flooding or HashDoS) is a hash function for which collisions cannot easily be created. Cryptographic hash functions would provide such a feature; however, they are computationally expensive, which makes them difficult to use efficiently in NetFlow probes. The implementation of such network monitoring elements with rigorous throughput may be challenging. Hardware acceleration of their crucial functions can be an aid here. Still, to the best of our knowledge, there is a lack of publications discussing hardware-accelerated network probes for network traffic analysis with dedicated hash functions that would be resilient to targeted attacks.

Our article aims at filling up this gap. In this work, we propose a hardware-accelerated network probe that accelerates extraction of network packet characteristics and calculation of the hash identifier. In addition, we describe the application of the cryptographic hash functions SHA-1 and SHA-3 to map a flow key to a flow cache location. The efficiency of our approach will be compared with the solutions discussed in [19].

Our article is organized as follows: First, in Section 2 we present the concept of a hardware-accelerated network probe and review different hashing algorithms. Next, in Section 3 we describe the experiments conducted. Their results are presented in Section 4, followed by discussion and conclusions in Section 5.

2. Materials and Methods

In this section, we outline the concept of a hardware-accelerated network probe (Section 2.1). Different hash algorithms that can produce hash table keys are discussed in Section 2.2. Details of hardware implementations and functional verification of the design are described in Section 2.3.

2.1. Hardware-Accelerated Network Probe

A network probe is a tool which acquires parameters from network traffic for traffic-analysis purposes. In this work, we used a hardware-accelerated version of the software network probe proposed in [22], which is also briefly presented here. The block diagram of the probe is presented in Figure 1. The network probe processes the traffic data in the following steps:

- capture network packets from a specific interface,
- analyze packets in chosen network stack layers,
- extract flow key and other features from the current packet,
- compute the hash value from the flow key,
- create or update a network flow record in the active flow cache,
- export inactive flows to the expired flow table,
- calculate flow parameters for the expired flows,
- store flow parameters in the output dataset.

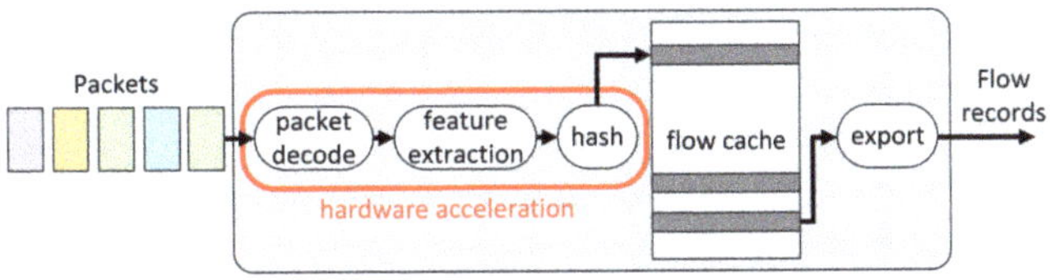

Figure 1. Block diagram of hardware-accelerated network probe.

The traffic captured from a network interface is analyzed and then a network flow record is created or an existing one is updated in the flow cache. Packet headers are analyzed in terms of second, third, and fourth ISO/OSI Reference Model layers. Assignment of new packets to flows is based on a hash function of the header parameters, which is calculated using the IP source address, the IP destination address, the source port number, the destination port number, and information on the transport layer protocol.

Considering the transport layer protocols, the conditions for classifying the stream as ended are RST or FIN flags in the case of TCP, and reaching a predefined inactivity time in the case of UDP. The flows considered as ended are statistically analyzed and their parameters extracted, as described in the next section. Expired flows are dumped to a file.

Captured packets are processed starting with the second ISO/OSI layer. From the data link layer, information about the timestamp and the packet length is fetched. The *Ether_type* field contains information about the higher-layer protocol used, which is, in the network probe's case, IPv4. After receiving the IP header, it is possible to decode the source and destination IP addresses, along with the transport layer protocol. Knowing the values of the headers of transport layer protocols, it is possible to decode the recipient's port, and the TCP flags, if applicable.

Current flows are stored in flow caches organized in *buckets*. For every incoming packet, a hash of the flow key is calculated and then checked against the existing flow keys in the appropriate bucket. If the hash does not exist, a new flow record is created in the given bucket, with parameters such as: source and destination IP addresses, source and destination port numbers, first packet timestamp, and transport layer protocol. If the hash already exists, the existing flow is updated. The packet count value is incremented, TCP flags are updated (if applicable), and a new timestamp and the packet size are added to the list.

In the case of the TCP protocol, the appearance of a FIN or RST flag means the end of the flow. Then, some of the flow's parameters are updated. Furthermore, the flow is moved from the active flows map to the expired flows list. Post-processing of the parameters consists of converting source and destination IP addresses to ASCII format; marking last timestamp; and calculating the flow's duration and total byte count, and its statistical parameters.

In the case of UDP packets, these are periodically checked by the application thread, which will be iterating through the active flows cache. The last packet's arrival time in a flow is compared to the last packet's arrival time on the network adapter, and if this exceeds the time difference by a predefined value (set in our case to 10 s), it is moved from the current flows cache to the expired flows list.

2.2. Hash Functions

Hashing is an extremely useful technique widely used to construct fast lookup methods to be able to quickly assign received packets to their corresponding flows. The hash functions used for mapping flow keys to hash values need to be chosen carefully to ensure optimal utilization of the hash table. Intuitively, a hash function is a function that maps every item to a hash value in a fashion that is somehow random. The most obvious model for a hash function is that it is fully random. Unfortunately, it is almost always impractical to construct fully random hash functions, as the space required to store such a function is essentially the same as that required to encode an arbitrary function as a lookupTable [23]. Thus, the hashing applied is usually a compromise between the randomness properties that are desired in a hash function and the computational resources needed to store and evaluate such a function.

Hash functions utilized in network monitoring devices should have the following features:

1. good performance—hash calculation cannot become a bottleneck in the network monitor;
2. uniform distribution—when this condition is fulfilled, buckets of the hash table which stores data describing monitored flows are randomly selected for traffic that is not manipulated, and none of them is likely to contain long list of packets (or to overflow);
3. collision resistance—when the hash function has this feature, it is extremely hard for an attacker to forge two packets with different flow characteristics that will end in the same hash table bucket, a situation that might eventually lead to bucket overflow.

Report [19] discusses hash algorithms used in two popular monitoring tools—nProbe [24] and Vermont [25]. The authors of the current paper have identified some flaws in both algorithms and proposed a modified version of Vermont. They also suggest that cryptographic hash functions might be best for such an application, if their implementations meet performance demands.

The network probe implements all three algorithms from [19] in hardware. In addition, two cryptographic hash functions were implemented—the cryptographically broken but still widely used SHA-1 and the state-of-the-art SHA-3. All of the algorithms are described in following subsections.

For the proposed network probe, a hash width of 32 bits was considered. If the result of a given algorithm was wider, this was reduced accordingly to 32 bits. The network probe considers source IP address, destination IP address, protocol, and protocol (TCP/UDP) source/destination port numbers as flow keys.

2.2.1. Sum Modulo 32—nProbe

The nProbe [24] monitoring tool utilizes simple sum modulo as its hash algorithm. For the proposed network probe, the calculation is presented as Equation (1):

$$h = (srcIP + dstIP + protocol + srcPort + dstPort) mod 32 \tag{1}$$

This algorithm is very simple; however, as the authors of [19] point out, after testing it with a captured network packet trace, it does not have a perfectly uniform distribution—a number of buckets contain considerably more entries than others. Another drawback is relative ease of generating collisions, because an attacker can freely manipulate the values of the flow keys provided that their sum is constant.

2.2.2. Nested CRC-32—Vermont

Cyclic redundancy checks or cyclic redundancy codes (CRC) have been utilized for error detection in computing for a long time. A digest is calculated from transmitted data and is appended to the frame. The same algorithm is applied to data upon frame reception, and when the result is the same as the code calculated by the transmitter, it means that the received packet is correct.

The actual algorithm can be described mathematically as polynomial division of binary data being interpreted as polynomial over GF(2) (every bit is a polynomial coefficient—zero or one) by generator polynomial G(x). The remainder of that division is treated as a check sequence, which is appended to the transmitted frame [26].

The CRC-32 implementation used in the proposed network probe is based on IEEE 802.3 [27] polynomial. Implementation parameters, according to [26], are presented in Table 1.

Table 1. The network probe hardware accelerator CRC-32's implementation parameters, following [26].

Parameter	Value
Polynomial	$x^{32} + x^{26} + x^{23} + x^{22} + x^{16} + x^{12} + x^{11} + x^{10} + x^8 + x^7 + x^5 + x^4 + x^2 + x^1 + 1$ (0x04C11DB7)
Data width	32
Initial value	0xFFFFFFFF
Reflect input	True
Reflect output	True
Final XOR	0xFFFFFFFF

Vermont [25] is built on nested CRC-32 invocations. The algorithm starts with a given initial seed, and Figure 2 presents how CRC-32 is invoked five times to include flow keys in the hash calculation. The result of the preceding CRC-32 function is utilized as seed for the next one.

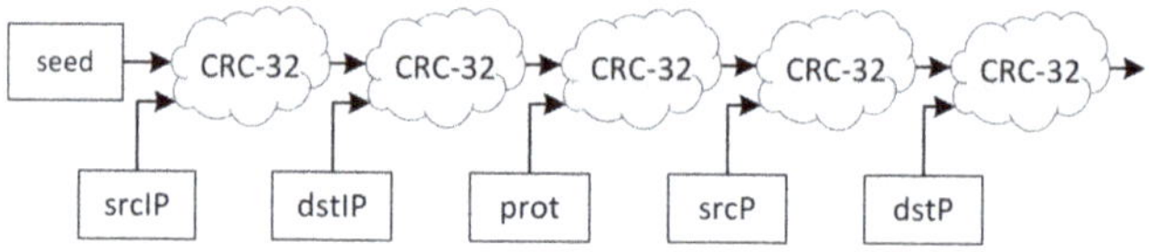

Figure 2. Illustration of Vermont hashing function.

The authors of [19] found that Vermont is computationally efficient and offers roughly uniform distribution; however, they also proved that an attacker is still able to create hash collisions on purpose.

2.2.3. Nested CRC-32 with w Constants—Modified Vermont

Report [19] proved that the CRC-based Vermont algorithm does not protect network monitoring devices from targeted collision attacks. The goal of the authors of this current paper was to design a function that does not have this flaw, but that offers the same statistical qualities. The result of their research is a modified Vermont algorithm, presented in Figure 3.

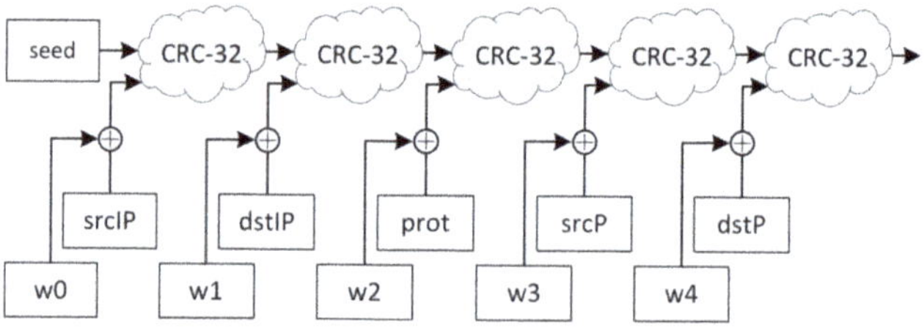

Figure 3. Illustration of modified Vermont hashing function, based on [19].

To ensure that an attacker cannot create collisions in a simple way, a unique secret random value (*w(i)*, initialized during network monitor activation) is added to every flow key before CRC-32 calculation. This significantly increases the cost of a targeted attack, but does not prevent it, since the CRC-32 scheme is still used.

2.2.4. SHA-1

SHA-1 is a cryptographic hash function created in 1995, described in [28,29]. In its cycle of life it is currently marked as deprecated, because it is prone to a variety of attacks. In 2015, a group of researchers was able to find a *freestart collision*, where the SHA-1 initialization vector was chosen by themselves [30], but soon the full SHA-1 algorithm was also cracked [31–33].

An organized crime syndicate in possession of tens of thousands of dollars can create an SHA-1 collision in about two months, and for instance, forge an SSL certificate. That is the reason famous brands such as Microsoft, Google, and Mozilla abandoned the SHA-1 algorithm; however, it still may be useful in real-time applications such as network monitoring.

The SHA-1 function produces a 160-bit hash. It is capable of hashing messages as long as $2^{64} - 1$ bits, which are divided into 512-bit blocks processed one by one.

The first step of the algorithm is *padding*, because the length of the message must be a multiple of 512 bits. During this process, the information about message length is encoded in 64 bits (hence the message length limit). This number is concatenated with exactly one "1" bit and an appropriate number of "0" bits, so when the padding bit string is appended to the message, the total length is a multiple of 512 bits. The temporary value of the hash is stored in five 32-bit variables H, initialized as in Listing 1.

Listing 1. Initial values of *H* variables in SHA-1 algorithm.

```
H_0(0) = 0x67452301
H_1(0) = 0xEFCDAB89
H_2(0) = 0x98BADCFE
H_3(0) = 0x10325476
H_4(0) = 0xC3D2E1F0
```

Every block of the message is processed through 80 rounds according to the scheme in Figure 4.

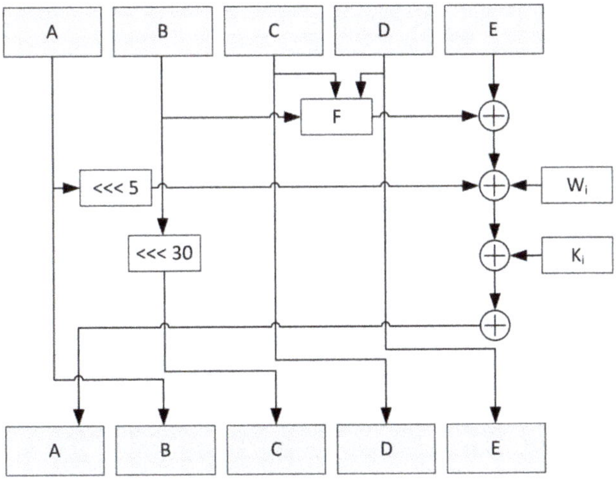

Figure 4. SHA-1 algorithm round scheme.

Variables *A* to *E* are assigned values of corresponding *H* registers from the previous block or *H(0)* for the first block. The *W* array is generated—the first 16 words are 32-bit chunks of the processed block and subsequent words are calculated with Equation (2).

$$W(i) = W(i-3) \oplus W(i-8) \oplus W(i-14) \oplus W(i-16) \tag{2}$$

Function *F* and the value of variable *K* depend on the current round number as in Equations (3) and (4).

$$F(i) = \begin{cases} (B \,\&\, C) \mid ((\sim B) \,\&\, D) & \text{for } 0 <= i <= 19 \\ B \oplus C \oplus D & \text{for } 20 <= i <= 39 \\ (B \,\&\, C) \mid (B \,\&\, D) \mid (C \,\&\, D) & \text{for } 40 <= i <= 59 \\ B \oplus C \oplus D & \text{for } 60 <= i <= 79 \end{cases} \tag{3}$$

$$K(i) = \begin{cases} 0x5A827999 & \text{for } 0 <= i <= 19 \\ 0x6ED9EBA1 & \text{for } 20 <= i <= 39 \\ 0x8F1BBCDC & \text{for } 40 <= i <= 59 \\ 0xCA62C1D6 & \text{for } 60 <= i <= 79 \end{cases} \tag{4}$$

After 80 rounds for the given block, the *H* registers are updated as in Listing 2. When all blocks of the message are processed, the hash can be read as a concatenation of *H* variables.

Listing 2. Update of *H* variables when block was processed in the SHA-1 algorithm.

```
H_0(i) = H_0(i-1) + A
H_1(i) = H_1(i-1) + B
H_2(i) = H_2(i-1) + C
H_3(i) = H_3(i-1) + D
H_4(i) = H_4(i-1) + E
```

In the proposed network probe, SHA-1 is applied to a 104-bit string that consists of 32-bit IP source and destination addresses, 8-bit IP protocol information, and 16-bit source and destination ports of TCP/UDP. The 160-bit hash is reduced to 32-bit words by XORing ($\oplus$) all *H* registers together.

2.2.5. SHA-3

SHA-3 [34] is the newest hash standard issued by NIST. Unlike previous SHA algorithms, it is based on *sponge construction* [35] instead of the Merkle–Damgård structure [36]. SHA-3 is in fact a slightly modified *Keccak* algorithm [37], the winner of the NIST contest. SHA-3, like SHA-2, is capable of four hash length generations: 224, 256, 384, and 512 bits, depending on the underlying sponge construction configuration.

Keccak has an internal state which is *b*-bit string *S*; this can be also presented as a three-dimensional array (named *A*, Figure 5) with mapping as in Equation (5). For SHA-3, $b = 1600$ and two more helper variables are derived from this value: $w = b/25 = 64$ and $l = log_2(w) = 6$.

$$A[x, y, z] = S[w(5y + x) + z] \tag{5}$$

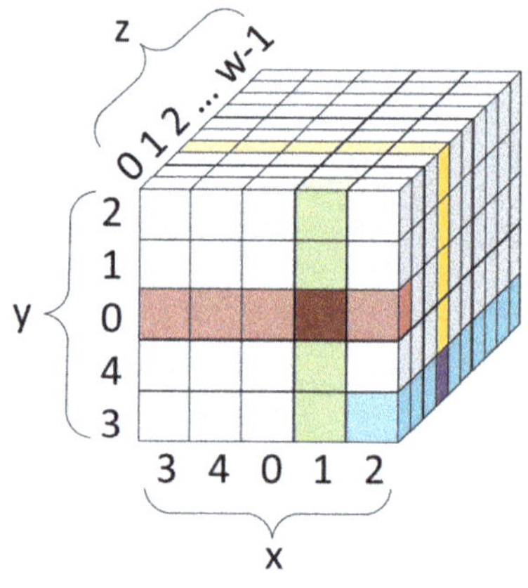

Figure 5. SHA-3 state as three-dimensional array *A*.

In Figure 5:

- the color green marks an example *column* of the state array ($x = 1, z = 0$),
- the color red marks an example *row* of the state array ($y = 0, z = 0$),
- the color blue marks an example *lane* of the state array ($x = 2, y = 3$),
- and the color yellow marks an example *slice* of the state array ($z = 3$).

An SHA-3 round consists of five step mappings denoted θ, ρ, π, χ, and ι (Equation (6)). Each of those mappings takes state array *A* as an input and returns an updated state array A'. The ι mapping also takes round index i_r as an argument.

$$Rnd(A, i_r) = \iota(\chi(\pi(\rho(\theta(A)))), i_r) \tag{6}$$

A detailed explanation of every step mapping can be found in [34], and the descriptions below will give a brief idea of how each of these works.

The effect of θ is to XOR ($\oplus$) each bit in the state with the parities of two columns in the array. The ρ operation result is modification of the z coordinate for every bit in each *lane* by an offset (modulo lane size), which depends on fixed x and y coordinates of this lane. The π operation effect is rearranged positions of lanes in every state array slice. In the χ operation, each bit of the state array is XORed ($\oplus$) with a non-linear function of two other bits in its row. The effect of the ι operation is to modify some of the bits in *Lane(0,0)* (the exact center of the state array slice) in a way that depends on the round index i_r. *Lane(0,0)* is XORed ($\oplus$) with a w-bit string, where most of the bits are "0", but a selected few are the result of *rc(x)* transformation dependent on round index i_r.

Before the message is fed into the sponge construction, a two-bit suffix "01" is appended to its end. It supports *domain separation* and allows us to distinguish the SHA-3 hash function from other algorithms. Now the message must be padded so its length is a multiple of *rate* (r) parameter, which essentially is the SHA-3 block width. SHA-3 utilizes a *pad10*1* padding scheme, which generates a bit string starting and ending with "1" and filled with an appropriate number of 0s (hence the asterisk, which in regular expression notation indicates *zero* or more).

Figure 6 presents the SHA-3 sponge construction's principle of operation. At the beginning, the SHA-3 state is initialized with a 1600-bit (b = 1600) string of zeros. In the phase called *absorption*, the padded message is divided into series of r-bit blocks and XORed ($\oplus$) into a state vector. Then f transformation, which consists of 24 SHA-3 rounds, is applied to the state. This process is repeated until the whole message is absorbed. In the second stage, the actual hash is *squeezed* from the sponge. For all SHA-3 hash lengths, the hash can be obtained without applying the f transformation again—an appropriate number of bits is taken directly from the state vector as r is always greater than the hash length (Table 2). Variable c is the *capacity* of the sponge, and for SHA-3 it is double the hash length ($c = 2d$). As variables r and c satisfy relation $r + c = b$, the selection of capacity determines the block width of the SHA-3 algorithm.

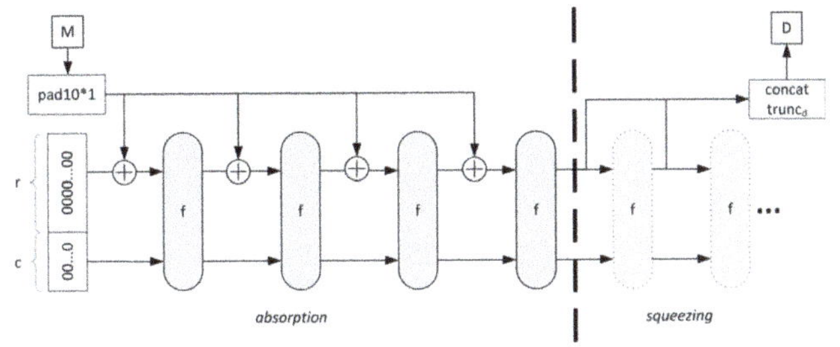

Figure 6. Sponge construction, which is the basis of SHA-3.

Table 2. Capacity c and rate r of SHA-3 algorithms in relation to hash length.

Hash Length d	Capacity $c = 2d$	Rate $r = b - c$
224	448	1152
256	512	1088
384	768	832
512	1024	576

In the network probe, SHA-3 is applied to a 104-bit string that consists of 32-bit IP source and destination addresses, 8-bit IP protocol information, and 16-bit source and destination ports for TCP/UDP. The SHA-3 digest is trimmed to the 32 most significant bits, which are considered the flow hash.

2.3. Implementation and Verification

2.3.1. Implementation

The proposed network probe hardware accelerator was implemented with the hardware description language Verilog [38]. The accelerator's top module is depicted in Figure 7. It has a 128-bit data path with two AXI4-Stream interfaces [39], Slave and Master, used for data flow. Packets are processed sequentially, and their order is not changed. Block *netprobe_top* consists of two submodules that implement the two main functions of the accelerator:

- *netprobe_parser_top*, where IP packet parsing and extraction of flow keys along with some other parameters (e.g., payload length, TCP flags) is performed,
- *netprobe_hash_top*, where calculation of the 32-bit hash over flow keys extracted from the IP packet header is carried out.

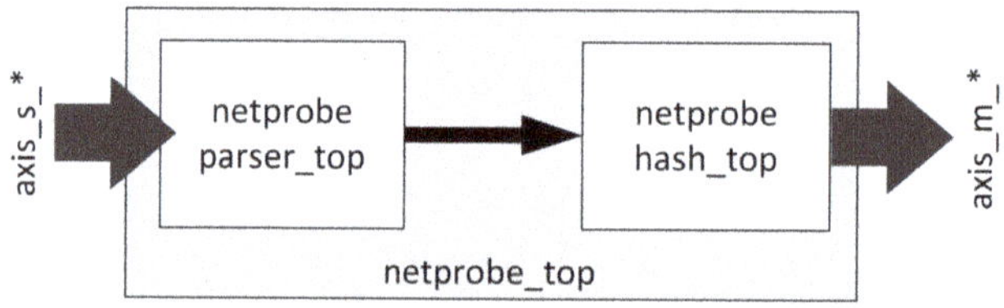

Figure 7. Block scheme of the network probe hardware accelerator's top module—*netprobe_top*.

Figure 8 presents the structure of the packet parser module. The first block in the data path is a protocol filter, responsible for dropping IP packets that contain a protocol other than TCP or UDP. Packets that pass this protocol check are distributed in a round-robin manner between two parallel parser engines which extract flow keys and other information from the packet header.

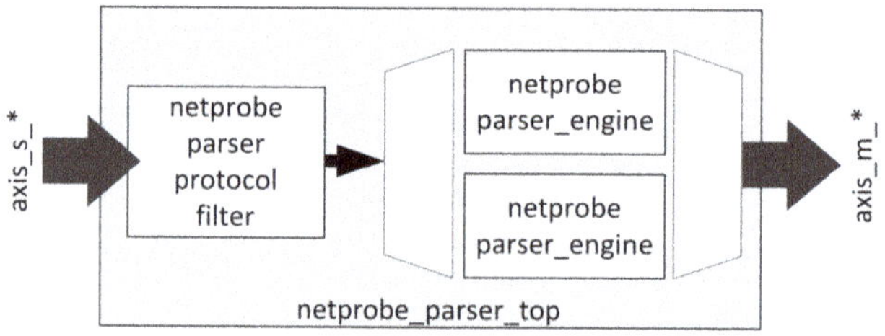

Figure 8. Block scheme of the network probe hardware accelerator packet parser module—*netprobe_parser_top*.

These modules were parallelized to avoid empty cycles on the Master interface due to the unfavorable header structure of processed IP packets, e.g., such as IP header length (IHL), and as a result the TCP header offset that causes the TCP port and TCP flag fields to be in different packet beats for the 128-bit data path width. Parser engines process the IP packet header, extract flow keys and the rest of the features, and forward data in an internal format (two beats in a 128-bit data path). A placeholder for the hash is included, although it is calculated later in the pipeline.

Module *netprobe_hash_top* is a block that wraps hash engines. It is parameterized with a HASH_ALGORITHM variable, which selects an appropriate algorithm submodule to be instantiated (Table 3).

Table 3. Values of HASH_ALGORITHM parameter for *netprobe_hash_top* module configuration.

HASH_ALGORITHM Value	Hash Algorithm
0	None—hash is not appended
1	Sum modulo 32—nProbe
2	Nested CRC-32—Vermont
3	Nested CRC-32 with w constants—modified Vermont
4	SHA-1
5	SHA-3

The module *netprobe_hash_top* also has a set of strap ports used for modified Vermont and SHA-3 algorithm configuration as in Table 4. In the network probe hardware accelerator, w constant straps were tied off to random integers and a 512-bit hash was selected for the SHA-3 algorithm.

Table 4. Module *netprobe_hash_top* strap ports for algorithm configuration.

Strap Input Port	Width	Description
strap_w_src_addr	32	32-bit constant w for modified Vermont algorithm to sum with Source IP Address key
strap_w_dst_addr	32	32-bit constant w for modified Vermont algorithm to sum with Destination IP Address key
strap_w_protocol	32	32-bit constant w for modified Vermont algorithm to sum with IP Protocol key
strap_w_src_port	32	32-bit constant w for modified Vermont algorithm to sum with Source Port key
strap_w_dst_port	32	32-bit constant w for modified Vermont algorithm to sum with Destination Port key
strap_hash_length	2	Selection of SHA-3 hash length—2′d0, 2′d1, 2′d2, 2′d3 mean 224, 256, 384, 512 bits, respectively.

The nProbe hash algorithm (for HASH_ALGORITHM = 1) was implemented as a simple 32-bit adder, whose inputs are flow keys extracted from the internal packet format and left-padded with zeros to 32-bit width if necessary.

The Vermont hash algorithm (for HASH_ALGORITHM = 2) was implemented as 5-stage pipeline, similarly to the diagram in Figure 2. Internal packet data are registered in parallel to CRC-32 logic, and at every stage an appropriate flow key is selected to be included in the hash.

The modified Vermont hash algorithm (for HASH_ALGORITHM = 3) was realized in a similar manner to regular Vermont. Flow keys are obfuscated with w constants before being used in CRC-32 calculations, as in Figure 3.

In the case of SHA-1 (for HASH_ALGORITHM = 4), concatenation of all flow keys forms a 104-bit word, which is considered input to the hash function. The length of the input word is less than 512 bits, which means that SHA-1 transformation (80 rounds) must be applied only to a single block. This makes pipelined algorithm implementation possible, as backpressure towards subsequent packets is not necessary.

Figure 9 presents an example of such a pipeline. Data with extracted flow keys are constantly fed to the input, and multiple packets are processed simultaneously. Since the internal packet format requires two cycles to be transmitted in a 128-bit data path, where only the first cycle carries valid flow keys, a valid hash is obtained at the final stage of the pipeline only for the first beat of this packet.

In regular SHA-1 implementation, the hash pipeline would have 80 stages—one per SHA-1 round. It is possible to reduce the number of stages by unfolding the algorithm loop and implementing two rounds between stage registers. This approach, however, leads to critical path extension of circuits and as a result decreases maximum clock frequency. The solution to this problem was proposed in [40], where the authors described a method with the SHA-1 algorithm loop unfolding using additional variables. This technique allows us to perform two algorithm rounds within one clock cycle and reduces the required number of stages by half. It was incorporated in the network probe hardware accelerator SHA-1 implementation; therefore, its pipeline had 40 stages.

For SHA-3 (for HASH_ALGORITHM = 5), as previously, concatenation of all flow keys creates a 104-bit input word. Again, this is less than the SHA-3 block length, so the approach illustrated in Figure 9 can be applied once more. The SHA-3 pipeline in the proposed network probe hardware accelerator has 24 stages, one per SHA-3 round.

In all cases, a 32-bit flow hash is inserted into the initial placeholder of the output accelerator packet.

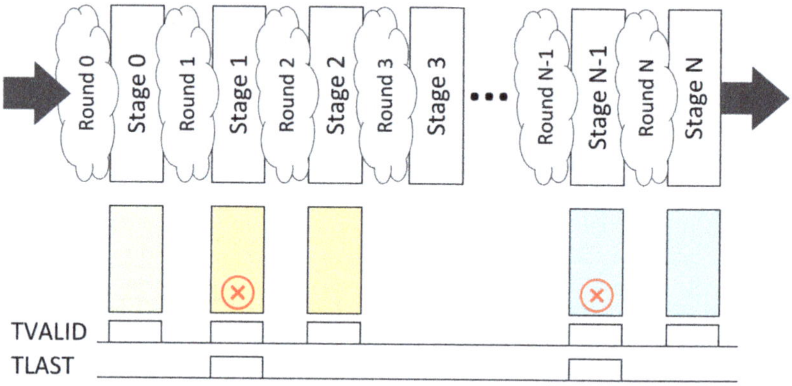

Figure 9. Pipelined hash algorithm implementation in the network probe hardware accelerator.

2.3.2. Functional Verification

Functional verification of the proposed network probe hardware accelerator was conducted using cocotb—an open source, Python-based testbench environment for VHDL/Verilog RTL [41]. It adopts the same concepts of constrained random verification as industry-standard UVM [42]; however, it is implemented in Python rather than SystemVerilog. This enables swift and productive construction of the verification environment, as Python scripting is simple, and additionally, a huge library of existing code is available (e.g., packet generation libraries and cryptographic algorithm implementations).

Figure 10 presents the structure of the *cocotb*-based verification environment. DUT (Design Under Test, here *netprobe_top*) was instantiated as top level in the simulator and was surrounded by verification environment components as drivers, monitors, and scoreboard, which were extended from infrastructure provided by cocotb. Ports of the tested module were stimulated directly from the Python function acting as a test case.

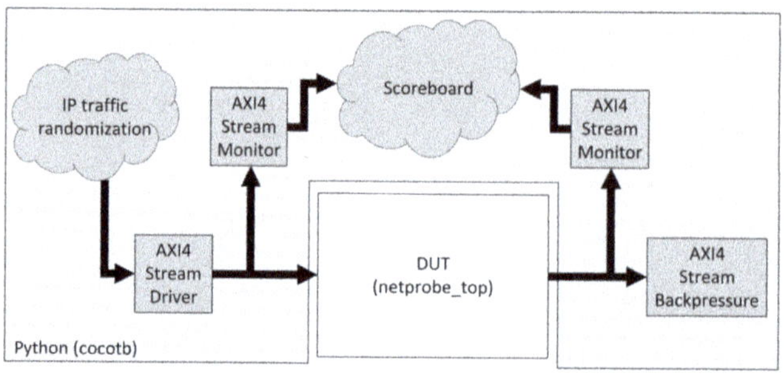

Figure 10. *cocotb*-based verification environment of *netprobe_top* module.

At the beginning, a number of transaction objects that mimic IP packets were created and randomized. The goal was to cover a broad space of possible network traffic, so multiple packet parameters were changed: packet length, addresses, encapsulated protocol, etc. These objects were passed to an AXI4-Stream driver, which transmitted them onto the Slave interface of the *netprobe_top* module. Both Slave and Master interfaces were watched by AXI4-Stream monitors, which were able to transform waveforms into transaction objects. Initial packets and those processed by DUT were fed to the scoreboard component. The DUT behavior model was applied to the stimulus packets there, and the result was compared with transactions processed by the *netprobe_top* module itself. They must be the same, and when this condition is not fulfilled, an error is reported.

Figure 11 is a screen capture from a simulation of *netprobe_top* module configured with the SHA-3 algorithm. The selected SHA-3 hash length was 512 bit (*strap_hash_length* equals 2'd3). The goal of the executed test case was to check the performance of the design. Signal *axis_m_tready* of the accelerator's Master interface was tied off to high value, which indicates no backpressure. DUT was flooded with a number of short IP packets—signal *axis_s_tvalid* went high at Cursor 1. After 32 clock cycles (latency for SHA-3 configuration), the first result packets were presented on the Master interface (Cursor 2, *axis_m_tvalid* goes high). Checks implemented in the testbench verified whether the *axis_s_tready* signal goes low. Module *netprobe_top* does not introduce backpressure on its own, and even in these harsh conditions, DUT behaved as expected.

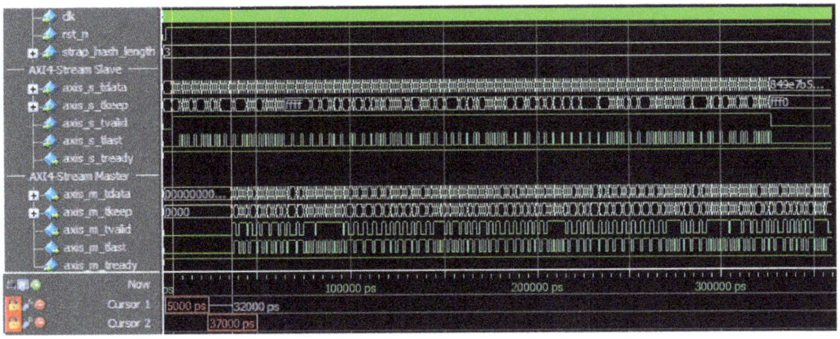

Figure 11. Simulation of *netprobe_top* module with the SHA-3 hash algorithm using the *cocotb*-based verification environment.

2.3.3. Synthesis Results

Synthesis of the network probe hardware accelerator was performed for Intel Stratix V GX FPGA (5SGXEA7N2F45C2), an element of the Terasic DE5-Net development kit [43], using Intel Quartus Prime 18.1 software.

Table 5 summarizes the synthesis results of the *netprobe_top* module for a range of hash algorithms. Since nProbe, Vermont, and modified Vermont are based on simple hashing schemes that use basic types of calculations (addition modulo 32 or CRC), hardware implementation of these algorithms requires little hardware resources (less than 1% of available resources of FPGA used in the experiment). Although SHA-1 and SHA-3 cryptographic functions are far more computationally expensive, the proposed implementation requires few enough resources to be efficiently used as a part of the hardware NetFlow probe. Even though SHA-1 and SHA-3 were optimized for performance, not for the area, the probe with the most complex SHA-3 algorithm utilized only 16.44% of resources, leaving enough of them to implement other functionalities of the NetFlow probe [17]. It is no surprise that straightforward hash algorithms (such as nProbe, Vermont, or modified Vermont) implementations can sustain multigigabit throughput, but realizations of cryptographic functions (SHA-1, SHA-3) definitely match this. All investigated hash algorithms offer throughput over 20 Gbit/s.

Table 5. Module *netprobe_top* implementation results in Stratix V FPGA.

Hash Algorithm	Logic Utilization	Maximum Clock Frequency F_{max} (MHz)	Throughput (Gbit/s)	Latency (Clock Cycles/ns)
nProbe	0.56%	206.48	26.43	9/43.59
Vermont	0.69%	207.51	26.56	13/62.65
modified Vermont	0.79%	197.63	25.30	13/65.78
SHA-1	6.17%	201.61	25.81	48/238.08
SHA-3	16.44%	175.04	22.41	32/182.82

It has been assumed that cryptographic hash functions such as SHA are computationally too expensive for efficient use in a flow monitor. The high bandwidth and low latency of the hardware accelerator based on the SHA-1 and SHA-3 functions definitely enables construction of a network probe working in a real-time manner—even when it is flooded with the smallest IP packets.

It is worth mentioning that the low percentage of logic utilization allows for further design optimization and parallelization [44]. Utilizing such techniques, it should be even possible to reach a 100 Gbit/s bandwidth limit.

3. Experiments

In our experiments, we wanted to verify the following research hypotheses:

- It is possible to realize a network traffic probe with a cryptographic hash function, working in a real-time regime.
- Cryptographic hash functions SHA-1 and SHA-3 provide comparable distribution of flows to the reference methods.

In the experiments, the NetFlow probe was supplied with selected traffic, and the distribution of flow records in the flow cache *buckets* was analyzed. We conducted tests for five hardware-accelerated probes implementing different hash functions. Each probe was supplied with three different types of network traffic to analyze the impact of traffic type on flow record distribution over buckets in the flow cache.

3.1. Experimental Testbed

Verification and performance tests of the NetFlow probe hardware-accelerator designs were carried out using a dedicated testbed. The hardware part of the probe was implemented in the DE5-Net FPGA development platform. A general-purpose PC containing 10 Gbps Ethernet interfaces (Intel 82599 10 Gigabit Ethernet card) was connected to the DE5-Net kit. The Ethernet connectivity between the DE5-Net FPGA platform and the PC was established by means of multi-mode fiber optics, with SFP+ transceivers. The PC was used as a traffic generator running the *tcreplay* network driver and as a network monitor implementing the software part of the NetFlow probe.

3.2. Network Traffic Used

In our experiments, we used the CICIDS 2017 dataset [45]. It contains the traffic captured during five days of activity in a simulated network. Both pcap and bidirectional flow formats have been published. These datasets cover various kinds of attack, such as botnets, (D)DoS, web application attacks, and SSH brute-force attempts. In total, 2,830,540 flows were collected over five days (from Monday to Friday).

In our experiments, we used the Monday, Wednesday, and Friday traffic. The traffic collected on Monday contained 496,943 flows, purely with benign network communication. The Wednesday traffic embraced 452,601 flows, which, apart from normal traffic, contained traffic captured during DoS, Heartbleed, slowloris, Slowhttptest, Hulk, and GoldenEye attacks. The Friday subset was the most numerous—it contained 792,487 flows with normal traffic and traffic with registered DDoS attacks, botnet communication, and various port scan attacks.

3.3. Metrics

The hardware-accelerated network flow probe was modified so that the flow cache stores records for all flows during a test session, i.e., flow records for terminated or expired flows, were not removed from the flow cache *buckets*. This allowed measurement of such values as:

- minimal number of flow records in a nonempty bucket (hereinafter named as *Min*),
- maximal number of flow records in a bucket (*Max*),
- mean number of flow records in a bucket (*Mean*),
- standard deviation (SD) of flow records in a bucket,
- number of nonempty buckets (*Buckets*).

This gave us an overview of the distribution of flow records over all buckets in the flow cache for a given hash computation scheme and for the selected traffic type.

4. Results

In our experiments, each hash function was used in 16-bit and 32-bit versions, which organized the flow cache into 2^{16} and 2^{32} buckets, respectively. Every probe was supplied with three types of traffic from the CICIDS 2017 dataset labeled *Normal (Monday)*, *Normal + attacks (Wednesday)*, and *Normal + attacks (Friday)* (see Section 3.2). For each traffic type, the number of flows it contains was given as N.

The results for hardware-accelerated probes using 16-bit hash functions have been presented in Table 6. For every traffic type used to supply, all probe metrics proposed in Section 3.3 were recorded. As can be seen, all hash functions except Mod32 yielded similar statistics over flow cache buckets. Mod32 function achieved noticeably worse Max, Mean, and SD values than the rest of the hash functions. We observed, however, that statistics for all functions were not much affected by anomalous traffic (DoS attacks, botnet communication, port scan attacks)—see the results for the Wednesday and Friday traffic. It can be noticed that traffic *Normal + attacks (Friday)* generated larger values of recorded parameters for all functions than the other two traffic types. However, this can be explained by the fact that it contains much more flows than the other two traffic types used. Graphical presentation of the distribution of flow records over the flow cache buckets for a hash function based on a simple modulo operation (Mod32), modified Vermont, or the SHA-3 cryptographic function is shown in Figure 12. It can be seen that the distribution produced by the simple modulo hash function is far from uniform. The modified Vermont hash function and that based on the cryptographic SHA-3 function offer much better distributions.

Table 6. Statistics of bucket occupation for various hash functions—2^{16} buckets used.

Hash Function	Normal (Monday) N = 496,943				Normal + Attacks (Wednesday) N = 452,601				Normal + Attacks (Friday) N = 792,487			
	Min	Max	Mean	SD	Min	Max	Mean	SD	Min	Max	Mean	SD
Mod32	0	38	7.58	5.45	0	34	6.91	5.18	0	52	12.09	7.62
Vermont	0	21	7.58	2.66	0	21	6.91	2.62	2	27	12.09	3.38
Modified Vermont	0	19	7.58	2.63	0	19	6.91	2.47	1	27	12.09	3.32
SHA-1	0	21	7.58	2.76	0	21	6.91	2.62	1	32	12.09	3.47
SHA-3	0	20	7.58	2.75	0	22	6.91	2.62	1	29	12.09	3.47

A more precise overview is given in Table 7, where the results for the 32-bit version of hash functions are presented. Such a hash size greatly increases the flow cache capacity (up to 2^{32} buckets). In this case, in addition to the metrics used in Table 6, the number of nonempty buckets is also given (*Buckets* column). Again, all hash functions, except Mod32, showed similar distribution over flow cache buckets, which was not affected by typical anomalous traffic. The Mod32 results significantly deviate from those obtained for the rest of the hash functions. It is worth noting that for Vermont, modified Vermont, and the two SHA hash functions, the mean value of flow records in a bucket was 1, and the number of nonempty buckets was almost equal to the number of all flows present in the traffic. This indicates that these functions put almost every flow record in a separate bucket, offering almost uniform distribution of flow records over flow cache buckets for normal traffic and typical anomalous traffic.

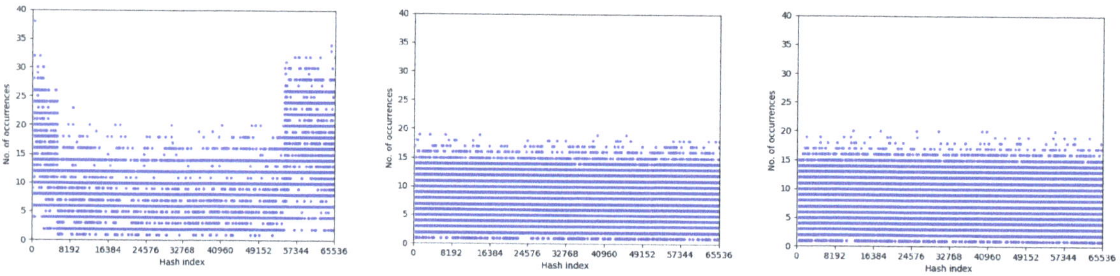

(a) (b) (c)

Figure 12. Visualization of bucket occupation for 2^{16} buckets. (**a**) Mod32; (**b**) modified Vermont; (**c**) SHA-3.

Table 7. Statistics of bucket occupation for various hash functions—2^{32} buckets used.

Hash	Normal (Monday) N = 496,943					Normal + Attacks (Wednesday) N = 452,601					Normal + Attacks (Friday) N = 792,487				
	Min	Max	Mean	SD	Buckets	Min	Max	Mean	SD	Buckets	Min	Max	Mean	SD	Buckets
Mod32	1	22	2.91	2.43	170,626	1	22	2.93	2.28	154,662	1	32	3.98	3.65	199,241
Vermont	1	2	1	0.01	496,919	1	2	1	0.01	452,577	1	2	1	0.01	792,465
Mod. Verm.	1	2	1	0.01	452,572	1	2	1	0.01	452,572	1	2	1	0.01	792,386
SHA-1	1	2	1	0.01	452,567	1	2	1	0.01	452,567	1	2	1	0.01	792,423
SHA-3	1	2	1	0.01	496,922	1	2	1	0.01	452,583	1	2	1	0.01	792,420

5. Discussion and Conclusions

A proper view of the statistics and the dynamics of a network is of great importance, since it enables us to detect network attacks. Thus, network monitors using the network flow concept are an important part of modern cybersecurity defense. As such, these devices themselves may be the targets of cyberattacks. One of the possible weak points of NetFlow probes is a network flow cache, which is usually implemented as a hash table. Due to the limited size of a hash table, it is inevitable that, sooner or later, two different flows will be mapped to the same hash bucket. It is essential that the hash function used for calculating the hash keys offers a uniform distribution of NetFlow records over available buckets, so that the lengths of all bucket lists would be almost equal. This makes it possible to use a reasonably sized hash data structure to make the flow lookup fast, because of minimal list lengths. The experiments conducted during this research show that even a relatively simple hash function may guarantee such characteristics.

However, nowadays, when components of cybersecurity systems themselves may be a targets of a cyberattack, a no less important feature of such systems is their resistance to attacks. In the case of a NetFlow probe, it should be impossible for an attacker to create directed collisions in the hash function. If an attacker is able to fabricate network traffic in such a way as to lead to a large number of collisions in the hash function, some buckets of the hash table may overflow, causing malfunction of the probe.

The results from Section 4 show that only very simple hash functions (i.e., Mod32) are susceptible to common malicious traffic, such as DDoS or port scan attacks. More complex methods, such as Vermont, based on CRC32, offer relatively uniform distribution of flow records over flow cache buckets for normal traffic, and typical anomalous traffic. However, as demonstrated in [19], it is possible to prepare a targeted attack exploiting a vulnerability of the implemented hash function.

Thus, it is crucial to select a hashing function that maps a small number of flow keys on to the same flow cache location. A hash function should therefore compute hash keys that are uniformly distributed, so that it should be impossible for an attacker to create directed collisions. At the same time, the hash function must be fast so that it does not become a bottleneck of the NetFlow probe.

The obvious countermeasure against hash collision-based attacks is the application of cryptographic hash functions, for which collisions cannot be created easily. The results presented in Section 4 prove that the use of the cryptographic functions SHA-1 and SHA-3 offers comparable distribution of flows in the flow cache to the dedicated methods (Vermont, modified Vermont) used as reference. The advantage of implementing a hash function based on cryptographic functions in a NetFlow probe is that it is very difficult (or even impossible) to prepare a targeted attack on such a probe by fabricating network traffic to overflow flow cache buckets through systematically creating packets that lead to hash collisions.

Cryptographic functions, however, have not usually been candidates for hash functions in NetFlow probes, since they are considered to be computationally too expensive for efficient use in flow monitoring. Our concept presented in Section 2.3 shows that it was possible to implement a hardware-accelerated network flow probe employing a cryptographic hash function that offered sufficient performance to construct a network probe working in real-time with multigigabit traffic, even when it was flooded with the smallest IP packets. Relatively low hardware resource utilization makes it possible to reach a 100 Gbit/s bandwidth limit by applying hardware-specific design optimization and parallelization.

It has to be emphasized that most available traffic datasets contain traffic with a relatively small number of flows. The set CICIDS 2017 used in our experiment contains, in total, 2,830,540 flows. Taking into account the fact that the flow cache of a probe that uses a 32-bit hash function contains 2^{32} buckets, the flow records fill only a small fraction of the flow cache. The use of datasets with significantly larger numbers of flows with normal and anomalous traffic might give a better view of possible differences in distribution of flow records over flow cache buckets for the evaluated hash functions. Such an approach, and the application of customized traffic containing flows intentionally constructed to produce hash collisions (which may not be a trivial task for some hash functions), could be the subject of future work.

To conclude, we can state that the resistance of cryptographic hash functions to collisions and the multigigabit efficiency of a hardware-accelerated implementation of hash computation allow the creation of an effective monitoring solution for modern cybersecurity systems while delivering a high level of resilience to targeted attacks.

Author Contributions: Conceptualization, M.K., M.R. and A.J.; methodology, M.K. and M.R.; software, M.K. and P.S.; validation, M.K. and P.S.; formal analysis, M.K. and M.R.; investigation, M.K., P.S., M.R. and A.J.; resources, M.K., P.S. and M.R.; data curation, P.S.; writing—original draft preparation, M.K., P.S., M.R. and A.J.; writing—review and editing, M.K., M.R. and A.J.; visualization, M.K.; supervision, M.R. and A.J.; project administration, A.J.; funding acquisition, A.J. All authors have read and agreed to the published version of the manuscript.

Funding: The study has been supported by the SIMARGL Project—Secure Intelligent Methods for Advanced RecoGnition of malware and stegomalware, with the support of the European Commission and the Horizon 2020 Program, under grant agreement number 833042. The publication was funded by the statutory activity subsidy from the Polish Ministry of Education and Science.

Conflicts of Interest: The authors declare no conflict of interest.

References

1. Al-Garadi, M.A.; Mohamed, A.; Al-Ali, A.K.; Du, X.; Ali, I.; Guizani, M. A Survey of Machine and Deep Learning Methods for Internet of Things (IoT) Security. *IEEE Commun. Surv. Tutor.* **2020**, *22*, 1646–1685. [CrossRef]
2. Fizza, K.; Banerjee, A.; Mitra, K.; Jayaraman, P.P.; Ranjan, R.; Patel, P.; Georgakopoulos, D. QoE in IoT: A vision, survey and future directions. *Discov. Internet Things* **2021**, *1*, 4. [CrossRef]
3. Federal Bureau of Investigations. The Cyber Threat. Available online: https://www.fbi.gov/investigate/cyber (accessed on 1 April 2022).
4. Caviglione, L.; Choraś, M.; Corona, I.; Janicki, A.; Mazurczyk, W.; Pawlicki, M.; Wasielewska, K. Tight Arms Race: Overview of Current Malware Threats and Trends in Their Detection. *IEEE Access* **2021**, *9*, 5371–5396. [CrossRef]
5. NETSCOUT. NETSCOUT Threat Intelligence Report. Available online: https://www.netscout.com/threatreport (accessed on 1 April 2022).
6. Morgan, S. Cybercrime To Cost The World $10.5 Trillion Annually By 2025. Special Report: Cyberwarfare In The C-Suite. 2020. Available online: https://cybersecurityventures.com/hackerpocalypse-cybercrime-report-2016/ (accessed on 1 April 2022).
7. AV-TEST Institute. Malware Statistics. 2022. Available online: https://www.av-test.org/en/statistics/malware/ (accessed on 1 April 2022).
8. Wagner, C.; François, J.; State, R.; Engel, T. Machine Learning Approach for IP-Flow Record Anomaly Detection. In Proceedings of the NETWORKING 2011, Valencia, Spain, 9–13 May 2011; Domingo-Pascual, J., Manzoni, P., Palazzo, S., Pont, A., Scoglio, C., Eds.; Springer: Berlin/Heidelberg, Germany, 2011; pp. 28–39.
9. Iglesias, F.; Ferreira, D.C.; Vormayr, G.; Bachl, M.; Zseby, T. NTARC: A Data Model for the Systematic Review of Network Traffic Analysis Research. *Appl. Sci.* **2020**, *10*, 4307. [CrossRef]
10. Krupski, J.; Graniszewski, W.; Iwanowski, M. Data Transformation Schemes for CNN-Based Network Traffic Analysis: A Survey. *Electronics* **2021**, *10*, 2042. [CrossRef]
11. Hofstede, R.; Čeleda, P.; Trammell, B.; Drago, I.; Sadre, R.; Sperotto, A.; Pras, A. Flow Monitoring Explained: From Packet Capture to Data Analysis With NetFlow and IPFIX. *IEEE Commun. Surv. Tutor.* **2014**, *16*, 2037–2064. [CrossRef]
12. Spognardi, A.; Villani, A.; Vitali, D.; Mancini, L.V.; Battistoni, R. Large-Scale Traffic Anomaly Detection: Analysis of Real Netflow Datasets. In *E-Business and Telecommunications*; Obaidat, M.S., Filipe, J., Eds.; Springer: Berlin/Heidelberg, Germany, 2014; pp. 192–208.
13. van der Steeg, D.; Hofstede, R.; Sperotto, A.; Pras, A. Real-time DDoS attack detection for Cisco IOS using NetFlow. In Proceedings of the 2015 IFIP/IEEE International Symposium on Integrated Network Management (IM), Ottawa, ON, Canada, 11–15 May 2015; pp. 972–977. [CrossRef]
14. Zadnik, M.; Pecenka, T.; Korenek, J. Netflow probe intended for high-speed networks. In Proceedings of the International Conference on Field Programmable Logic and Applications, Tampere, Finland, 24–26 August 2005; pp. 695–698. [CrossRef]
15. Novotný, J.; Čeleda, P.; Žádník, M. Hardware-Accelerated Framework for Security in High-Speed Networks. In *Information Assurance for Emerging and Future Military Systems*; NATO Science and Technology Organization: Brussels, Belgium, 2008. [CrossRef]
16. Forconesi, M.; Sutter, G.; Lopez-Buedo, S.; Aracil, J. Accurate and flexible flow-based monitoring for high-speed networks. In Proceedings of the 2013 23rd International Conference on Field programmable Logic and Applications, Porto, Portugal, 2–4 September 2013; pp. 1–4. [CrossRef]
17. Trzepiński, M.; Skowron, K.; Korona, M.; Rawski, M. FPGA Implementation of Memory Management for Multigigabit Traffic Monitoring. In *Man–Machine Interactions 5*; Gruca, A., Czachórski, T., Harezlak, K., Kozielski, S., Piotrowska, A., Eds.; Springer: Berlin/Heidelberg, Germany, 2018; pp. 555–565.
18. Sonchack, J.; Michel, O.; Aviv, A.J.; Keller, E.; Smith, J.M. Scaling Hardware Accelerated Network Monitoring to Concurrent and Dynamic Queries With *Flow. In Proceedings of the 2018 USENIX Annual Technical Conference (USENIX ATC 18), USENIX Association, Boston, MA, USA, 11–13 July 2018; pp. 823–835.
19. Eckhoff, D.; Limmer, T.; Dressler, F. Hash tables for efficient flow monitoring: Vulnerabilities and countermeasures. In Proceedings of the 2009 IEEE 34th Conference on Local Computer Networks 2009, Zurich, Switzerland, 20–23 October 2009; pp. 1087–1094. [CrossRef]
20. Kang, M.S.; Lee, S.B.; Gligor, V.D. The Crossfire Attack. In Proceedings of the 2013 IEEE Symposium on Security and Privacy, Berkeley, CA, USA, 19–22 May 2013; pp. 127–141. [CrossRef]
21. Zhao, Z.; Shi, X.; Wang, Z.; Li, Q.; Zhang, H.; Yin, X. Efficient and Accurate Flow Record Collection With HashFlow. *IEEE Trans. Parallel Distrib. Syst.* **2022**, *33*, 1069–1083. [CrossRef]
22. Szumełda, P.; Orzechowski, N.; Rawski, M.; Janicki, A. VHS-22—A Very Heterogeneous Set of Network Traffic Data for Threat Detection. In Proceedings of the European Interdisciplinary Cybersecurity Conference (EICC 2022), Barcelona, Spain, 15–16 June 2022. [CrossRef]
23. Kirsch, A.; Mitzenmacher, M.; Varghese, G. Hash-Based Techniques for High-Speed Packet Processing. In *Algorithms for Next Generation Networks*; Springer: Berlin/Heidelberg, Germany, 2010.
24. Deri, L. nProbe: An Open Source NetFlow Probe for Gigabit Networks. In Proceedings of the TERENA Networking Conference 2003, Zagreb, Croatia, 21 May 2003.

25. Lampert, R.T.; Sommer, C.; Münz, G.; Dressler, F. Vermont—A Versatile Monitoring Toolkit for IPFIX and PSAMP. In Proceedings of the IEEE/IST Workshop on Monitoring, Attack Detection and Mitigation (MonAM 2006), Tübingen, Germany, 28–29 September 2006.
26. Williams, R.N. A Painless Guide to CRC Error Detection Algorithms. Available online: http://ross.net/crc/download/crc_v3.txt (accessed on 24 April 2022).
27. *IEEE Std 802.3-2018*; IEEE Standard for Ethernet. Revision of IEEE Std 802.3-2015. IEEE: Piscataway, NJ, USA, 2018; pp. 1–5600. [CrossRef]
28. Dang, Q. *Secure Hash Standard (SHS)*; National Institute of Standards and Technology: Gaithersburg, MD, USA, 2015. [CrossRef]
29. Eastlake, D.E., 3rd; Jones, P. US Secure Hash Algorithm 1 (SHA1); RFC 3174. Available online: https://www.rfc-editor.org/info/rfc3174 (accessed on 27 April 2022).
30. Stevens, M.; Karpman, P.; Peyrin, T. Freestart Collision for Full SHA-1. In Proceedings of the Annual International Conference on the Theory and Applications of Cryptographic Techniques, Vienna, Austria, 8–12 May 2016; pp. 459–483.
31. Stevens, M.; Bursztein, E.; Karpman, P.; Albertini, A.; Markov, Y. The First Collision for Full SHA-1. In Proceedings of the Advances in Cryptology—CRYPTO 2017, Santa Barbara, CA, USA, 20–24 August 2017; pp. 570–596.
32. Leurent, G.; Peyrin, T. From Collisions to Chosen-Prefix Collisions—Application to Full SHA-1. Cryptology ePrint Archive, Report 2019/459. 2019. Available online: https://ia.cr/2019/459 (accessed on 27 April 2022).
33. Leurent, G.; Peyrin, T. SHA-1 Is a Shambles—First Chosen-Prefix Collision on SHA-1 and Application to the PGP Web of Trust. Cryptology ePrint Archive, Report 2020/014. 2020. Available online: https://ia.cr/2020/014 (accessed on 27 April 2022).
34. Dworkin, M. *SHA-3 Standard: Permutation-Based Hash and Extendable-Output Functions*; National Institute of Standards and Technology: Gaithersburg, MD, USA, 2015. [CrossRef]
35. Bertoni, G.; Daemen, J.; Peeters, M. *Cryptographic Sponge Functions*; Citeseer: University Park, PA, USA, 2011.
36. Merkle, R.C. Secrecy, Authentication, and Public Key Systems. Ph.D. Thesis, Stanford university, Stanford, CA, USA, 1979.
37. Bertoni, G.; Daemen, J.; Peeters, M.; Van Assche, G. Keccak. In *Advances in Cryptology–EUROCRYPT 2013*; Johansson, T., Nguyen, P.Q., Eds.; Lecture Notes in Computer Science; Springer: Berlin/Heidelberg, Germany, 2013; Volume 7881. [CrossRef]
38. *IEEE Std 1364-2005*; IEEE Standard for Verilog Hardware Description Language. Revision of IEEE Std 1364-2001. IEEE: Piscataway, NJ, USA, 2006; pp. 1–590. [CrossRef]
39. ARM. AMBA® 4 AXI4-Stream Protocol Version 1.0 (ARM IHI 0051A). 2010. Available online: https://documentation-service.arm.com/static/60d5e2510320e92fa40b4788 (accessed on 27 April 2022).
40. Lee, E.H.; Kim, S.M.; Lee, J.H.; Cho, K. Design of a High Speed SHA-1 Architecture Using Unfolded Pipeline for Biomedical Applications. In Proceedings of the International Multi-Conference on Society, Cybernetics and Informatics (IMSCI 2009), Orlando, FL, USA, 10–13 July 2009.
41. Various. Cocotb's Documentation. Available online: https://docs.cocotb.org/en/stable (accessed on 24 April 2022).
42. Accellera. Universal Verification Methodology. Available online: https://www.accellera.org/community/uvm (accessed on 24 April 2022).
43. Terasic. *DE5-Net FPGA Development Kit. User Manual*; Terasic: Hsinchu, Taiwan, 2018.
44. Korona, M.; Skowron, K.; Trzepiński, M.; Rawski, M. High-performance FPGA architecture for data streams processing on example of IPsec gateway. *Int. J. Electron. Telecommun.* **2018**, *64*, 351–356.
45. Sharafaldin, I.; Lashkari, A.H.; Ghorbani, A.A. Toward Generating a New Intrusion Detection Dataset and Intrusion Traffic Characterization. In Proceedings of the 4th International Conference on Information Systems Security and Privacy (ICISSP 2018), Funchal, Portugal, 22–24 January 2018.

electronics

MDPI

Article

An Exploratory Study of Cognitive Sciences Applied to Cybersecurity

Roberto O. Andrade [1], Walter Fuertes [2], María Cazares [3], Iván Ortiz-Garcés [4,*] and Gustavo Navas [3]

1 Facultad de Ingeniería en Sistemas, Escuela Politécnica Nacional, Quito 170525, Ecuador; roberto.andrade@epn.edu.ec
2 Department of Computer Sciences, Universidad de las Fuerzas Armadas ESPE, Sangolquí P.O. Box 17-15-231B, Ecuador; wmfuertes@espe.edu.ec
3 IDEIAGEOCA, Universidad Politécnica Salesiana, Cuenca 010102, Ecuador; mcazares@ups.edu.ec (M.C.); gnavas@ups.edu.ec (G.N.)
4 Facultad de Ingeniería y Ciencias Aplicadas, Escuela de Ingeniería en Tecnologías de la Información, Universidad de las Américas, Quito 170125, Ecuador
* Correspondence: ivan.ortiz@udla.edu.ec

Abstract: Cognitive security is the interception between cognitive science and artificial intelligence techniques used to protect institutions against cyberattacks. However, this field has not been addressed deeply in research. This study aims to define a Cognitive Cybersecurity Model by exploring fundamental concepts for applying cognitive sciences in cybersecurity. For achieving this, we developed exploratory research based on two steps: (1) a text mining process to identify main interest areas of research in the cybersecurity field and (2) a valuable review of the papers chosen in a systematic literature review that was carried out using PRISMA methodology. The model we propose tries to fill the gap in automatizing cognitive science without taking into account the users' learning processes. Its definition is supported by the main findings of the literature review, as it leads to more in-depth future studies in this area.

Keywords: cognitive security; cybersecurity; cyberattacks

Citation: Andrade, R.O.; Fuertes, W.; Cazares, M.; Ortiz-Garcés, I.; Navas, G. An Exploratory Study of Cognitive Sciences Applied to Cybersecurity. *Electronics* **2022**, *11*, 1692. https://doi.org/10.3390/electronics11111692

Academic Editor: Krzysztof Szczypiorski

Received: 25 April 2022
Accepted: 23 May 2022
Published: 26 May 2022

1. Introduction

Cybersecurity attacks have been relevant since the appearance of the first computers. However, their evolution due to the level of techniques and tools has converted them into the world's main risk. The World Economic Forum [1] has classified cyberattack as one of the top ten worldwide risks. Its impact is considered more significant than a food crisis due to its scope in modern society and its probability of occurrence. Reactive solutions focus mainly on attack alleviation processes, while proactive solutions could predict possible cyberattacks and generate self-protection systems. This scenario has motivated companies and researchers in the cybersecurity field to look for alternatives for replacing reactive solutions with proactive ones. One approach used by specialized firms and researchers is to establish anomaly detection processes that discover possible attack patterns and identify attackers' behaviors. In the last three years (2019–2021), several contributions to anomaly detection have been developed in different domains such as SCADA systems, smart grids, smart cities, critical infrastructures, and Cyber-Physical Systems (CPS) [2].

The anomaly detection process requires identifying features or components that differ from typical behaviors [3]. In the initial phase of this anomaly detection process, modeling cybersecurity expert knowledge and cognitive processes are relevant for building better proactive solutions. However, the large volume of data generated by the different interconnected devices in the digital world makes the identification process more challenging to implement [4]. Several alternatives have been defined for supporting analysts' cognitive processes (i.e., augmented cognition) by using computational models that simulate the cognitive processes performed by cybersecurity experts. The identification of security risk patterns based on the analysts' cognitive processes can be approached through the Observe–Orient–Decide–Act model (OODA) or the Monitor–Analyze–Plan–Execute model (MAPE-K) [5].

Researchers have proposed the automation and support of the cognitive processes defined in the OODA and MAPE-K models through different machine learning techniques [6]. In the same research line, we found that several works from 2019 to 2021 used convolution networks, K-means, or deep learning for detecting phishing, ransomware, and even attacks against smart grids [7].

Researchers have identified that the possible actions or strategies of adversaries can be studied using game theory models with incomplete information based on Stackelberg's proposals [8]. This approach could support identifying a possible future attack and the possible strategies used by the adversary. In this way, cybersecurity research's central objective is to expand security analysts' cognitive capacity through data analysis, machine learning techniques, and game theory in cybersecurity [9].

Researchers have proposed a more in-depth approach to improve the cybersecurity proposals, focused on the adversary to identify their behavioral characteristics that lead them to decide on a specific attack strategy [10]. Furthermore, this allows for identifying the techniques that the adversary could select and how to use them. This approach could enable cybersecurity analysts to anticipate and establish a more optimal defense mechanism. Research has included the psychological perspective to analyze the adversaries' behavior [11]. Incorporating Artificial Intelligence, Machine Learning, data analytics, and psychology, among other fields related to cognitive sciences in cybersecurity, has generated a new cybersecurity approach called cognitive security [12]. This approach goes one step ahead of security intelligence to propose the best defensive strategies and take advantage of both cognitive processes: cybersecurity analysts and adversaries [13].

This study aims to identify the fundamental concepts related to the application of cognitive sciences in cybersecurity for establishing defense strategies to minimize the impact of cyberattacks. For this reason, we developed an exploratory study based on two stages:

- A text mining process to identify challenges in the field of cybersecurity and analyze the impact of cyberattacks and the future direction of cybersecurity solutions based on cognitive science;
- A Systematic Literature Review (SLR) to identify the contributions of applied cognitive sciences in cybersecurity as alternatives for proactive strategies. The main contribution of this study is the definition of a cognitive cybersecurity model supported by the findings of a literature review in this research area based on the PRISMA methodology.

This study is structured as follows. Section 2 introduces and describes the theory that explains the components of the research problem under research. Section 3 provides the methodological procedure applied to judge the validity of the results of this study. Section 4 presents a proposal for a cognitive cybersecurity model. Finally, the Section 6 describes the main findings and the lines of future work.

2. Background

2.1. Adversarial and User Analysis

In cyberattack scenarios, a competitive advantage by the adversary could exist in the first instance. Table 1 shows the adversary has valuable information such as personal user information, type of operating system, and user applications. Additionally, the adversary has information about the types of security vulnerabilities that can be exploited. The adversary has been trained in several cybersecurity areas, such as ethical hacking, vulnerability analysis, and reverse engineering. In this context, a user has a clear disadvantage, and from the perspective of game theory, we are faced with a game scenario with incomplete information from the user's side. The user does not know information related to the adversary, such as the type of cyberattack it could perform, which techniques will be used to execute the attack, and which kind of resources are available. Establishing an optimal defense/security attack strategy requires more information from a user perspective [14].

Table 1. Comparative of resources adversarial versus user.

Role	Techniques	IT Resources	Information
User	Empirical Knowledge	Office or Home Desktop	No information related to the adversaries
Organization	Tactics, Techniques, and Procedures (TTP) Ofensive/Defensive approaches	Perimetral security (Firewall, IPS, IDS) Security Event Management (SIEM)	No or low information related to adversaries. Adversaries could use VPN or deep network to hide their information and maintain anonymity.
Adversaries	Offensive approaches (hacking, vulnerability scans, deep network) MITRE ATT&CK defines 245 techniques of attacks, distributed in 14 categories.	Vulnerability tools Exploit tools Obfuscation tools Lateral Movement Frameworks Remote access trojans	Data from Social networks (Facebook, Instagram, twitter) Data from personal or enterprise blogs or web pages. Data for deep network.

Alternatively, another drawback for the user is the stimulus that affects his/her decision criteria. For example, the COVID-19 pandemic has created a scenario where adversaries interact with web pages with drug procurement for the virus or access to free entertainment platforms [15]. In this context, the response time window in which the user must decide between clicking or abstaining from clicking is critical. For gathering information related to the adversary, pattern recognition techniques are used [7]. Meanwhile, decision-making models based on Bayesian networks [16] and diffusion models [17] are used for modeling user response time. Simmons et al. [18] propose the characterization of cyberattacks based on five major classifiers: attack vector, operational impact, attack target, defense, and informational impact. The adversary's characterization is based on two aspects: Risk adverseness and Experience level. Venkatesan et al. [19] propose that the modeling of the adversary behavior considers at least the following aspects:

- Cultural characteristics;
- Behavior patterns;
- Types of attacks.

At this point, incorporating cognitive sciences can improve the development of proactive cybersecurity solutions.

2.2. *Cognitive Sciences*

Research on cognitive sciences applied to cybersecurity acknowledges the importance of the human factor in cybersecurity; this is particularly relevant with the challenges generated by the growth of technologies such as cloud, mobile, IoT, and social networks [20,21]. Cognitive science could enhance the processes of perception, comprehension, and projection used by cybersecurity analysts to detect cyberattacks and establish future defense actions [9].

2.3. *Cognitive Process*

Currently, information is increasing fast, and the availability of processing data surpasses human capacities. According to [22], cognitive architectures and models have primarily been developed using Artificial Intelligence to serve as decision aids to human users. Analyzing the rational cognitive process can allow the design of the computational level of cognitive prediction. Cassenti et al. [23] mention that by using technology based on adaptive aids, the user's cognitive state can be obtained and difficulties detected at any stage of cognition. Additionally, Cassenti mentions that one missing element in technology models concerns the human learning process, providing feedback that allows technology to adapt to the user and accomplish goals. According to Cameron [24], cognitive strategies are mental processes developed by humans to regulate the thought processes inside the mind to achieve goals or solve problems (See, Figure 1).

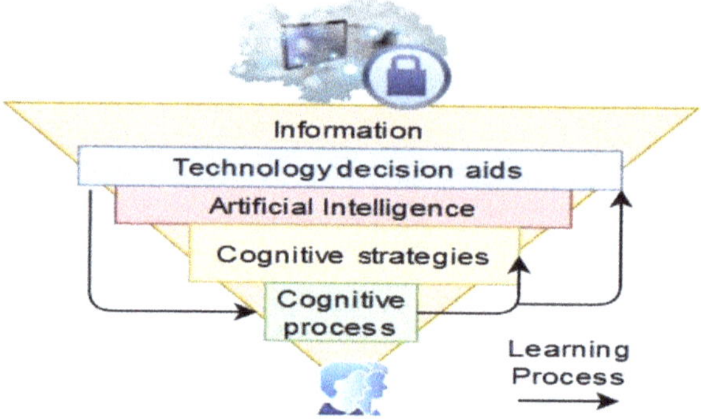

Figure 1. Relation between Information, Technology aids, and Cognitive Processes.

2.4. Cognitive Security

Cognitive security is the ability to generate cognition for efficient decision-making in real-time by modeling human thought processes to detect cybersecurity attacks and develop defense strategies. Specifically, it responds to the need to build situational awareness of cybersecurity related to the environment of technology systems and the insights about itself. In addition, cognitive security allows programmers to develop defense actions by analyzing structured or unstructured information using cognitive sciences approaches, for instance, by incorporating Artificial Intelligence techniques such as data mining, machine learning, natural language processing, human-computer interaction, data analytics, big data, stochastic processes, and game theory. These emulate the human thought process for generating continuous learning, decision making, and security analysis [5].

2.5. Prisma Methodology

The PRISMA methodology is divided into four stages: identification, screening, eligibility analysis, and inclusion [25]. The identification stage includes the development of the following phases: study selection, inclusion and exclusion criteria, manual search, and duplicate removal. The screening stage consists of choosing papers according to relevant titles and abstracts. Next, the eligibility analysis stage includes the process of reading the full texts that accomplished the screening criteria. Finally, the inclusion stage consists of the relevant data extraction from full papers [26].

2.6. Text Mining

In this work, we applied text mining to execute the data analysis of selected papers. Text mining can be defined as mathematical analysis to deduce patterns and trends in the data. A classic exploration can detect these patterns because the relationships are very complex or large amounts of text where repetitive patterns, trends, or rules that explain the text's behavior are discovered. Text Mining's objective, an essential part of Data Science, is to help understand the content of a set of texts through statistics and search algorithms related to Artificial Intelligence [27]. In the text mining process, we obtain information from large amounts of text, with unstructured information and the context in which it was written, intending to extract non-obvious information. Text mining could conduct a qualitative research project with a large sample size similar to a quantitative research study [28].

3. Methods

Cognitive sciences applied to cybersecurity; an exploration based on PRISMA.

The methodology used in this study was the development of a systematic literature review based on the PRISMA methodology, which includes four stages: identification, screening, eligibility analysis, and inclusion (see Figure 2). Study selection was based on a systematic review following the Prisma Guidelines [21]. In the identification stage, we found works in the following databases: Springer, Scopus, IEEE, Association for Computing Machinery (ACM), Web of Science, and Science Direct, in the last three years, 2019 to 2020, to identify the trends in cybersecurity. The search queries established were the following:

- "Cybersecurity" AND "Attacks" AND "Trends";
- "Cybersecurity" AND "Trends" AND "Challenges".

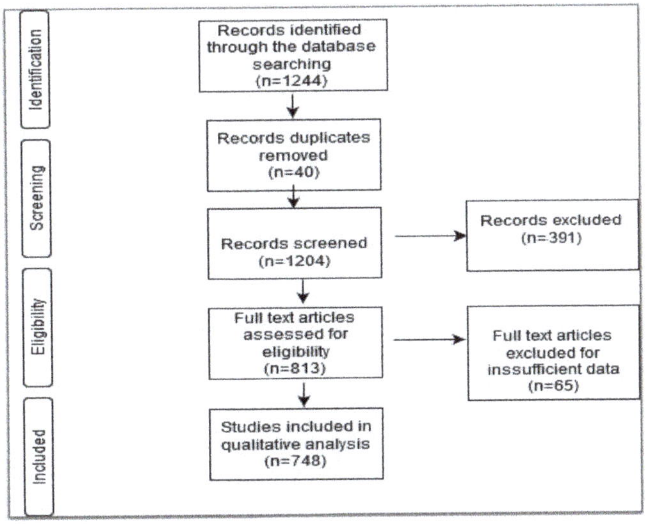

Figure 2. SLR according to Prisma methodology.

The inclusion criteria were: (i) documents published on the scientific database from 2019 to 2021. The exclusion criteria included: (i) documents not related to cybersecurity and (ii) documents out of the research period (2019–2021). Figure 3 shows the screening and eligibility process of the 1244 studies. Then, based on the review of papers' titles and abstracts using a web application, Rayyan, created for the systematic review process, we removed the papers that did not comply with the inclusion criteria. At the end of the screening process, 813 articles were selected for full-text reading. Finally, we removed studies without clear proposals in the cybersecurity field, excluding 748 papers.

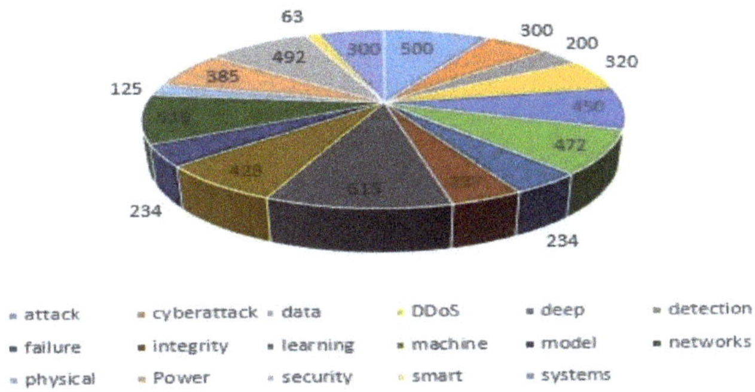

Figure 3. General topics in cybersecurity between 2019 to 2021.

Qualitative analysis using text mining technique.

Text mining, which is considered another field in cognitive science, is essential for qualitative cybersecurity research. However, text mining requires text cleaning and tokenization as prerequisites. In this way, the cleaning process of text, within the scope of text mining, consists of eliminating everything that does not provide information on its subject, structure, or content from the corpus. It should be noted that there is no single way to do this step. It depends on the purpose of the analysis and the text source. We applied a text mining analysis using R software to all 748 studies obtained in the included stage of PRISMA methodology. Thus, we eliminated non-informative patterns (web page URLs), punctuation marks, and single characters. We generated the text tokenization, which divides the text into the units for the analysis in question. We proceeded to store the tokenized text. Each element of the tokenized_text column is a list with a character vector containing the generated tokens. However, there has been a significant change when doing the tokenization process. Before the text's division, the study elements (observations) were the titles and keywords of selected papers. Each one was in a row, thus fulfilling the condition of tidy data: one observation per row. When performing the tokenization, the study element has become each token (word), thus violating the condition of tidy data. Thus, each token list must be expanded to recover the ideal structure, doubling the other columns' value as many times as necessary [29]. We carried out the analysis for the years 2020–2021, obtaining the results in Table 2 and Figure 3.

Table 2. General topics in cybersecurity between 2019 to 2021.

5G	Cloud Security	Microgrid
AC microgrids	Data integrity attack (DIA.)	Multistate model
Advanced metering	Deep learning	Offensive Security
Artificial intelligence (AI.)	Distributed resilient control	Open software
Battery pack	DevOps Security	Plug-in electric vehicles
Blockchain	Digital transformation	Robust algorithm
Call detail record (CDR.)	Event-triggered mechanism	Smart contract
Cybersecurity awareness	Energy security	Smart sensor
Cyberattacks	False data injection attacks	Smart meters
Cyberattack detection	Human Security	Smart City
Cyber-physical systems (CPS.)	Internet of Things (IoT)	Software-defined architecture
Cyber power network	Machine learning	Supply chain management

We included the studies of all the works that evidenced the development of strategies and structures in cybersecurity. Furthermore, we considered articles referring to models developed for learning defense against a cyberattack.

Then, we developed a word cloud process to obtain more detail on scientific studies' contributions in the cybersecurity domain. Algorithm 1 shows the R script used to determine the cybersecurity topics, and Figure 4 shows the word cloud results.

Algorithm 1: Pseudo-code of R script to word cloud process

```
wordcloud ⇐ function(group, df)
print(group)
wordcloud(words = dftoken, freq = dffrequency
max.words = 400, random.order = FALSE,
rot.per = 0.35, color = brewer.pal(8,"Dark2"))
```

Figure 4. Results of a word cloud of 748 papers.

Cybersecurity attacks and their impact

According to the World Economic Forum [1], cyberattacks were considered a fifth of the worldwide risks above food crisis in 2020. Adversaries developed several cyberattack scenarios. For instance, the United States Department of Justice discloses public information about scams perpetrated through websites, social networks, emails, and robocalls, among other means. All these related to fake news about COVID-19 vaccines, treatments, protective equipment, and obviously, about criminals who conducted fake businesses to steal identities or file fraud cite the USDJ. Another cybersecurity scenario covered by the United States Department of Justice is when adversaries send text messages using fictitious phone numbers and social media accounts to harass, intimidate, cyberstalk, and attempt to sex-extort women [30]. On the other hand, CISA mentions that adversaries use Bots to conduct credential harvesting, mail exfiltration, crypto mining, point-of-sale data exfiltration, and the deployment of ransomware [31]. According to the FBI, from 2015 to 2019, reports about fraud in the FBI's Internet Crime Complaint Center (IC3) went from USD 1.1 billion to USD 3.5 billion [32]. Establishing the most appropriate cybersecurity defense solution is necessary to identify the characteristics of cyberattacks [33]. There are currently a great variety of cyberattacks [34]; Table 3 shows those that are the most recurrent among the selected papers for this study.

Table 3. Cybersecurity attacks detected in text mining process.

Type	Description	Reference
Phishing	Is a malicious and deliberate attempt of sending fake messages that appear to come from a reputable source	[35–40]
Insider threats	Are encountered in international security, geopolitics, business, trade, and cybersecurity. Insider threats could be considered more damaging than outsider attacks	[41–43]
APT	Are designed to steal corporate or national secrets.	[44–46]
Cybercrime	Criminal activity either targets or uses a computer, a computer network, or a networked device.	[47,48]
Malware	Malicious software designed to infiltrate a device without knowledge	[49,50]
DDoS	Denial of service attack on a computer system or network that causes a service or resource tobe inaccessible to legitimate users	[51–55]
Ransomware	A type of malicious program that restricts access to certain parts of files of the infected operating system and demands a ransom in exchange for removing this restriction	[56–60]
Mobile malware	Its name suggests malicious software that targets explicitly the operating systems on mobile phones	[61–64]
Watering hole	Refers to a tactic used during targeted attack campaigns where the APT is distributed through a trusted website that is usually visited by employees of the target company or entity	[65,66]

We contrasted this result with an international organization related to cybersecurity. We found that some of them were considered the most relevant cyberattacks in the year 2020, according to The European Union Agency for Cybersecurity (ENISA) [34]. Additionally, we compared this result with the report of a specialized cybersecurity firm. We found that four out of nine attacks documented in our study had a growth rate of between 7 and 25 percent in 2020 in America, Europe, and Asia (see Table 4). According to [67], a classification of cyberattacks is based on the effects they cause against a system or its architecture: misuse of resources; user access compromise; root access compromise; web access; malware; and denial of service.

Table 4. Growth rate percentage of cyberattack 2020.

Attack	Americas	Europa	Asia
DDoS	13%	17%	23%
Ransomware	3%	4%	6%
Mobile malware	15%	15%	25%
Phishing	7%	11%	14%

Other cyberattacks use machines as attack vectors [68], while others focus on human behaviors [69]. In the case of phishing, attackers seek to exploit human vulnerabilities resulting from factors such as solidarity, desperation, or authority control to carry out their attack [70]. In contrast, Ransomware attacks exploit vulnerabilities in operating systems or applications to encrypt users' or organizations' sensitive information [71]. Within this context, Watering hole attackers use exploit kits with stealth features and seek to compromise a specific group of end-users by infecting websites [65]. A malicious URL attacker defines a link created to distribute malware or facilitate a scam [72]. Form hacking is a type of cyberattack where hackers inject malicious JavaScript code into legitimate website payment forms [73]. Table 5 shows a classification of attacks based on an adversary's resource (machine or human).

Table 5. Attacks adversary's targets (machine or human).

Type	Human	Machine
Phishing	X	-
Insider threats	X	X
Web Based Attacks	-	X
Advanced persistent threat (APT)	-	X
Spam	X	X
Identity theft	X	X
Data breach	X	X
Botnets	-	X
Physical manipulation	X	-
Cybercrime	X	X
Malware	X	X
DDoS	-	X
Ransomware	-	X
Mobile malware	X	X
Watering hole	X	X
Information leakage	X	X

X represents the affectation of target due to attack.

Another way to classify cyberattacks could be based on the target, such as energy, healthcare, and transportation [74,75]. Table 6 shows some services considered targets by adversaries. An exciting fact obtained from text mining analysis is that most research works focus on cybersecurity in the energy domain. False data injection is the most famous attack in energy services because it focuses on modifying forecasted demand data [76]. The main issue with energy services, such as smart grids, is connected to network infrastructure and smart meters, which could have some vulnerabilities. This aspect increases the probability of cyberattacks on smart grid infrastructures [77]. Research focuses on preventing and overcoming cyberattacks by using machine learning techniques, such as artificial neural networks, to solve cybersecurity challenges, especially with the considerable volume of data on power systems [74].

Table 6. Classification of cybersecurity attacks based on target services.

Services	Description	Reference
Financial services	Financial institutions are exposed due to their network dependence. Financial services include payment systems or trading platforms. An example of an attacker on financial services is accessing SWIFT credentials to send fraudulent payment orders.	[38]
The energy	The energy sector is vulnerable to attacks because they need real-time operations. Cyberattacks can generate failure or breakdown of generation, transmission, distribution, or substation systems	[51]
Healthcare	The prime target is the theft of medical information. Cyber-criminals' medical information is more valuable than personal financial information. Ransomware attacks are growing on medical devices	[52]

Table 7 shows topics related to cybersecurity in energy facilities. Healthcare is another domain of interest for adversaries for sensitive and personal information [75]. In healthcare, one relevant issue is legacy software [78]. It is difficult for some hospitals or medical centers to migrate their medical records to new systems, e.g., for factors such as budget, data format, or time; this could be a disadvantage from a cybersecurity perspective. Some research is focused on improving authentication methods to reduce this gap [79], following the topics related to healthcare cybersecurity:

- Physical security, two-way authentication, security protocol, and privacy;
- Security medical devices and legacy software.

Adversary takes advantage of vulnerabilities in different domains, such as [80]:

- Hardware failure;
- Software failure;
- Data encryption;
- Loss of backup power;
- Accidental user error;
- External security breach;
- Physical security;
- Accidental user error;
- External security breach.

Table 7. Cybersecurity topics related to energy facilities.

Energy Systems	Cybersecurity Scope	Applied Mechanism
Cyber-physical power system (CPPS)	Intrusion detection	Temporal-topological correlation
Distribution systems	Anomaly detection	Multi-agent system
Electric drive system	Attack pattern	Fuzzy feature analysis
Industrial system	Cyberattack monitoring and detection	Frequent pattern tree
Smart distribution networks	Situation awareness	State estimation
Networked control system	Resilience control	Markov chain
Steam turbines	Active defense	k-connected graph
Microgrids	Quantization effect	Ruin probability
Smart grid	Sequential false data injection attacks	
	Power outage	
	Stealthy attack	
	False data injection (FDI)	
	Denial-of-service (DoS)	

The growth of new electronic services and technologies such as IoT, big data, and artificial intelligence have allowed the development of new attack vectors [81,82]. IoT has generated interest by adversaries in carrying out security attacks due to its lack of advanced security and great coverage [83]. IoT solutions are very attractive for attackers because of the variety of attacks that can be performed on different components of IoT, among which we can mention the following [84]:

- Mobile devices;
- Embedded systems;
- Consumer technologies;
- Operational systems.

The growth of crypto-currency and distributed authentication architecture is driving the use of blockchain architecture [85]. Another use of blockchain is in healthcare organizations to improve data integrity, authentication, and privacy issues, especially those with sensitive features such as medical records [86]. On the other hand, IoT is growing in different domains such as healthcare, smart city, and smart home [26]. Establishing authentication such as PKI architectures for IoT ecosystems could be expensive for many IoT devices, so smart contracts based on blockchain architecture are an alternative [87]. Following, we outline the topics related to blockchain and cybersecurity in papers selected in this work, which were developed between 2019 to 2021:

- Energy trading;
- Cryptocurrency, crypto-jacking, money laundering;
- Public organization;
- Decentralized consensus decision-making (DCDM);
- Fuzzy static Bayesian game model (FSB-GM);
- Internet of Things, smart contracts;
- Electronic health records.

Some cyberattacks take advantage of new technologies such as 5G, IoT, and the cloud to perform DDoS attacks [88]. The growth of IoT devices with limited computational resources and lack of security configurations make them vulnerable to different cyberattacks. For instance, Mirai Botnet malware exploited the vulnerabilities of an estimated 600,000 IoT devices, resulting in massive Distributed Denial of Service (DDoS) attacks [89]. Cloud computing services are used to launch Distributed Denial of Service (DDoS) attacks. However, adversaries are focusing on low-rate DDoS attacks because they are more challenging to detect due to their stealthy and low-rate traffic [58].

On the other hand, using the hijacked Connection-less Lightweight Directory Access Protocol, an attacker could perform DDoS attacks at 2.3 terabytes per second [90]. Social media platforms have achieved relevance for interaction and social information exchange. However, the attackers have used them to deceive people and make them victims of attacks [91]. An adversary has found a striking attack target in humans because they can be deceived through persuasion techniques [15]. Attacks based on human vulnerabilities, called social engineering, have grown in recent years [66]. Figure 5 shows a word cloud of topics related to social engineering. We can observe that human factors are relevant in this kind of attack. The pandemic has created tremendous pressure on cybersecurity aspects. During the COVID-19 pandemic, the social engineering attacks carried out were phishing, spamming, and scamming. These attacks were combined with socio-technical methods such as fake emails, websites, and mobile apps [92]. The need to work remotely has changed the attack surface of organizations. Attacks on VPNs, hijacking of video meetings, fake news campaigns, and phishing attacks have increased during the COVID-19 pandemic [15]. According to the text mining process, we identified the following topics related to COVID-19 and cybersecurity:

- Malicious web pages;
- Malicious Mobil Apps;
- Malicious Emails messages;

- Misinformation and fake news;
- Security and privacy.

Figure 5. Word cloud of topics related to Social Engineering.

Challenges in cybersecurity solutions

To face cyberattacks, organizations have established cybersecurity mechanisms that could be physical, software-oriented, or procedural. Below, we show some of the most common defense mechanisms:

- Security intelligence systems;
- Perimeter controls;
- Encryption technologies;
- Data loss prevention;
- Governance risk;
- Automated policy management.

The mechanisms described above are the most common solutions for cyberattacks. However, it is possible to define specific defense mechanisms for each type of cyberattack in some cases. For instance [67], the two defense techniques against phishing attacks are:

- Software-based defense approaches;
- User education.

However, MITRE [93] has defined 245 techniques that the attacker could use for executing cyberattacks. The techniques are distributed in 14 stages; each stage is associated with the attackers' process of executing cyberattacks. Figure 6 shows the number of techniques associated with each stage. Figure 7 shows the frequency of MITRE techniques included in the works selected in this study, which were developed between 2019 and 2021. Our text mining analysis found that the most relevant techniques are reconnaissance, discovery, lateral movement, collection, command-control, and impact. On this point, it is important to mention that the absence of frequency in other techniques, such as initial access or privilege escalation, is not an indicator that these techniques are not used in cyberattacks. The information shown in Figure 8 reveals that researchers are more focused on the result of one specific technique in their study. However, for the review made, we can observe that not all selected works considered the cycle of a cyberattack; this aspect is relevant for developing a good defense strategy. We found that most of the techniques mentioned in the selected studies focused on gathering information, such as reconnaissance, discovery, and collection.

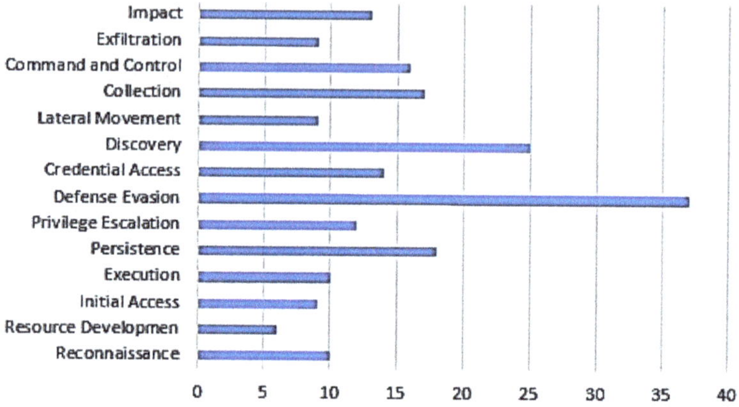

Figure 6. Cybersecurity Techniques according to MITRE.

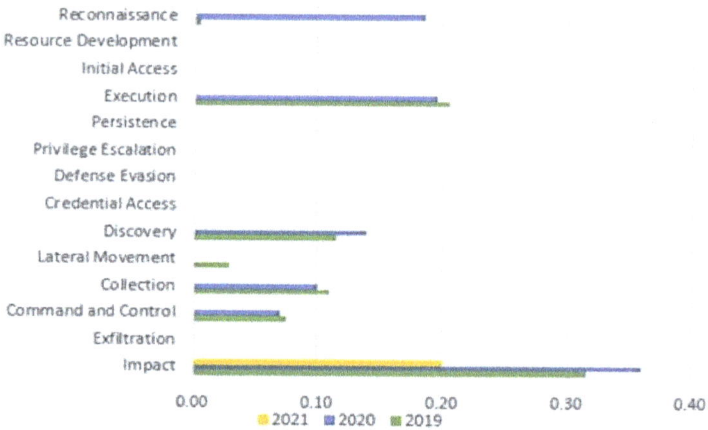

Figure 7. Techniques MITRE identified in works selected in this study.

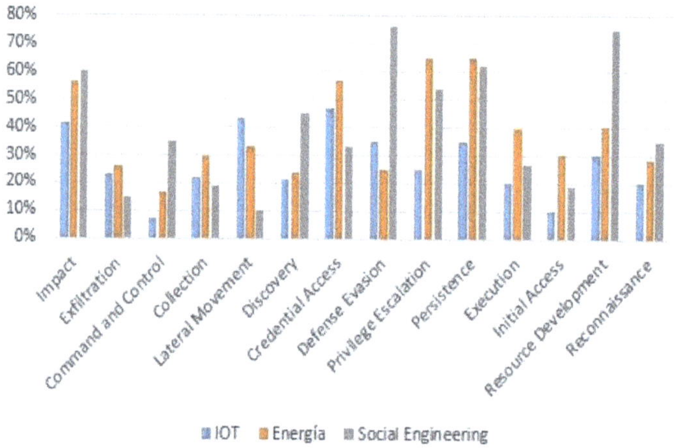

Figure 8. Techniques MITRE used in vertical domains such as Energy, Social Engineering, and IoT.

Figure 8 shows the relation between cybersecurity techniques and domain attacks: energy, IoT, and social engineering. We observed that the relevance of a specific technique depends on the type of cyberattack. For instance, the most relevant techniques in social engineering are reconnaissance, resource development, persistence, and defense evasion. On the other hand, the most relevant techniques in IoT attacks are credential access, lateral movement, and collection. This number of techniques could be a challenge because cybersecurity analysts need to have the capability to detect them in real-time when they are used in cyberattacks to select the best defense strategy.

Figure 9 shows some variants of cybersecurity attacks based on social engineering, which show the incredible versatility of attacks, which can vary depending on the attack techniques used digitally, in person, or by phone.

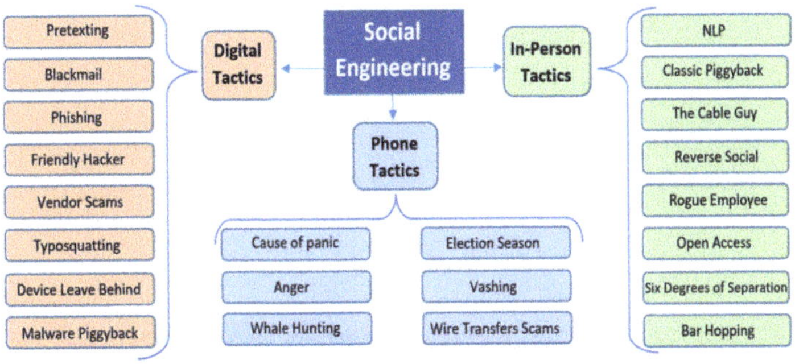

Figure 9. Classification of Social Engineering attacks.

Cybersecurity solutions require adapting to new challenges:

- The heterogeneity of IoT solutions;
- The expansion of the attack surface by IoT and Machine Learning;
- Attacks on Cloud infrastructures;
- Cognitive hacking.

Cybersecurity firms and researchers have been developing some alternatives by mainly focusing on anomaly detection. Inside the anomaly detection process, the objective is to detect some pattern, behavior, or component used by attackers [94]. Table 8 shows topic development from 2019 to 2021 related to anomaly detection. Cybersecurity companies and researchers in the field have moved on from reactive solutions to proactive ones [95].

Table 8. Cybersecurity topics related to IoT.

The Mechanism Applied Based on	IoT Cybersecurity Context
Data analysis	IoT attack classification
ANN	Attack–defense trees
Graph neural nets	DDoS attacks
Cognitive packet network	Botnet
Random neural networks	Attack countermeasures
	Home security threat
	Identification anomaly

Cybersecurity research is trying to stay one step ahead and take advantage of cybersecurity analysts' cognitive capabilities to define proactive cybersecurity defense strategies. So, several research types are focused on incorporating cognitive models to generate these proactive solutions. In the selected period (2019–2021), several studies included artificial intelligence and machine earning concepts applied to cybersecurity (See Table 9).

Table 9. Topics related to anomalies.

Cybersecurity Context	Scope	Applied Mechanism
Cyber-physical power system (CPPS)	Behavior pattern	Multiagentsystems (MASs)
Internet of Things	Attack pattern	Honeypots
Connected and automated vehicles (CAVs)	Anomaly detection	Convolution neural network (CNN)
Smart home	Anomaly identification	Dimensionality reduction
Intelligent transportation system (ITS)	Attack pattern	Principal component analysis

The use of supervised machine learning such as Decision Tree (DT), Support Vector Machine (SVM), Naïve Bayes (NB), Random Forest (RF), and unsupervised algorithms such as K-nearest neighbor (kNN) and Artificial Neural Network (ANN's) for building intrusion detection systems (IDS), or anomaly pattern detection, are the most exciting topics in cybersecurity. A relevant fact observed in the selected papers was the growing number of studies related to deep learning applications. Researchers have considered deep learning a good alternative for facing different cybersecurity issues. How can deep learning be applied to detect IoT attacks, APT, DDoS, malware, and anomaly detection? An interesting fact is that there are three variants of deep learning:

1. Deep learning;
2. Deep reinforcement learning;
3. Deep transfer learning.

Table 10 shows topics identified from papers in the text mining process related to security in IoT. Research focus on defense solutions to face DDoS include the use of cognitive sciences approaches such as [88]:

- Deep learning;
- Machine learning;
- Deep Convolutional Neural Network (CNN);
- Genetic algorithms;
- Game theory;
- PCA;
- Large-Scale System (LSS.)

Table 10. Machine learning applied to cybersecurity.

Learning Techniques	Cybersecurity Application Context
Decision Tree	Cryptojacking
k-nearest neighbors	Internet of things
Random Forest	Advanced persistent threat
Naive Bayes	Collaborative attacks
Recurrent Neural Networks (RNNs)	Traffic flow monitoring
Generative adversarial networks	Distributed denial-of-service attacks
Deep learning	Malicious javascript detection
Deep reinforcement learning	Intruder detection
Deep transfer learning	

Below there are some approaches of studies between 2019 and 2021 with solutions based on machine learning and deep learning for identifying malicious URLs or sentiment analysis in social media:

- Deep learning for word embedding;
- Natural language processing and sentiment analysis on online social networks.

Game theory is another alternative of cognitive sciences applied to cybersecurity. Its objective is to try to guess the next step for adversaries during cyberattacks. Figure 10 shows a word cloud with topics related to game theory. We identified that game theory could be applied to different domains such as energy, investment, cyber-physical systems, and computer security. Additionally, game theory research shows approaches in defense mechanisms, information dissemination, and decision making. Game theory uses computational modeling to take advantage of security analysts' cognitive processes and adversaries to improve decision-making based on information analysis to face attacks [96]. Game theory is mostly used in the economy field, which is responsible for studying optimal decisions and strategies for given situations. According to the definition of Nash equilibrium, the strategy or set of strategies of each player responds to the other players' actions to maximize each player's profit. The player's strategy is a specific action at a particular moment of the game [96]. A game is defined as interacting with two or more participants seeking a reward. During the game, participants develop strategies to maximize their profit. Players do not necessarily represent people; they can be organizations or groups. There are two classic games in-game theory: cooperative games and non-cooperative games. There are two ways for the mathematical representation of a game: a standard form using matrices and an extensive form using decision trees. A cooperative game is based on the players' interaction reaching agreements to establish the decision-making that each player will carry out, achieving the objective of reaching coalitions, and determining how to distribute the rewards [97]. However, in non-cooperative games, each player must decide what decision to make without knowing the rest's decisions. These are more subject to the reality of what happens in the cybersecurity domain. Complete information games are those in which each player knows all the events in the game's course from the beginning, especially when making a decision. A classic example of a complete information game is the game of chess. Incomplete information games, in most cases, are simultaneous decision-making games, so each player knows something that the others do not. Interactions between an adversary and the user could be modeled based on two players' stochastic game. Using a non-linear program is possible to compute Nash equilibrium to define the best response strategies for players [98]. Developing games that consider cost, time, reward, and performance could define effective game strategies.

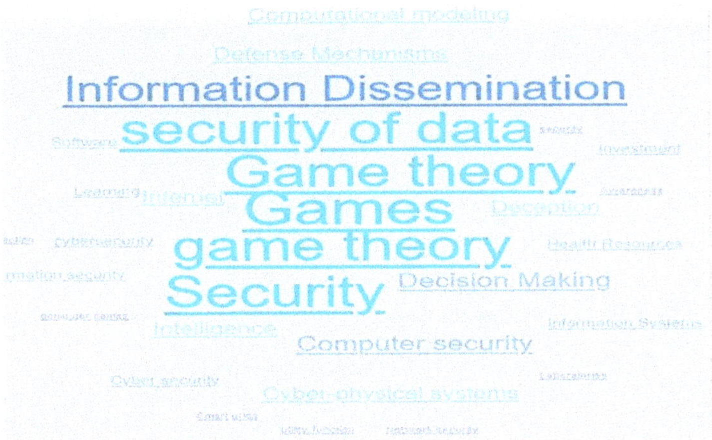

Figure 10. Game theory research topics.

4. Results and Discussion

Cognitive Cybersecurity Model

Our text mining process found that the works selected in this study do not consider indirect cognitive processes or cognitive models such as OODA or MAPE-K. Including game theory in cybersecurity can lead to strategies to minimize cyberattacks from a cognitive perspective. A complete information model is the most appropriate to obtain the best decision from the game theory approach. Big network environments are very complex scenarios for developing detection and protection cybersecurity solutions. The integration of machine learning and deep learning with game theory techniques could improve proactive security solutions. Concerning Figure 2, Cassenti et al. [23] mention that technology does not consider the user learning processes. From our perspective, the game theory approach could be a solution to this because it validates the user's decision-making processes based on a set of experiences and patterns. From the game theory perspective, if the user (player) improves the learning process or the decision-making process based on cognitive processes, the probability of winning the game increases. In this sense, we propose in Figure 11, a cognitive cybersecurity model based on integrating cognitive process and machine learning, deep learning, and game theory approach applied in cybersecurity. As shown in Figure 11, we structured the model into three layers. The first layer of the cognitive model addresses the aspects of perception related to the cognitive processes. It associates them with sources of information that can be analyzed to establish patterns of anomalies based on space-time criteria. The second layer establishes the association of the understanding processes with machine learning (ML) techniques or deep learning (DL) that can be used for the anomaly detection processes. This association must have bi-directional feedback between analysts and technology to improve ML or DL algorithms' training processes. The way towards the analysis allows us to generate perspicacity about cybersecurity situation awareness. Additionally, this feedback should support the improvement in the analyst's cognitive processes to detect cyberattacks.

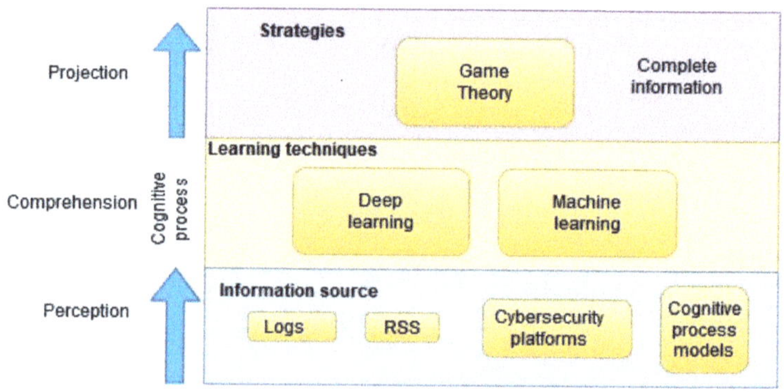

Figure 11. Proposal for a Cognitive Cybersecurity model.

Finally, the third layer associates the cognitive projection process with game theory techniques. At this level, the decision-making processes to establish the best defense strategy must be supported by the information obtained from the ML and DL processes carried out in the lower layer. The bidirectional relation, in one sense, is the computational model of the game theory component. In another sense, it should improve the cognitive decision-making processes. However, establishing the proposed model is complex without obtaining all the information from the analyst and adversary (See Figure 12). Modeling the adversary's characteristics would allow analysts to have a complete vision to establish a better decision. For instance, they knew that the adversary could use a combination of tools (T), techniques (Th), and procedures (C2F). However, the list of tools and procedures can be extensive and varied. Below is a list of the most widespread RATs:

- OSSEC is an open-source HIDS for data gathering;
- Snort is an intrusion Prevention System (IPS) to detect malicious network activity;
- Suricata is an open-source system for real-time intrusion detection (IDS) and intrusion prevention (IPS);

- Security Onion is open-source used for threat hunting, security monitoring, and log management;
- OpenWIPS-NG is an intrusion prevention system (IPS), preferred for wireless packet tracking;
- Fail2ban is a software that scans log files and bans IPs that show malicious activity. Procedures used for adversaries could be based on Command and Control (C2) frameworks. Following, we list some C2 frameworks:
    - FudgeC2 is a campaign-orientated Powershell C2;
    - Callidus is an open-source C2 framework that leverages Outlook, OneNote, Microsoft Teams for command and control;
    - APfell is a cross-platform, OPSEC aware, red teaming, post-exploitation C2 framework;
    - DaaC2: is an open-source C2 framework that makes use of Discord as a C2;
    - Koadic is an open-source for post-exploitation;
    - TrevorC2 is a client/server model for masking command and control through web browsers

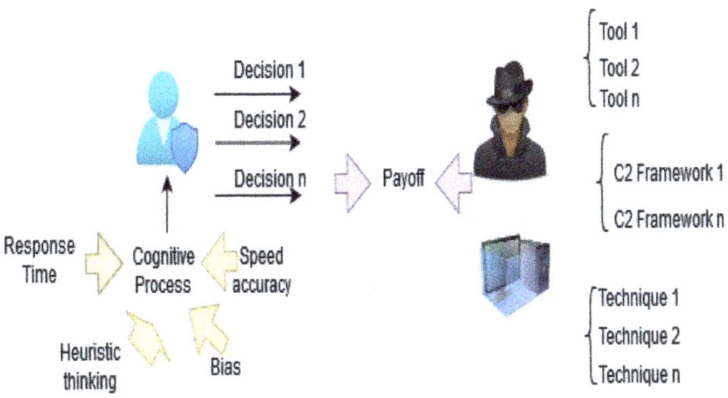

Figure 12. Attack and defense components.

We represent in Equation (1) the attack as the combination of tools (T), procedures (C2F), and techniques (Th), where w represents the weights based on the tool, procedure, and technique used by the adversary.

$$\text{Attack} = w(T) + w(C2F) + w(Th) \quad (1)$$

On the other hand, cybersecurity analysts developed a set of cognitive processes to establish the defense process. From a macro vision, the analyst must decide if a possible event could be an attack or not, based on the cognitive process of perception. Bitzer et al. [99] mention that perceptual decision making is applied to two-alternative forced-choice tasks to judge perceptual feature differences. According to Bitzer, drift-diffusion models have been used to quantitatively analyze behavioral data, i.e., reaction times and accuracy. In the same vein, Dale et al. [100] mention that the cognitive analysis of vast amounts of data requires the application of the heuristics process and that people often mistakenly judge the likelihood of a situation by not taking all relevant data into account. However, according to Nikolić et al. [101], the application of heuristics as mental strategies and certain deformations in the thoughts and perceptions of decision makers affect their attitudes and approach to problem solving. Trueblood et al. [102] mention that we need to understand how people make perceptual decisions to improve training to minimize misdiagnoses in the medical field. So, let us adapt this approach to cybersecurity: the defense strategy must be oriented toward the factors associated with the cognitive process; this is described in Equation (2), where: R.T is the Response Time associated with the time for executing a defense action by a cybersecurity analyst; H.T is the heuristic thinking associated with the process of selecting a decision; B is the Bias related to human thinking, and S.A is the speed accuracy in the decision-making process.

$$\text{Cognitive Process} = w1(R.T) + w2(H.T) + w3(B) + w4(S.A) \quad (2)$$

where wi is the weight assigned to each variable.

Once the cognitive process has been carried out, the best decision is made considering the weight of each variable in the cognitive process, expressed in Equation (3).

$$\text{Decision(j)} = (\Delta P\ j)\ \text{Cognitive process} \quad (3)$$

where Delta P j is the variation due to weights in cognitive processes.

Therefore, the defense strategy is expressed as Equation (4).

$$\text{Defense} = (\text{Decision } j) + \text{Error} \quad (4)$$

However, analysts in the cybersecurity decision-making process could be affected by factors such as Bias and speed accuracy. Bias (B) effects and speed-accuracy effects are ubiquitous in experimental psychology. Bias effects arise when the two stimulus alternatives occur with unequal frequency or have unequal rewards attached to them. Speed-accuracy (SA) effects arise as the result of explicit instructions emphasizing speed or accuracy [103]. Computational models of decision making present a solution to this problem. In particular, we choose Response Time (RT) models such as the drift-diffusion model (D.D.M.), proposed by Ratcliff [103], and the linear ballistic accumulator (LBA) model, proposed by Brown [104]. Accumulator models assume that evidence is accumulated over time until a threshold amount is reached for a commitment to that response option. These models contain four primary parameters related to different psychological components of simple decisions: caution, Bias, stimulus processing, and motor sense.

5. Discussion

In this study, a literature review for the period 2019 to 2021 was carried out. Text-mining was used to determine the most addressed topics in chosen papers in the area of cybersecurity. This exploratory analysis focused on the most relevant used words in the content. The words we found included security, attack, detection, networks, machine learning, and power. This result made us deduce that cybersecurity research has been related to detecting cyberattacks on electricity grids through machine learning in recent years. Another finding in our literature review was that the mainstream research has been dedicated to implementing proactive cybersecurity. Cognitive science is being applied for this purpose. We actually found relevant contributions in which machine learning and deep learning-based solutions were proposed. Figure 13 shows the percentage of works that use machine learning and deep learning, respectively, from the papers included in the literature review that we carried out.

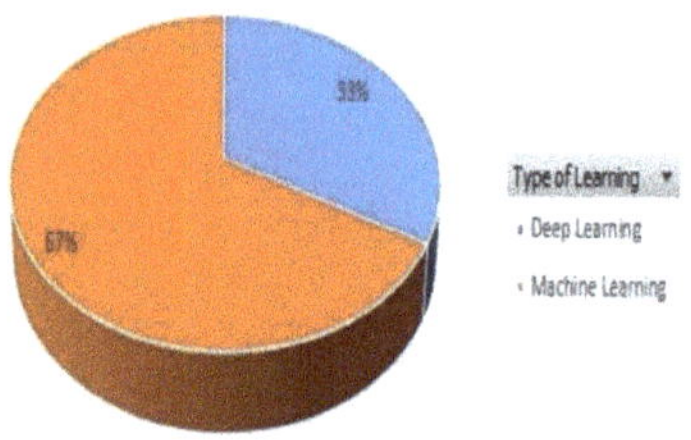

Figure 13. Deep Learning vs. Machine Learning.

The period 2019 to 2021 was atypical in the way some activities were carried out worldwide due to it being in the context of a pandemic driven by COVID-19. In this context, the reasoning in some sectors has to consider the greater use of technological resources, such as tele-education, tele-health, government, and private electronic services. From the perspective of the digital transformation of organizations and cities, the pandemic was an essential accelerator in the adoption of technologies in specific sectors. It made organizations and people more dependent on technological resources. However, this context generated the need to address essential aspects of cybersecurity. For example, children increased their availability of internet connections, increasing their exposure to online risks [105]. Organizations based their logistics and supply chain processes on internet-based technologies, expanding the attack surface [106]. The inclusion of IoT for data collection and process automation increases the need to acquire an end-to-end secure IoT environment [107]. The use of social engineering attacks based on the human need to obtain information about the pandemic increased their probability of accessing fake news or being a victim of social engineering attacks [108].

During the same period, 2019–2021, even within the context of the pandemic, there was no reduction in attacks on organizations' information systems or the impact on people through social engineering attacks. The literature review carried out for the perid 2019–2021 showed that the financial, energy, and healthcare services were the most attacked, and the fastest-growing attacks were DDoS, Ransomware, Mobile malware, and Phishing. This context highlighted the need for organizations to strengthen their cybersecurity strategies concerning:

- Security intelligence systems;
- Perimeter controls;
- Encryption technologies;
- Data loss prevention;
- Governance risk;
- Automated policy management.

While from a user perspective, it highlighted the need to generate more awareness concerning:

- Malicious web pages;
- Malicious Mobile Apps;
- Malicious Email messages;
- Misinformation and fake news;
- Security and privacy.

Faced with this continuous growth of cybersecurity attacks and the need to improve security strategies to protect people and organizations, the literature review carried out shows that research has promoted the use of learning techniques as a resource to strengthen their security strategies, specifically to automate activities such as behavior pattern, attack pattern, anomaly detection, and anomaly identification. The most-used learning techniques in the cybersecurity domain correspond to Decision Tree, k-nearest neighbors, Random Forest, Naive Bayes, Recurrent Neural Networks (RNNs), generative adversarial networks, deep learning, deep reinforcement learning, and deep transfer learning, and you can see a growing interest in what corresponds to deep learning. Although game theory is not new in its application to cybersecurity, it has had significant growth in recent years, especially in improving decision-making processes related to cybersecurity in the financial, energy, and critical infrastructure sectors.

This finding encourages future work to understand how security organizations and specialists are preparing to adopt cognitive techniques based on learning as a security strategy. It has also proposed a possible future analysis of how our organizations can have their learning capacity (situational awareness and self-awareness) capable of establishing that it is being attacked and can establish a level of resilience. From the user's perspective, it highlights how these learning techniques can be used to strengthen cognitive processes in detecting security attacks, especially those based on social engineering techniques.

The design of cognitive models applied in cybersecurity compared to traditional security methods is based on obtaining or abstracting information from the user's cognitive processes, organization, and adversary roles, for which a cognitive model could define the following steps:

1. Implementation of infrastructure for handling a large volume of data;
2. Incorporation of cognitive sciences in security strategies such as artificial intelligence, machine learning, data analytics, and psychology;
3. Cognitive model design based on:
    A. Cognitive processes Observe–Orient–Decide–Act model (OODA);
    B. the Monitor–Analyze–Plan–Execute model (MAPE-K).
4. Identification of cognitive processes:
    A. Users' or analysts' cognitive processes;
    B. The adversary's behavioral characteristics.

6. Conclusions

The literature review found that much attention has been paid to proactive cybersecurity solutions, acceptable cybersecurity practices, and cybersecurity hygiene strategies for mitigating cyberattacks. In this context, the use of cognitive science techniques has grown significantly. Answers in this area are being proposed, and they mainly look for the improvement of the response time of cyberattacks' countermeasures that work in real-time.

In general, cognitive science is being used to understand the behavior of adversaries to minimize the impact of cyberattacks. In this context, machine learning and deep learning are the techniques that are used the most. The model we propose tries to fill the gap that exists in automatizing cognitive science without considering the users learning processes. Our opinion is that incorporating game theory represents a significant contribution to bringing cognitive sciences to decision-making processes. A set of heuristic, Bias, and quantitative perception measures was defined as part of the cognitive cybersecurity model we have proposed. These measures make it possible to integrate machine learning and deep learning techniques with game theory. We conclude that social and psychological analysis in cybersecurity may improve the process of obtaining information that helps in the decision-making processes.

The present work, investigating the period 2019–2021, understands the evolution of cybersecurity under an atypical context such as a pandemic. Work carried out during the year 2022 has not been considered because it is a period still in progress and has had a change based on the progressive return of activities. Therefore, we believe that future complementary work would be to analyze how this new change has affected cybersecurity processes.

This work was based on the literature review of scientific bases. It would be interesting to extend it with a study of different organizations and their perspective on the inclusion or management of cognitive techniques applied to cybersecurity, including understanding how these techniques can provide security in the requirements analysis, and by performing security configurations in the context of DevOps [109] and Digital transformation [110], in addition to how cognitive techniques tie in with Open-source tools, which are widely used to maintain network security, endpoint security, and system security [111]. Although our literature review does not show them explicitly, these are very relevant topics in cybersecurity today. This leads us in future work to propose new search strings that allow us to expand our study to these topics.

Author Contributions: Conceptualization, R.O.A. and W.F.; methodology, R.O.A.; validation, W.F.; formal analysis, R.O.A.; investigation, M.C.; data curation, I.O.-G.; writing—review and editing, I.O.-G. and G.N.; funding acquisition, I.O.-G. and G.N. All authors have read and agreed to the published version of the manuscript.

Funding: This research received no external funding.

Acknowledgments: The authors would like to thank the Universidad de las Fuerzas Armadas ESPE of Sangolquí, Ecuador, for the resources granted for the development of the research project entitled: "Detection and Mitigation of Social Engineering attacks applying Cognitive Security, Code: PIC-ESPE-2020-SE". The author also acknowledges the Universidad de Las Américas of Ecuador and his Engineer degree in Information Technology for support in this work. Additionality, we want to thank the members of the group IDEIAGEOCA for all their support in the research.

Conflicts of Interest: The authors declare no conflict of interest.

References

1. WEF: Word Economic Forum. The Global Risks Report 2021. Available online: https://www.weforum.org/reports/the-global-risks-report-2021 (accessed on 21 May 2021).
2. Donevski, M.; Zia, T. A survey of anomaly and automation from a cybersecurity perspective. In Proceedings of the 2018 IEEE Globecom Workshops (GC Wkshps), Abu Dhabi, United Arab Emirates, 9–13 December 2018; pp. 1–6. [CrossRef]
3. Yang, Q.; Jia, X.; Li, X.; Feng, J.; Li, W.; Lee, J. Evaluating Feature Selection and Anomaly Detection Methods of Hard Drive Failure Prediction. *IEEE Trans. Reliab.* **2020**, *70*, 749–760. [CrossRef]
4. Alrashdi, I.; Alqazzaz, A.; Aloufi, E.; Alharthi, R.; Zohdy, M.; Ming, H. AD-IoT: Anomaly Detection of IoT Cyberattacks in Smart City Using Machine Learning. In Proceedings of the IEEE 9th Annual Computing and Communication Workshop and Conference (CCWC), Las Vegas, NV, USA, 7–9 January 2019; pp. 0305–0310. [CrossRef]
5. Andrade, R.; Torres, J. Self-awareness as an enabler of cognitive security. In Proceedings of the 2018 IEEE 9th Annual Information Technology, Electronics and Mobile Communication Conference (IEMCON), Vancouver, BC, Canada, 1–3 November 2018; pp. 701–708. [CrossRef]
6. Leung, H. An integrated decision support system based on the human ooda loop. In Proceedings of the 2018 IEEE 17th International Conference on Cognitive Informatics Cognitive Computing (ICCI*CC), Berkeley, CA, USA, 16–18 July 2018; p. 1. [CrossRef]
7. Shaukat, K.; Luo, S.; Varadharajan, V.; Hameed, I.A.; Xu, M. A Survey on Machine Learning Techniques for Cyber Security in the Last Decade. *IEEE Access* **2020**, *8*, 222310–222354. [CrossRef]
8. Brückner, M.; Scheffer, T. Stackelberg games for adversarial prediction problems. In Proceedings of the 17th ACM SIGKDD International Conference on Knowledge Discovery and Data Mining, New York, NY, USA, 21–24 August 2011; pp. 547–555. [CrossRef]
9. Andrade, R.; Torres, J. Enhancing intelligence soc with big data tools. In Proceedings of the 2018 IEEE 9th Annual Information Technology, Electronics and Mobile Communication Conference (IEMCON), Vancouver, BC, Canada, 1–3 November 2018; pp. 1076–1080. [CrossRef]
10. Le, D.C.; Zincir-Heywood, N. Exploring adversarial properties of insider threat detection. In Proceedings of the 2020 IEEE Conference on Communications and Network Security (CNS), Avignon, France, 29 June–1 July 2020; pp. 1–9. [CrossRef]
11. Rajivan, P.; Gonzalez, C. Creative Persuasion: A Study on Adversarial Behaviors and Strategies in Phishing Attacks. *Front. Psychol.* **2018**, *9*, 135. [CrossRef] [PubMed]
12. Andrade, R.; Yoo, S.G. Cognitive security: A comprehensive study of cognitive science in cybersecurity. *J. Inf. Secur. Appl.* **2019**, *48*, 102352. [CrossRef]
13. Alqahtani, H.; Kavakli-Thorne, M. Exploring factors affecting user's cybersecurity behaviour by using mobile augmented reality app (cybar). In *Proceedings of the 2020 12th International Conference on Computer and Automation Engineering. ICCAE, Sydney, NSW, Australia, 14–16 February 2020*; Association for Computing Machinery: New York, NY, USA; pp. 129–135. [CrossRef]
14. Kakkad, V.; Shah, H.; Patel, R.; Doshi, N. A Comparative study of applications of Game Theory in Cyber Security and Cloud Computing. *Procedia Comput. Sci.* **2019**, *155*, 680–685. [CrossRef]
15. Andrade, R.O.; Ortiz-Garces, I.; Cazares, M. Cybersecurity attacks on smart home during covid-19 pandemic. In Proceedings of the Fourth World Conference on Smart Trends in Systems, Security and Sustainability (WorldS4), London, UK, 27–28 July 2021; pp. 398–404. [CrossRef]
16. Orunsolu, A.; Sodiya, A.; Akinwale, A. A predictive model for phishing detection. *J. King Saud Univ. Comput. Inf. Sci.* **2022**, *34*, 232–247. [CrossRef]
17. Schubert, A.-L.; Frischkorn, G.T.; Hagemann, D.; Voss, A. Trait Characteristics of Diffusion Model Parameters. *J. Intell.* **2016**, *4*, 7. [CrossRef]
18. Simmons, C.; Ellis, C.; Shiva, S.; Dasgupta, D.; Wu, C. Avoidit: A cyber attack taxonomy. In Proceedings of the 9th Annual Symposium on Information Assurance (ASIA'14), Albany, NY, USA, 3–4 June 2014.
19. Venkatesan, S.; Sugrim, S.; Izmailov, R.; Chiang, C.J.; Chadha, R.; Doshi, B.; Hoffman, B.; Allison Newcomb, E.; Buchler, N. On detecting manifestation of adversary characteristics. In Proceedings of the MILCOM 2018—2018 IEEE Military Communications Conference (MILCOM), Los Angeles, CA, USA, 29–31 October 2018; pp. 431–437. [CrossRef]
20. Andrade, R.O.; Yoo, S.G.; Tello-Oquendo, L.; Ortiz-Garces, I. A Comprehensive Study of the IoT Cybersecurity in Smart Cities. *IEEE Access* **2020**, *8*, 228922–228941. [CrossRef]

21. Zambrano, P.; Torres, J.; Tello-Oquendo, L.; Jacome, R.; Benalcazar, M.E.; Andrade, R.; Fuertes, W. Technical Mapping of the Grooming Anatomy Using Machine Learning Paradigms: An Information Security Approach. *IEEE Access* **2019**, *7*, 142129–142146. [CrossRef]
22. Lebiere, C.; Morrison, D.; Abdelzaher, T.; Hu, S.; Gonzalez, C.; Buchler, N.; Veksler, V. Cognitive models of prediction as decision aids. In Proceedings of the 14th International Conference on Cognitive Modeling, University Park, PA, USA, 4–6 August 2016.
23. Cassenti, D.; Veksler, V. Using cognitive modeling for adaptive automation triggering. In Proceedings of the AHFE 2017 International Conference on Human Factors in Simulation and Modeling, Los Angeles, CA, USA, 17–21 July 2017; pp. 378–390. [CrossRef]
24. Cameron, L.; Jago, L. *Cognitive Strategies*; Springer: New York, NY, USA, 2013; p. 453. [CrossRef]
25. Mengist, W.; Soromessa, T.; Legese, G. Method for conducting systematic literature review and meta-analysis for environmental science research. *MethodsX* **2019**, *7*, 100777. [CrossRef]
26. Andrade, R.O.; Yoo, S.G. A Comprehensive Study of the Use of LoRa in the Development of Smart Cities. *Appl. Sci.* **2019**, *9*, 4753. [CrossRef]
27. Antons, D.; Grünwald, E.; Cichy, P.; Salge, T.O. The application of text mining methods in innovation research: Current state, evolution patterns, and development priorities. *R&D Manag.* **2020**, *50*, 329–351. [CrossRef]
28. Lee, C.; Cheng, C.; Zeleke, A. Can text mining technique be used as an alternative tool for qualitative research in education? In Proceedings of the 15th IEEE/ACIS International Conference on Software Engineering, Artificial Intelligence, Networking and Parallel/Distributed Computing (SNPD), Las Vegas, NV, USA, 30 June–2 July 2014; pp. 1–6. [CrossRef]
29. Ceron, J.C.A.; Gomez, L.J.P.; Ceballos, H.G.; Cantu-Ortiz, F.J. Twitter data analysis on the topic: Tec de monterrey. In Proceedings of the 2020 3rd International Conference on Computer Applications Information Security (ICCAIS), Riyadh, Saudi Arabia, 19–21 March 2020; pp. 1–6. [CrossRef]
30. USDJ. New York Man Pleads Guilty to Cyberstalking after Harassing and Sextorting Multiple Victims. 2021. Available online: https://www.justice.gov/usao-mdfl/pr/new-york-man-pleads-guilty-cyberstalking-after-harassing-and-sextorting-multiple (accessed on 25 May 2021).
31. CISA. Ransomware Activity Targeting the Healthcare and Public Health Sector. 2021. Available online: https://us-cert.cisa.gov/ncas/alerts/aa20--302a (accessed on 25 May 2021).
32. FBI. Fraudsters Prey on Emotions and Bank Accounts in Money Mule Schemes. 2021. Available online: https://www.fbi.gov/contact-us/field-offices/elpaso/news/press-releases/fraudsters-prey-on-emotions-and-bank-accounts-in-money-mule-schemes (accessed on 25 May 2021).
33. Al-Mohannadi, H.; Mirza, Q.; Namanya, A.; Awan, I.; Cullen, A.; Disso, J. Cyber-attack modeling analysis techniques: An overview. In Proceedings of the 2016 IEEE 4th International Conference on Future Internet of Things and Cloud Workshops (FiCloudW), Vienna, Austria, 22–24 August 2016; pp. 69–76. [CrossRef]
34. ENISA. Threat Landscape. 2020. Available online: https://www.enisa.europa.eu/topics/threat-risk-management/threats-and-trends (accessed on 12 February 2021).
35. Singh, C. Meenu: Phishing website detection based on machine learning: A survey. In Proceedings of the 2020 6th International Conference on Advanced Computing and Communication Systems (ICACCS), Coimbatore, India, 6–7 March 2020; pp. 398–404. [CrossRef]
36. Athulya, A.; Praveen, K. Towards the detection of phishing attacks. In Proceedings of the 2020 4th International Conference on Trends in Electronics and Informatics (ICOEI) (48184), Tirunelveli, India, 15–17 June 2020; pp. 337–343. [CrossRef]
37. Chapla, H.; Kotak, R.; Joiser, M. A machine learning approach for url based web phishing using fuzzy logic as classifier. In Proceedings of the 2019 International Conference on Communication and Electronics Systems (ICCES), Coimbatore, India, 17–19 July 2019; pp. 383–388. [CrossRef]
38. Zubair Hasan, K.M.; Hasan, M.Z.; Zahan, N. Automated prediction of phishing websites using deep convolutional neural network. In Proceedings of the 2019 International Conference on Computer, Communication, Chemical, Materials and Electronic Engineering (IC4ME2), Rajshahi, Bangladesh, 11–12 July 2019; pp. 1–4. [CrossRef]
39. Kunju, M.V.; Dainel, E.; Anthony, H.C.; Bhelwa, S. Evaluation of phishing techniques based on machine learning. In Proceedings of the 2019 International Conference on Intelligent Computing and Control Systems (ICCS), Madurai, India, 15–17 May 2019; pp. 963–968. [CrossRef]
40. Zheng, K.; Wu, T.; Wang, X.; Wu, B.; Wu, C. A Session and Dialogue-Based Social Engineering Framework. *IEEE Access* **2019**, *7*, 67781–67794. [CrossRef]
41. Joshi, C.; Aliaga, J.R.; Insua, D.R. Insider Threat Modeling: An Adversarial Risk Analysis Approach. *IEEE Trans. Inf. Forensics Secur.* **2020**, *16*, 1131–1142. [CrossRef]
42. Duncan, A.; Creese, S.; Goldsmith, M. A combined attack-tree and kill-chain approach to designing attack-detection strategies for malicious insiders in cloud computing. In Proceedings of the 2019 International Conference on Cyber Security and Protection of Digital Services (Cyber Security), Oxford, UK, 3–4 June 2019; pp. 1–9. [CrossRef]
43. Khan, A.Y.; Latif, R.; Latif, S.; Tahir, S.; Batool, G.; Saba, T. Malicious Insider Attack Detection in IoTs Using Data Analytics. *IEEE Access* **2019**, *8*, 11743–11753. [CrossRef]
44. Alshamrani, A.; Myneni, S.; Chowdhary, A.; Huang, D. A Survey on Advanced Persistent Threats: Techniques, Solutions, Challenges, and Research Opportunities. *IEEE Commun. Surv. Tutor.* **2019**, *21*, 1851–1877. [CrossRef]

45. Su, Y. Research on apt attack based on game model. In Proceedings of the 2020 IEEE 4th Information Technology, Networking, Electronic and Automation Control Conference (ITNEC), Chongqing, China, 12–14 June 2020; Volume 1, pp. 295–299. [CrossRef]
46. Khosravi, M.; Ladani, B.T. Alerts Correlation and Causal Analysis for APT Based Cyber Attack Detection. *IEEE Access* **2020**, *8*, 162642–162656. [CrossRef]
47. Sajal, S.Z.; Jahan, I.; Nygard, K.E. A survey on cyber security threats and challenges in modem society. In Proceedings of the 2019 IEEE International Conference on Electro Information Technology (EIT), Brookings, SD, USA, 20–22 May 2019; pp. 525–528. [CrossRef]
48. Park, J.; Cho, D.; Lee, J.K.; Lee, B. The Economics of Cybercrime. *ACM Trans. Manag. Inf. Syst.* **2019**, *10*, 1–23. [CrossRef]
49. Cao, M.; Badihi, S.; Ahmed, K.; Xiong, P.; Rubin, J. On benign features in malware detection. In Proceedings of the 35th IEEE/ACM International Conference on Automated Software Engineering ASE'20, Virtual Event, Australia, 21–25 December 2020; Association for Computing Machinery: New York, NY, USA; pp. 1234–1238. [CrossRef]
50. Samantray, O.P.; Tripathy, S.N.; Das, S.K. A study to understand malware behavior through malware analysis. In Proceedings of the 2019 IEEE International Conference on System, Computation, Automation and Networking (ICSCAN), Pondicherry, India, 29–30 March 2019; pp. 1–5. [CrossRef]
51. Vishwakarma, R.; Jain, A.K. A survey of DDoS attacking techniques and defence mechanisms in the IoT network. *Telecommun. Syst.* **2019**, *73*, 3–25. [CrossRef]
52. Liang, X.; Znati, T. An empirical study of intelligent approaches to ddos detection in large scale networks. In Proceedings of the 2019 International Conference on Computing, Networking and Communications (ICNC), Honolulu, HI, USA, 18–21 February 2019; pp. 821–827. [CrossRef]
53. Priya, S.S.; Sivaram, M.; Yuvaraj, D.; Jayanthiladevi, A. Machine learning based ddos detection. In Proceedings of the 2020 International Conference on Emerging Smart Computing and Informatics (ESCI), Pune, India, 12–14 March 2020; pp. 234–237. [CrossRef]
54. Nandi, S.; Phadikar, S.; Majumder, K. Detection of ddos attack and classification using a hybrid approach. In Proceedings of the 2020 Third ISEA Conference on Security and Privacy (ISEA-ISAP), Guwahati, India, 27 February–1 March 2020; pp. 41–47. [CrossRef]
55. Rohit, M.H.; Fahim, S.M.; Khan, A.H.A. Mitigating and detecting ddos attack on iot environment. In Proceedings of the 2019 IEEE International Conference on Robotics, Automation, Artificial-intelligence and Internet-of-Things (RAAICON), Dhaka, Bangladesh, 29 November–1 December 2019; pp. 5–8. [CrossRef]
56. Agrawal, N.; Tapaswi, S. Defense Mechanisms Against DDoS Attacks in a Cloud Computing Environment: State-of-the-Art and Research Challenges. *IEEE Commun. Surv. Tutor.* **2019**, *21*, 3769–3795. [CrossRef]
57. Bijitha, C.V.; Sukumaran, R.; Nath, H.V. A survey on ransomware detection techniques. In *Secure Knowledge Management in Artificial Intelligence Era*; Sahay, S.K., Goel, N., Patil, V., Jadliwala, M., Eds.; Springer: Singapore, 2020; pp. 55–68.
58. Alzahrani, A.; Alshahrani, H.; Alshehri, A.; Fu, H. An intelligent behavior-based ransomware detection system for android platform. In Proceedings of the 2019 First IEEE International Conference on Trust, Privacy and Security in Intelligent Systems and Applications (TPS-ISA), Los Angeles, CA, USA, 12–14 December 2019; pp. 28–35. [CrossRef]
59. Adamov, A.; Carlsson, A. Reinforcement learning for anti-ransomware testing. In Proceedings of the 2020 IEEE East-West Design Test Symposium (EWDTS), Varna, Bulgaria, 4–7 September 2020; pp. 1–5. [CrossRef]
60. Bahrani, A.; Bidgly, A.J. Ransomware detection using process mining and classification algorithms. In Proceedings of the 2019 16th International ISC (Iranian Society of Cryptology) Conference on Information Security and Cryptology (ISCISC), Mashhad, Iran, 28–29 August 2019; pp. 73–77. [CrossRef]
61. Shahpasand, M.; Hamey, L.; Vatsalan, D.; Xue, M. Adversarial attacks on mobile malware detection. In Proceedings of the 2019 IEEE 1st International Workshop on Artificial Intelligence for Mobile (AI4Mobile), Hangzhou, China, 24 February 2019; pp. 17–20. [CrossRef]
62. Tahtaci, B.; Canbay, B. Android malware detection using machine learning. In Proceedings of the 2020 Innovations in Intelligent Systems and Applications Conference (ASYU), Istanbul, Turkey, 15–17 October 2020; pp. 1–6. [CrossRef]
63. Mbaziira, A.V.; Diaz-Gonzales, J.; Liu, M. Deep learning in detection of mobile malware. *J. Comput. Sci. Coll.* **2020**, *36*, 80–88.
64. Diaz-Gonzalez, J.; Mbaziira, A.V.; Liu, M. An exploratory deep learning approach to mobile malware detection. *J. Comput. Sci. Coll.* **2019**, *35*, 219.
65. Allen, J.; Yang, Z.; Landen, M.; Bhat, R.; Grover, H.; Chang, A.; Ji, Y.; Perdisci, R.; Lee, W. Mnemosyne: An effective and efficient postmortem watering hole attack investigation system. In Proceedings of the 2020 ACM SIGSAC Conference on Computer and Communications Security, CCS'20, Virtual Event, USA, 9–13 November 2020; Association for Computing Machinery: New York, NY, USA, 2020; pp. 787–802. [CrossRef]
66. Research CP. CYBER ATTACK TRENDS: 2020 MID-YEAR REPORT—Check Point Research. 2021. Available online: https://research.checkpoint.com/2020/cyber-attack-trends-2020-mid-year-report/ (accessed on 21 May 2021).
67. Dasgupta, D.; Akhtar, Z.; Sen, S. Machine learning in cybersecurity: A comprehensive survey. *J. Déf. Model. Simul. Appl. Methodol. Technol.* **2020**, *19*, 57–106. [CrossRef]
68. Brewer, J.N.; Dimitoglou, G. Evaluation of attack vectors and risks in automobiles and road infrastructure. In Proceedings of the 2019 International Conference on Computational Science and Computational Intelligence (CSCI), Las Vegas, NV, USA, 5–7 December 2019; pp. 84–89. [CrossRef]

69. Li, T.; Wang, K.; Horkoff, J. Towards effective assessment for social engineering attacks. In Proceedings of the 2019 IEEE 27th International Requirements Engineering Conference (RE), Jeju, Korea, 23–27 September 2019; pp. 392–397. [CrossRef]
70. Tandale, K.D.; Pawar, S.N. Different types of phishing attacks and detection techniques: A review. In Proceedings of the 2020 International Conference on Smart Innovations in Design, Environment, Management, Planning and Computing (ICSIDEMPC), Aurangabad, India, 30–31 October 2020; pp. 295–299. [CrossRef]
71. Badami, C.; Kettani, H. On Malware Detection in the Android Operating System. In Proceedings of the 4th International Conference on Algorithms, Computing and Systems, Rabat, Morocco, 6–8 January 2020; Association for Computing Machinery: New York, NY, USA, 2020; pp. 45–50. [CrossRef]
72. George, R.; Jalal, R.; Raju, R.M.; Sunny, S.S.; Hari, M. High responsive plug-in for malicious url detection. In Proceedings of the 2019 3rd International Conference on Trends in Electronics and Informatics (ICOEI), Tirunelveli, India, 23–25 April 2019; pp. 357–359. [CrossRef]
73. Song, X.; Chen, C.; Cui, B.; Fu, J. Malicious JavaScript Detection Based on Bidirectional LSTM Model. *Appl. Sci.* **2020**, *10*, 3440. [CrossRef]
74. Dogaru, D.I.; Dumitrache, I. Cyber security of smart grids in the context of big data and machine learning. In Proceedings of the 2019 22nd International Conference on Control Systems and Computer Science (CSCS), Bucharest, Romania, 28–30 May 2019; pp. 61–67. [CrossRef]
75. Spanakis, E.G.; Bonomi, S.; Sfakianakis, S.; Santucci, G.; Lenti, S.; Sorella, M.; Tanasache, F.D.; Palleschi, A.; Ciccotelli, C.; Sakkalis, V.; et al. Cyber-attacks and threats for healthcare—A multi-layer thread analysis. In Proceedings of the 2020 42nd Annual International Conference of the IEEE Engineering in Medicine Biology Society (EMBC), Montreal, QC, Canada, 20–24 July 2020; pp. 5705–5708. [CrossRef]
76. Pilz, M.; Naeini, F.B.; Grammont, K.; Smagghe, C.; Davis, M.; Nebel, J.-C.; Al-Fagih, L.; Pfluegel, E. Security attacks on smart grid scheduling and their defences: A game-theoretic approach. *Int. J. Inf. Secur.* **2019**, *19*, 427–443. [CrossRef]
77. Canaan, B.; Colicchio, B.; Abdeslam, D.O. Microgrid Cyber-Security: Review and Challenges toward Resilience. *Appl. Sci.* **2020**, *10*, 5649. [CrossRef]
78. Tervoort, T.; De Oliveira, M.T.; Pieters, W.; Van Gelder, P.; Olabarriaga, S.D.; Marquering, H. Solutions for Mitigating Cybersecurity Risks Caused by Legacy Software in Medical Devices: A Scoping Review. *IEEE Access* **2020**, *8*, 84352–84361. [CrossRef]
79. Fatima, K.; Nawaz, S.; Mehrban, S. Biometric authentication in health care sector: A survey. In Proceedings of the 2019 International Conference on Innovative Computing (ICIC), Lahore, Pakistan, 1–2 November 2019; pp. 1–10. [CrossRef]
80. Kazemi, Z.; Fazeli, M.; Hely, D.; Beroulle, V. Hardware security vulnerability assessment to identify the potential risks in a critical embedded application. In Proceedings of the 2020 IEEE 26th International Symposium on On-Line Testing and Robust System Design (IOLTS), Napoli, Italy, 13–15 July 2020; pp. 1–6. [CrossRef]
81. Kotenko, I.; Saenko, I.; Kushnerevich, A.; Branitskiy, A. Attack detection in iot critical infrastructures: A machine learning and big data processing approach. In Proceedings of the 2019 27th Euromicro International Conference on Parallel, Distributed and Network-Based Processing (PDP), Pavia, Italy, 13–15 February 2019; pp. 340–347. [CrossRef]
82. Sen, S.; Jayawardena, C. Analysis of cyber-attack in big data iot and cyber-physical systems—A technical approach to cybersecurity modeling. In Proceedings of the 2019 IEEE 5th International Conference for Convergence in Technology (I2CT), Bombay, India, 29–31 March 2019; pp. 1–7. [CrossRef]
83. Neshenko, N.; Bou-Harb, E.; Crichigno, J.; Kaddoum, G.; Ghani, N. Demystifying IoT Security: An Exhaustive Survey on IoT Vulnerabilities and a First Empirical Look on Internet-Scale IoT Exploitations. *IEEE Commun. Surv. Tutor.* **2019**, *21*, 2702–2733. [CrossRef]
84. Gressl, L.; Steger, C.; Neffe, U. Consideration of security attacks in the design space exploration of embedded systems. In Proceedings of the 2019 22nd Euromicro Conference on Digital System Design (DSD), Kallithea, Greece, 28–30 August 2019; pp. 530–537. [CrossRef]
85. Chauhan, V.; Arora, G. A review paper on cryptocurrency portfolio management. In Proceedings of the 2019 2nd International Conference on Power Energy, Environment and Intelligent Control (PEEIC), Greater Noida, India, 18–19 October 2019; pp. 60–62. [CrossRef]
86. Shahnaz, A.; Qamar, U.; Khalid, A. Using Blockchain for Electronic Health Records. *IEEE Access* **2019**, *7*, 147782–147795. [CrossRef]
87. Gong, X.; Liu, E.; Wang, R. Blockchain-based iot application using smart contracts: Case study of m2m autonomous trading. In Proceedings of the 2020 5th International Conference on Computer and Communication Systems (ICCCS), Shanghai, China, 15–18 May 2020; pp. 781–785. [CrossRef]
88. Chen, Y.-W.; Sheu, J.-P.; Kuo, Y.-C.; Van Cuong, N. Design and implementation of iot ddos attacks detection system based on machine learning. In Proceedings of the 2020 European Conference on Networks and Communications (EuCNC), Dubrovnik, Croatia, 15–18 June 2020; pp. 122–127. [CrossRef]
89. Ahmed, Z.; Danish, S.M.; Qureshi, H.K.; Lestas, M. Protecting iots from mirai botnet attacks using blockchains. In Proceedings of the 2019 IEEE 24th International Workshop on Computer Aided Modeling and Design of Communication Links and Networks (CAMAD), Limassol, Cyprus, 11–13 September 2019; pp. 1–6. [CrossRef]
90. Sambangi, S.; Gondi, L. A machine learning approach for ddos (distributed denial of service) attack detection using multiple linear regression. *Proceedings* **2020**, *63*, 51.
91. Sai, A.M.V.V.; Li, Y. A Survey on Privacy Issues in Mobile Social Networks. *IEEE Access* **2020**, *8*, 130906–130921. [CrossRef]

92. Hijji, M.; Alam, G. A Multivocal Literature Review on Growing Social Engineering Based Cyber-Attacks/Threats During the COVID-19 Pandemic: Challenges and Prospective Solutions. *IEEE Access* **2021**, *9*, 7152–7169. [CrossRef] [PubMed]
93. MITRE. Enterprise Attacks Techniques MITRE. 2021. Available online: https://attack.mitre.org/techniques/enterprise/ (accessed on 25 May 2021).
94. Wankhede, S.B. Anomaly detection using machine learning techniques. In Proceedings of the 2019 IEEE 5th International Conference for Convergence in Technology (I2CT), Bombay, India, 29–31 March 2019; pp. 1–3. [CrossRef]
95. Wollaber, A.; Peñna, J.; Blease, B.; Shing, L.; Alperin, K.; Vilvovsky, S.; Trepagnier, P.; Wagner, N.; Leonard, L. Proactive cyber situation awareness via high performance computing. In Proceedings of the 2019 IEEE High Performance Extreme Computing Conference (HPEC), Waltham, MA, USA, 24–26 September 2019; pp. 1–7. [CrossRef]
96. Iqbal, A.; Gunn, L.J.; Guo, M.; Babar, M.A.; Abbott, D. Game Theoretical Modelling of Network/Cybersecurity. *IEEE Access* **2019**, *7*, 154167–154179. [CrossRef]
97. Zhou, X.; Cheng, S.; Liu, Y. A Cooperative Game Theory-Based Algorithm for Overlapping Community Detection. *IEEE Access* **2020**, *8*, 68417–68425. [CrossRef]
98. Kwiatkowska, M.; Norman, G.; Parker, D.; Santos, G. Prism-games 3.0: Stochastic game verification with concurrency, equilibria and time. In *Computer Aided Verification*; Lahiri, S.K., Wang, C., Eds.; Springer: Cham, Switzerland, 2020; pp. 475–487.
99. Bitzer, S.; Park, H.; Blankenburg, F.; Kiebel, S.J. Perceptual decision making: Drift-diffusion model is equivalent to a Bayesian model. *Front. Hum. Neurosci.* **2014**, *8*, 102. [CrossRef]
100. Dale, S. Heuristics and biases: The science of decision-making. *Bus. Inf. Rev.* **2015**, *32*, 93–99. [CrossRef]
101. Nikolić, J. Biases in the decision-making process and possibilities of overcoming them. *Èkon. Horiz.* **2018**, *20*, 45–59. [CrossRef]
102. Trueblood, J.S.; Holmes, W.R.; Seegmiller, A.C.; Douds, J.; Compton, M.; Szentirmai, E.; Woodruff, M.; Huang, W.; Stratton, C.; Eichbaum, Q. The impact of speed and bias on the cognitive processes of experts and novices in medical image decision-making. *Cogn. Res. Princ. Implic.* **2018**, *3*, 28. [CrossRef]
103. Smith, P.; Ratcliff, R. An introduction to the diffusion model of decision making. In *An Introduction to Model-Based Cognitive Neuroscience*; Springer: New York, NY, USA, 2015; pp. 49–70. [CrossRef]
104. Nishiguchi, Y.; Sakamoto, J.; Kunisato, Y.; Takano, K. Linear Ballistic Accumulator Modeling of Attentional Bias Modification Revealed Disturbed Evidence Accumulation of Negative Information by Explicit Instruction. *Front. Psychol.* **2019**, *10*, 2447. [CrossRef]
105. Quayyum, F.; Cruzes, D.S.; Jaccheri, L. Cybersecurity awareness for children: A systematic literature review. *Int. J. Child Comput. Interact.* **2021**, *30*, 100343. [CrossRef]
106. Cheung, K.-F.; Bell, M.G.; Bhattacharjya, J. Cybersecurity in logistics and supply chain management: An overview and future research directions. *Transp. Res. Part E: Logist. Transp. Rev.* **2021**, *146*, 102217. [CrossRef]
107. Hassija, V.; Chamola, V.; Saxena, V.; Jain, D.; Goyal, P.; Sikdar, B. A Survey on IoT Security: Application Areas, Security Threats, and Solution Architectures. *IEEE Access* **2019**, *7*, 82721–82743. [CrossRef]
108. Wang, Z.; Zhu, H.; Sun, L. Social Engineering in Cybersecurity: Effect Mechanisms, Human Vulnerabilities and Attack Methods. *IEEE Access* **2021**, *9*, 11895–11910. [CrossRef]
109. Rahman, A.U.; Williams, L. Software Security in DevOps: Synthesizing Practitioners' Perceptions and Practices. In Proceedings of the 2016 IEEE/ACM International Workshop on Continuous Software Evolution and Delivery (CSED), Austin, TX, USA, 14–15 May 2016; pp. 70–76.
110. Maglaras, L.; Drivas, G.; Chouliaras, N.; Boiten, E.; Lambrinoudakis, C.; Ioannidis, S. Cybersecurity in the Era of Digital Transformation: The case of Greece. In Proceedings of the 2020 International Conference on Internet of Things and Intelligent Applications (ITIA), Zhenjiang, China, 27–29 November 2020; pp. 1–5. [CrossRef]
111. Sharma, R.; Dangi, S.; Mishra, P. A Comprehensive Review on Encryption based Open Source Cyber Security Tools. In Proceedings of the 2021 6th International Conference on Signal Processing, Computing and Control (ISPCC), Solan, India, 7–9 October 2021; pp. 614–619. [CrossRef]

electronics

Article

Threat Matrix: A Fast Algorithm for Human–Machine Chinese Ludo Gaming

Fuji Han and Man Zhou *

School of Cyber Science and Engineering, Huazhong University of Science and Technology, Wuhan 430074, China; u201911105@hust.edu.cn
* Correspondence: zhou_man1125@hust.edu.cn

Abstract: Chinese Ludo, also known as Aeroplan Chess, has been a very popular board game for several decades. However, there is no mature algorithm existing for human–machine gambling. The major challenge is the high randomness of the dice rolls, where the algorithm must ensure that the machine is smarter than a human in order to guarantee that the owner of the game machines makes a profit. This paper presents a fast Chinese Ludo algorithm (named "Threat Matrix") that we have recently developed. Unlike from most chess programs, which rely on high performance computing machines, the evaluation function in our program is only a linear sum of four factors. For fast and low-cost computation, we innovatively construct the concept of the threat matrix, by which we can easily obtain the threat between any two dice on any two positions. The threat matrix approach greatly reduces the required amount of calculations, enabling the program to run on a 32-bit 80 × 86 SCM with a 100 MHz CPU while supporting a recursive algorithms to search plies. Statistics compiled from matches against human game players show that our threat matrix has an average win rate of 92% with no time limit, 95% with a time limit of 10 s, and 98% with a time limit of 5 s. Furthermore, the threat matrix can reduce the computation cost by nearly 90% compared to real-time computing; memory consumption drops and is stable, which increases the evaluation speed by 58% compared to real-time computing.

Keywords: game software; threat matrix computing; evaluation function; data modeling

Citation: Han, F.; Zhou, M. Threat Matrix: A Fast Algorithm for Human–Machine Chinese Ludo Gaming. *Electronics* **2022**, *11*, 1699. https://doi.org/10.3390/electronics11111699

Academic Editor: Krzysztof Szczypiorski

Received: 17 April 2022
Accepted: 23 May 2022
Published: 26 May 2022

1. Introduction

Board games are a form of intellectual confrontation invented over the course of a long history of production activities [1,2]; in addition, they are important applications that can help to explore and promote research into artificial intelligence [3]. The selection of pieces depends on a perfect evaluation function with a computation complexity of approximately b^d, where b is the number of legal moves per position and d is the number of positions of the chessboard. Because different types of board games have different b and d, the total amount of calculations varies greatly [4,5]. The evaluation algorithms of chess programs involve various artificial intelligence (AI) techniques, including machine learning, expert systems, neural networks, search trees, etc. [6,7]. The critical jobs that most game software focuses on consist of two aspects: (1) how to define and obtain comprehensive evaluation factors; and (2) how to construct a perfect evaluation function.

The advance of AI chess can be attributed to two factors: (1) high performance computing devices [8]; and (2) powerful evaluation algorithms [9]. In recent years, with the rapid development of high-performance computing and network technologies, as long as a suitable evaluation algorithm can be constructed a computer will be able process a very large number of calculations. With the recent rapid development of high-performance computer and network technologies, as long as the evaluating algorithm is properly constructed the huge amount of computation can be handled by machines [10,11]. Because machines are accurate, fast, and not influenced by emotion, they often beat human players.

At present, computer players of classic board games such as chess, Go, Shogi [4], and Checkers [12] can beat expert human players. R. Ruben et al. [13] have described a general video game AI (GVGAI) that provides a way to benchmark AI algorithms for games written in specific description languages for many application domains. The continuous promotion of AI algorithms from games to reality is of great significance for the development of both robot swarm intelligent collaboration and gaming [14,15].

The world chess champion Garry Kasparov was defeated by IBM's Deep Blue computer, which has become a milestone in the advance of artificial intelligence over human beings. Deep Blue typically searches to a depth of between six and eight moves up to a maximum of twenty moves, or even more in certain positions [16]. This brute force type of search can benefit to a degree from advances in computer performance. However, moving from brute force to informed search is a great endeavour in AI board game research. A recent study by Ansk involved an idea to solve two-player zero-sum and extensive-form games [17]. They were able obtain perfect information and simultaneous moves using the Double-Oracle Method and Serialized Alpha-Beta Search. In addition, Branislav et al. [18] are devoted to developing synchronous move game algorithms and evaluating exact backward induction methods with efficient pruning and sampling algorithms under different settings. David B. and his group focus on evolutionary algorithms, while Sebastian presents a rules learning method for general game playing. Thus, this great endeavour requires a combination of methods from a variety of subdisciplines [19,20].

Due to the enormous search space needed for the combination of large moves, the development of a program that can beat humans at the ancient Chinese chess-like game Go has been considered as one of the last great challenges [21,22]. Tesauro's approach is already considered to outperforms the best human players in playing backgammon. AlphaGo [3], developed by Google DeepMind, has beaten a top professional human Go player without handicaps on a full-sized 1919 board. AlphaGo's algorithm uses a combination of machine learning and search tree techniques combined with extensive training on both human and computer play [23].

Similar to chess, Go, and backgammon, Chinese Ludo gaming, known as Aeroplan Chess, was invented in China by reference to other ancient board games to commemorate the Flying Tigers in World War II; the game's rules and paths are derived from Lufbery circles in air warfare [24]. To encourage students to master chemical equations in the game, a new flying chess game called CHEMTrans has been developed.

In recent years, many online programs for Chinese Ludo gaming have been developed. These simply provide a platform for interaction between game players, and the computer does not engage in intelligent competition [25]. Developing a human–machine Chinese Ludo gambling game, while an attractive idea, represents a highly challenging task. Apart from the high randomness of the dice rolling, the algorithm must ensure that the machine is smarter than humans in order to guarantee that the owner of the game machines makes a profit. As a game requires 2–4 players and random dice rolling, no such gaming machine has been developed [26,27].

It is an interesting idea and a challenging task to adapt Chinese Ludo to the gambling industry. Despite the randomness of the game, the game machines must be "smarter" than people to ensure that the game owners make money. The primary contributions in this paper are as follows.

- We develop a human–machine gambling Chinese ludo game which is unlike most other chessboard programs; while these involve brute-force algorithms that require high-performance machines, our evaluation function is a simple linear sum of four variables.
- For high-frequency threat computation between any two dice, we innovatively propose the concept of a 'threat matrix'. Using the threat matrix, our program can instantly acquire the threat value between any two positions, rather than relying on brute-force computing based on the current positions. As a result, our program can reduce the real-time (immediate) computation cost by nearly 88%.

- Threat matrix can run on an 80 × 86 SCM (Single-Chip Microcomputer), and has achieved a 97% winning rate against human players.

The rest of this paper is organized as follows. Section 2 briefly introduces the rules of the Chinese Ludo game and reviews related existing research works. Section 3 provides details on the evaluation function, threat matrix, and our proposed algorithms. To clearly show how to calculate the evaluation function, Section 4 presents a concrete example. Section 5 analyses and compares the complexity of the main algorithms. Section 6 presents experimental validations and shows comparisons of various strategies. Section 7 concludes the paper.

2. Chinese Ludo Game

In this section, for convenience in understanding our algorithms we first briefly introduce the rules of improved man–machine gambling Chinese Ludo by the Lanhai Technology Company. Note that while the rules may be slightly different among different versions, the main rules are always the same.

1. Composition
   (a) Dice: One traditional dice, with each of its six faces showing a different number of dots from 1 to 6.
   (b) Board: One board, as shown in Figure 1, with 64 square positions marked with different colors. Each player is assigned a color, either red or blue. The top left corner and the right bottom corner are staging areas ('hangars') and the two closed rectangles colored with red and blue in the center are the ending squares corresponding to each game player.
   (c) Planes: Each player has four planes, in colors matching those of the board. There are two players, one human and one machine, and eight planes in total.

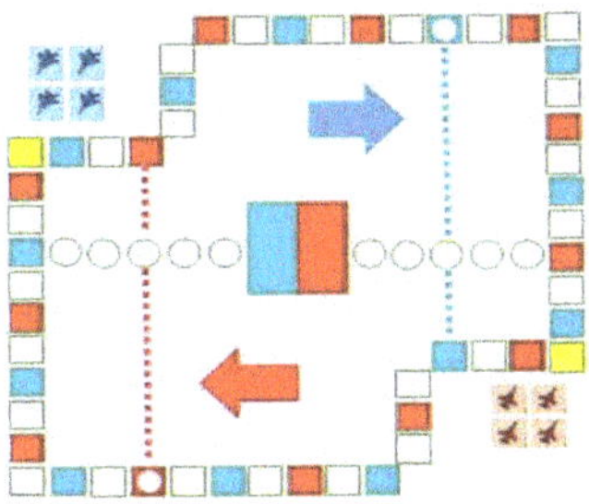

Figure 1. The chessboard of the Chinese Ludo game.

2. Game rules
   (a) Launch: At the start of the game, all of the planes of the two players are in the staging area. After randomly determine a player to roll the dice first, the players take turns rolling the dice. Only when a player rolls a 6 can s/he select one of her/his planes from its staging area to prepare to fly (in its starting square); the planes fly in the clockwise direction.
   (b) Move: When a player has one or more planes in play, s/he selects one to move (fly) over squares of the number shown on the dice. If the landing square is the same color as the plane's color, the plane can fly directly to the next square in the same color; however, this cannot be done with circles, which is called a 'jump'. Furthermore, if the landing square has the same color and is connected with a dotted line, the plane can fly along the dotted line, which is called a 'dotfly'. The rolling of a 6 earns the player an additional roll in that turn, which can repeat until the roll is not 6.

(c) Hit: If the landing square is occupied by the opponent's plane(s), the opponent's plane(s), either one or more, is hit and forced to return to the staging area. Planes of the same color are not hit.

(d) Win: The first player who is able to fly all of her/his planes to the ending square wins the game.

3. Evaluation Function

According to the rules of the Chinese Ludo game, the number that the dice shows after rolling is random, and the player has to decide which of her/his planes at different positions to select. Thus, in addition to luck, for a given position winning or losing depends on the selection of planes.

Compared to traditional chess games, the randomness of the dice in Chinese Ludo brings uncertainty to the evaluation of potential moves. Additionally, multi-layered nesting judgement resulting from the rules can introduce looping and crashing during the game. These two key difficulties may explain why there is no mature algorithm for Chinese Ludo gaming suitable for human–machine gaming.

Without loss of generality, let us suppose the computer is blue and the opponent is red. In the rest of the paper, we use blue for computers and red for their human opponents.

3.1. The Description and Definition of the Chess Position

To facilitate the evaluation of the planes, we first define the chess position. First, the squares in the chessboard are numbered as shown in Figure 2. For planes of the same player, the only difference is their position (i.e., different squares). Thus, we can use the number to represent each plane's position on the game board. The position of each of the four blue planes is denoted by X_i with $(i = 1, 2, 3, 4)$. Similarly, the position of each of the red planes is denoted by Y_j with $(j = 1, 2, 3, 4)$. We use $S(X_1, X_2, X_3, X_4, Y_1, Y_2, Y_3, Y_4)$ to represent the eight planes' positions on the board and R to represent the number that the dice shows after rolling. Thus, (S, R) can determine a specific Chinese Ludo position.

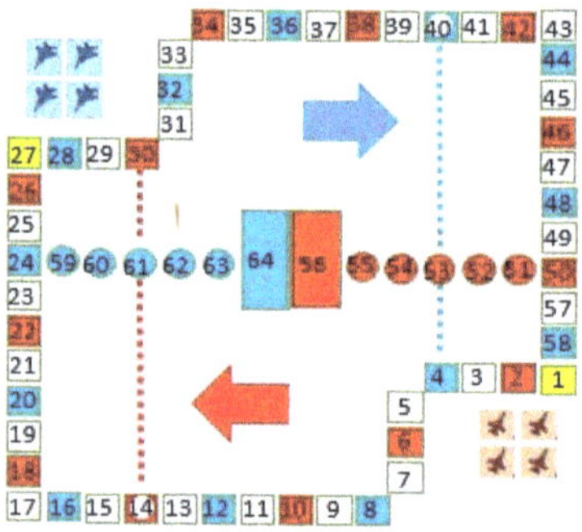

Figure 2. The definition of a chess position on the chessboard.

3.2. Evaluation Function

Next, we define the value of a position, denoted as $I(X_i)$, which is the number of squares from the starting square to the current position; $I(X_i)$ varies with the change of X_i, and the closer it is to the ending square, the greater it is. Note that the position of blue restarts from 1 after a modulo operation of 26.

At a chess position (S, R) we need to define an evaluation function, f, in order to determine which plane to move. Obviously, f is determined by position (S, R); thus, we use $f_i(S, R)(i = 1, 2, 3, 4)$ to represent the evaluation function of the i-th blue plane at position (S, R). By computing and comparing $f_i(i = 1, 2, 3, 4)$, we can make our decision as to which plane to move. Certainly, $f_i(S, R)$ can be used to evaluate the red/opponent's planes as well.

Considering the game rules, we analyze the four key factors that affect $f_i(S,R)$, as follows.

First, according to its rules (see Section 2), Chinese Ludo allows Jump and Dotfly moves. Therefore, planes at different squares may move different distances with the same dice roll. This is referred to as 'Incremented Distance', denoted by V_1. Without considering other factors, players intend to choose the plane that can move the longest distance.

Second, we consider Hits, i.e., whether, after moving, a blue plane lands on a square occupied by a red plane and the red plane is hit back to its staging area. We call this the 'Opponent's Hit Value', denoted by V_2. Clearly, V_2 greatly reduces the chances of the opponent/red player winning and increases the chance that the blue player wins.

Third, if after moving a plane does not hit the opponent's plane on the current roll, it may hit the opponent's plane on the next roll. This means that a plane's threat to the opponent's planes changes with each move. Certainly, players intend to choose the plane which offers the largest increase in the threat, which we call the 'Increased Threat Value of Blue on Red', denoted by T_1.

In the same way, the threat represented by the opponent's plane changes after each move. We aim to choosing the plane which will result in the smallest increase in the opponent's threat, which we call the 'Increased Threat Value of Red on Blue', denoted by T_2.

Through the analysis of the above four factors, it can be seen that all four values actually reflect changes in position, and position is the ultimate determination of which player wins or loses the game. Thus, the four values are linearly correlated with the evaluation function, $f_i(S,R)(i=1,2,3,4)$, where the evaluation function is defined as

$$f_i(S,R) = V_1 + V_2 + T_1 - T_2.(i = 1,2,3,4) \tag{1}$$

By calculating each $f_i(S,R)(i=1,2,3,4)$ according to Equation (1), we are able choose the plane with the greatest $f_i(S,R)$. Next, we provide details on how to calculate the four values. For the convenience of the reader, we list the main symbols used in this paper below in Table 1.

Table 1. Notations of the main symbols.

X_i	The position of the i-th blue plane
Y_j	The position of the opponent's j-th plane in red
$I(X_i)$	The number of squares from the starting square to the current position of X_i
$p(X_i,Y_j)$	Possibility of plane X_i to hit plane Y_j, that is, threat of X_i to Y_j
P	The threat matrix, $P=(p_{ij})_{64\times 64}$, where $p_{ij}=p(X_i,Y_j)$
V_1	Incremented distance
V_2	Opponent's hit value
T_1	Increased threat value of blue on red
T_2	Increased threat value of red on blue
O	Computational complexity

3.2.1. Incremented Distance after Moving (V_1)

According to the above definition of V_1, at position (S,R), if blue's i-th plane is on X_i and the number on the dice is R, according to the game rules, there are three cases for X_i to move:

1. Regular Move: this is the most common case, where the i-th plane moves forward for R steps; therefore, $V_1 = R$.
2. Jump: if $X_i + R = 4, 8, 12, 16, 20, 24, 28, 32, 36, 40, 44, 48$, or 58, that is, squares which have the same color (here, blue), X_i can move four more steps and jump to the next square of the same color, i.e., $V_1 = R + 4$.
3. Dotfly: if $X_i + R = 36$ or 40, i.e., two special positions, it can fly along the dotted line, i.e.,

$$V_1 = R + 4 + 16 = R + 20, \tag{2}$$

thus,

$$V_1 = R||(R+4)||(R+20) \tag{3}$$

where "||" means the logical operation "OR".

3.2.2. Opponent's Hit Value (V_2)

According to the rules, a plane can hit an opponent's plane if the square on which it lands is occupied by the opponent's plane.

1. If $X_i + R = Y_j$, i.e., the current position plus the dice number happens to be the current position of the opponent, the opponent's plane is hit back to its staging area;
2. If $X_i + V_1 = Y_j$, i.e., after the jump the blue plane can hit the opponent's plane, red loses the position, i.e., V_2.

For the above two cases, V_2 is the position value of the opponent's plane that is lost by the hit; otherwise, the opposing plane has no value to lose:

$$V_2 = \begin{cases} I(Y_j), & \text{if } X_i + R = Y_j \text{ or } X_i + V_1 = Y_j \\ 0, & \text{otherwise.} \end{cases} \tag{4}$$

The reason we calculate V_1 and V_2 separately from T_1 and T_2 is that the action of V_2 is a definite effect; the valuation of T_1 and T_2 is based on the randomness of the dice, which is a probabilistic event belonging to the threat, which we introduce below.

3.2.3. Increased Threat Value of Blue on Red (T_1)

In addition to the specific threat of blue to red V_2, we must consider the change in the overall threat due to moving.

At a (S, R), assume a red plane at Y_j has the position value $I(Y_j)$ and the positions of a blue plane before and after moving are X_i and $X_i + V_1$, respectively. We use $p(X_i, Y_j)$ and $p(X_i + V_1, Y_j)$ to represent the respective threat of the blue plane to the red plane before and after the move. Then, their difference, $p(X_i + V_1, Y_j) - p(X_i, Y_j)$, can denote the increasing threat of X_i to Y_j. If we let the difference multiply the position value of Y_j, we obtain the increased threat value of of X_i to Y_j, i.e., $T_1 = (p(X_i + V_1, Y_j) - p(X_i, Y_j)) \times I(Y_j)$.

Each player has four planes; thus, before moving, the respective overall threat of the blue plane compared to the four red planes is the sum of the product of $p(X_i, Y_j)$ and $I(Y_j)$,

$$\sum_{j=1}^{4} p(X_i, Y_j) \times I(Y_j). \tag{5}$$

After moving, the respective overall threat of the blue plane compared to the four red planes is the sum of the product of $p(X_i + V_1, Y_j)$ and $I(Y_j)$,

$$\sum_{j=1}^{4} p(X_i + V_1, Y_j) \times I(Y_j). \tag{6}$$

Thus, the sum of the increased value of the threat of X_i to all $Y_j (j = 1, 2, 3, 4)$ T_1 can be computed as

$$\begin{aligned} T_1 &= \sum_{j=1}^{4} p(X_i + V_1, Y_j) \times I(Y_j) - \sum_{j=1}^{4} p(X_i, Y_j) \times I(Y_j) \\ &= \sum_{j=1}^{4} (p(X_i + V_1, Y_j) - p(X_i, Y_j)) \times I(Y_j). \end{aligned} \tag{7}$$

When $T_1 \geq 0$, which indicates the threat of the blue plane to the red plane, is increasing, it is beneficial for blue; similarly, if $T_1 < 0$, which indicates the threat of the blue plane to the red plane, is decreasing, this is disadvantageous for blue. Thus, we should choose the plane with the largest T_1.

3.2.4. Increased Threat Value of Red on Blue (T_2)

Similar to the above T_1, before and after the move of the blue plane, the threat of the red plane to the blue plane changes. We call the difference between the product of the blue threat to red and the position value of the blue plane before and after the move as the "increased threat value of red on blue", denoted by T_2.

At position (S, R), if the red plane is on Y_j and the blue plane is on X_i and $X_i + V_1$ before and after the blue plane move, the position values of the blue plane before and after the move will be $I(X_i)$ and $I(X_i + V_1)$. Thus, the respective threat of the red plane to the blue plane before and after the move is $p(Y_j, X_i)$ and $p(Y_j, X_i + V_1)$.

Then, the total threat value of the four red planes to the blue plane can be calculated as the product of $p(Y_j, X_i)$ and $I(X_i)$ before the move

$$\sum_{j=1}^{4} p(Y_j, X_i) \times I(X_i). \tag{8}$$

and as the product of $p(Y_j, X_i + V_1)$ and $I(X_i + V_1)$ after the move

$$\sum_{j=1}^{4} p(Y_j, X_i + V_1) \times I(X_i + V_1). \tag{9}$$

Therefore, according to its definition, the threat increment, T_2, is

$$T_2 = \sum_{j=1}^{4} p(Y_j, X_i + V_1) \times I(X_i + V_1) - \sum_{j=1}^{4} p(Y_j, X_i) \times I(X_i). \tag{10}$$

Intuitively, $T_2 \geq 0$ indicates that the threat of the red plane to the blue plane increases and is disadvantageous for blue, while $T_2 < 0$ indicates that the threat to the blue plane decreases and is advantageous to blue.

Based on the above analysis, when entering Equations (3)–(10) into Equation (1), we can obtain $f_i(S, R)$ as follows:

$$\begin{aligned} f_i(S,R) &= V_1 + V_2 + T_1 - T_2 \\ &= (R\|(R+4)\|(R+20)) + (I(Y_j)\|0) + \sum_{j=1}^{4}(p(X_i + V_1, Y_j) - p(X_i, Y_j)) \times I(Y_j) \\ &+ \sum_{j=1}^{4} p(Y_j + X_i + V_1) \times I(X_i + V_1) - p(Y_j, X_i) \times I(X_i). \end{aligned} \tag{11}$$

By computing and comparing each f_i with $i = 1, 2, 3, 4$, we can select the plane which has the largest f_i to move.

Until now, we have not yet shown how to calculate p. As p is the most frequently used variable in the evaluation function, the key to improving the efficiency of this algorithm is calculating p rapidly. Next, we address this problem, then focus on the calculation of p in a separate section.

3.3. *Calculating Threat and Threat Matrix*

3.3.1. Calculating Threat

From a given position, $p(X_i, Y_j)$ is computable according to the rules and the definition of p.

At a position (S, R), suppose the blue plane is on X_i and the red plane is on Y_j. If a hit happens, i.e., $(Y_j = X_i)$, we have $p(Y_j, X_i) = 1$. Sometimes, after moving a plane cannot hit an opponent's plane because the moving distance is too short. Then, we must consider the case of rolling the dice continuously until it is not 6. Next, we use an example to explain how to calculate $p(Y_j, X_i)$.

Assume the opponent's red plane is on Y_j, a blue plane is on X_i, and the number on the die is R. According to the rules, there are three cases we must consider in order to calculate $p(Y_j, X_i)$.

1. If $Y_j < X_i \leq Y_j + 6$, the possibility of the opponent's plane on Y_j hitting blue's plane on X_i by rolling once is
$$p(Y_j, X_i) = 1/6. \tag{12}$$
2. If $Y_j + 6 < X_i \leq Y_j + 12$, the opponent's red plane on Y_j cannot hit blue's plane on X_i by rolling once. For it to be possible to hit, the red player must roll 6 the first time, landing on position $Y_j + 6$, then roll again. Thus, the probability is $p(Y_j, X_i) = 1/6 \times 1/6 = 1/36$. By analogy to more general cases, if rolling 6 up to seven times continuously the opponent's red plane on Y_j does not threaten the blue plane on X_i according to the board. Thus, if $Y_j + 6k < X_i \leq Y_j + 6(k+1)$ $(k = 0, 1, 2, 3, 4, 5, 6, 7)$, then
$$p(Y_j, X_i) = (1/6)^{k+1}. \tag{13}$$
3. Calculate the probabilities for special positions. According to the rules, a plane can jump if the landing square is the same color, such as 2, 6, 10, 14, 18, 22, 26, 30, 34, 38, 42, 46, and 50 for red, which can add the extra possibility of the plane on Y_j jumping and hitting X_i, in addition to the above two cases.
Squares of the same color are separated by three squares. Therefore, after each roll it is possible to land on the same color square. There are thus two cases, as follows.
    (a) When $Y_j < X_i \leq Y_j + 6$ and X_i is on the red square (e.g., $Y_j = 1$ and $X_i = 6$), if the dice roll is 5, the plane on Y_j hits the plane on X_i, thus there is no jump; if the dice roll is 2, the plane on Y_j first moves onto square 2, then jump onto square 6, then hits the plane on X_i. The probability is
    $$p(Y_j, X_i) = 1/6 + 1/6 = 1/3. \tag{14}$$
    (b) When $Y_j + 6k < X_i \leq Y_j + 6(k+1)$ and X_i is on a red square, then
    $$p(Y_j, X_i) = (1/6)^{k+1} + (1/6)^k. \tag{15}$$

With the above analysis and formulae, we can calculate any position's threat to another position and calculate the evaluation function, then make the choice of which plane to move.

3.3.2. Threat Matrix

Based on the above discussion, we have two methods to calculate the threat p:

1. Real-time computing
Real-time computing calculates p for current position (S, R) using Equations (12)–(15) each time the dice is rolled.
However, in order to reduce the manufacturing cost of the game machine, the program runs on an 80 × 86 SCM, which has limited computing and storage power. Although the computing load decreases with respect to certain specific chess positions (such as all eight planes moving into the circle), the computation takes about 3.7 s in extreme cases, which is intolerable to players. We therefore sought faster algorithms to perform the same job.
Through our observations, we found that no matter how the game changes, the threat is only ever determined by the positions. This naturally triggers a new idea, i.e., to compute the threat between any two positions and store them in a lookup table that is independent of any single position, herein called the 'Threat Matrix'.
2. Using the Threat Matrix to store pre-calculated threat values

There are 64 positions on a chessboard. Therefore, we built a 64×64 matrix, $P = (p_{ij})_{64\times 64}$, where $p_{ij} = p(X_i, Y_j)$ is the threat of a plane located on X_i to a plane located on Y_j. From the above discussion, the threat can be calculated using Equations (12)–(15). Part of the threat matrix is as shown in Table 2.

Table 2. An illustration of our threat matrix.

	1	**2**	**...**	**29**	**30**	**31**	**...**	**63**	**64**
1	1	0.1667	...	0.00012	0.00025	0.000128	...	0	0
2	0	1	...	0	0.05556	0	...	0	0
⋮	⋮	⋮		⋮	⋮	⋮		⋮	⋮
29	0	0	...	1	0.1667	0.1667	...	0	0
30	0	0	...	0	1	0.1667	...	0	0
31	0	0	...	0	0	1	...	0	0
⋮	⋮	⋮		⋮	⋮	⋮		⋮	⋮
63	0	0	...	0	0	0	...	0	0
64	0	0	...	0	0	0	...	0	0

4. Threat Matrix-Based Movement Strategy

In order for readers to better understand our algorithm, we use the following chess position (as shown in Figure 3) as an example to illustrate how our algorithm operates. In this example, it is assumed that $S = (3, 5, 0, 0, 1, 11, 0, 0)$ and $R = 3$ and it is blue's turn to make the choice of its plane.

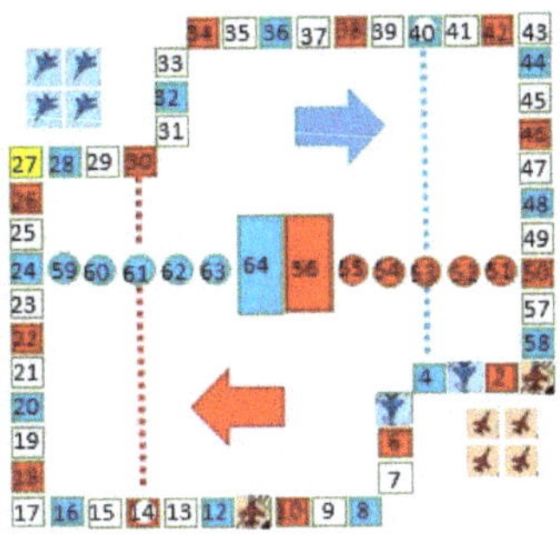

Figure 3. An example with chess position (S, R), where $S = (3, 5, 0, 0, 1, 11, 0, 0)$ and $R = 3$.

In this example, there are two blue planes (shown as the third and fifth squares in Figure 3) and two red planes (shown as the first and eleventh squares in Figure 3) on the circle, and all other planes are in their staging areas. The threat matrix-based movement strategy for the plane is shown in Algorithm 1.

Algorithm 1: Threat matrix-based movement strategy for plane.

Input: Positions of the blue plane, f_1, and opponent plane, f_2: X_1,X_2,Y_1,Y_2
Output: Movement strategy

```
Initialize position values of blue plane and opponent plane: I(X1) = 29,
  I(X2) = 31, I(X3) = I(X4) = 0, I(Y1) = 1, I(Y2) = 11, and I(Y3) = I(Y4) = 0;
for plane f_i (i = 1,2) do
    Compute incremented distance V1 of f_i;
    Compute opponent's hit value V2 of f_i;
    Compute increased threat value T1 of f_i;
    Compute increased threat value T2 of f_i;
    Compute the evaluation value f_i: f_i = V1 + V2 + T1 − T2;
end
return Choose suitable number to move based on evaluation values;
```

5. Recursion and Complexity Analysis for Multiple Moves

5.1. Recursive for n Plies

In the algorithm and example described above, we have evaluated only one move for the plane based on the current position, which is somewhat short-sighted. For most chess games, in order to win the game multiple moves need to be considered, and this is true for Chinese Ludo.

As the game rules are the same for both sides, the evaluation function applies to both blue's planes and red's planes. The plies of (S, R) are limited because there are only four planes and the number of R can only be 1 to 6, which makes it possible to look multiple moves ahead. This means that we can evaluate the opponent's planes based on their hypothetical choices and then consider the opponent's hypothetical assessment.

Let (S^n, R^n) represent the hypothetical position of n moves ahead; we can obtain the evaluation of $n+1$ steps based on $f_i(S^n, R^n)$, denoted by $f_i(S^{n+1}, R^{n+1})$. Thus, based on the Equations (1)–(15), we have

$$f_i(S^{n+1}, R^{n+1}) = V_1 \times f_i(S^n, R^n) + V_2 \times f_i(S^n, R^n) + T_1 \times f_i(S^n, R^n) + T_2 \times f_i(S^n, R^n). \tag{16}$$

It is apparent that Equation (16) is a recursive formula; the initial computation is expressed as Equation (11), with which we can look any number of moves ahead. Although the more moves we look ahead, the more beneficial it is to the blue, the computation cost increases exponentially with the increase in the number of moves. Next, we analyze the relationship between the complexity and the number of moves ahead.

5.2. Complexity Analysis

5.2.1. Computation Complexity of One Move

Let us first look at the computation complexity of evaluating $f_i(S, R)$ based on the current position.

First, according to the evaluation functions in Equations (3) and (4), the calculation of V_1 and V_2 does not need to consider the opponent's threat; X_i and Y_j, the positions of planes, are both integers and not more than 64, and only judgement and a simple addition operation of the positions are needed. For each side, there are only thirteen special positions. Therefore, to calculate V_1 + V_2, there are at most 13 + 13 = 26 comparisons of special positions.

Next, according to Equations (7) and (10), the calculation of T_1 and T_2 mainly involves computing $p(X_i, Y_j)$, which is essentially the comparison of positions.

As the real-time computing and threat matrix are two different methods, we discuss them separately in order to compare their respective advantages and disadvantages.

1. Complexity of Real-Time Computing

Considering the most complicated case, where each of Equations (12) and (13) need to be calculated, Equation (12) includes one comparison and Equation (13) includes eight comparisons, while Equations (14) and (15) include two special cases with special positions; thus, there are at most $(1+8)\alpha$ comparisons. The number of all comparisons is at most $(1+8)+(1+8)\alpha$.
There are at most four opposing planes in play, and the number of all comparisons is therefore $\beta \times (1+8+\alpha \times (1+8))$.
The above represents the calculation of one blue plane, and the blue player has four planes; the number of comparisons is therefore

$$\beta \times (2\alpha + \beta \times (1+8+\alpha \times (1+8))) = 9\alpha\beta^2 + 9\beta^2 + 2\alpha\beta. \tag{17}$$

2. Computing Complexity Using Threat Matrix
According to the definition of the threat matrix, there is no need to calculate $p(X_i, Y_j)$ using Equations (12)–(15) each time, as we can simply look up the pre-calculated matrix. According to Equation (7), we look up the threat matrix twice for each opposing plane. As there are β opposing planes, there are at most 2β search instances. Similarly, according to Equation (10), there are at most 2β search instances for T_2. Then, considering the calculation of V_1 and V_2, which is 2α, the full computation for one of blue's planes is $2\alpha + 4\beta$.
As there are at most four blue planes in play, the maximum computation is

$$\beta \times (2\alpha + 4\beta) = 4\beta^2 + 2\alpha\beta. \tag{18}$$

Comparing Equation (18) with Equation (17), the computation required for the threat matrix is obviously only $(4\beta^2 + 2\alpha\beta)/(9\alpha\beta^2 + 9\beta^2 + 2\alpha\beta) = 7.92\%$ with real-time computing, showing that the computation of the threat matrix is only one-eighth that required by real-time computing. In this game, we use the threat matrix instead of real-time computing.

5.2.2. Complexity of More Moves Ahead/per n Plies

According to Equation (11), each player has four planes; thus, there are at most four cases of S. R takes a value from 1 to 6; thus, the complexity of computing $f_i(S^{n+1}, R^{n+1})$ is 6β instances required to compute $f_i(S^n, R^n)$. Therefore, if we look k moves ahead, $(6\beta)k$, i.e., the computation complexity is $O((6\beta)k)$.

Using the real-time computing method, the complexity is $9\alpha\beta^2 + 9\beta^2 + 2\alpha\beta$, according to Equation (17). If we look one move ahead, the computation is $(9\alpha\beta^2 + 9\beta^2 + 2\alpha\beta) \times 6\beta$, and if we look k steps forward, the computation complexity will be $O((54\alpha\beta^3 + 54\beta^3 + 2\alpha\beta^2)k)$;

When using the threat matrix, according to Equation (18) the computation is $4\beta^2 + 2\alpha\beta$. If we look one-step forward, the computation is $(4\beta^2 + 2\alpha\beta) \times 6\beta$, and if we look k steps forward, the computation complexity will be $O((24\beta^3 + 12\alpha\beta^2)k)$;

It is apparent that there is a clear advantage when using the threat matrix, especially in multiple recursions, which we validate in the following experimental section.

6. Experiment Validation

The development of the Chinese Ludo game used an 80 × 86 SCM 32-bit single board computer with a 100 MHz CPU. Because the computing power was very limited, we used the threat matrix and evaluated only one move ahead of the current position. For more advanced game players, we have developed a more complex two-layer recursive algorithm which can evaluate two moves ahead. We used the 32-bit Windows XP OS as the development platform because the game was eventually expected to run on the 80 × 86 32-bit microprocessor. In addition, we chose Microsoft Visual C as the programming language for easier adoption.

To verify the effectiveness of our algorithms, we carried out three groups of experiments to contrast human play versus the machine, evaluation of one move ahead versus two moves ahead, and real-time computing versus using the threat matrix.

6.1. Contrasting Results for Human versus Machine Play

When the development of the game had finished, a team of 50 players was assigned to play against the machine. The whole test lasted for about one month during which there were more than 15,000 matches played, including 5000 matches with no time limit, 5000 matches with a 10 s limit, and 5000 matches with a 5 s limit. The time limit is the maximum time allowed before reacting after the dice are rolled. It was proposed by the company to increase fun and accelerate the game speed. Figure 4 shows the match results. Figure 5 shows the contrast in reaction times between human players and the machine.

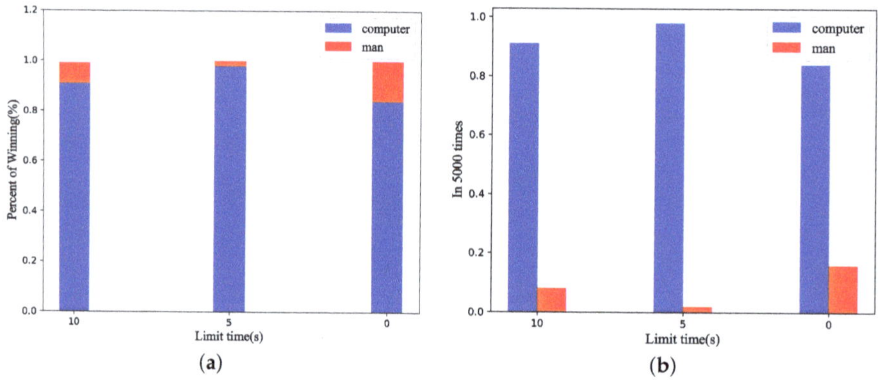

Figure 4. Contrast in match results between humans and machine: (**a**) statistics on the match results; (**b**) comparison of wins.

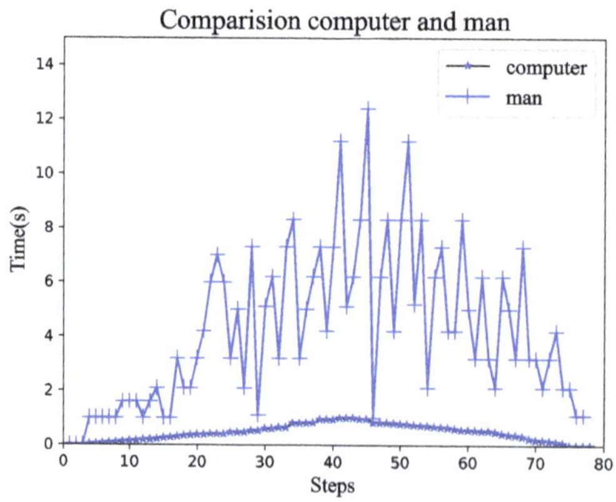

Figure 5. Comparison of reaction time between humans and machine.

From Figure 5, the following can be seen:

1. Aside from the randomness of dice rolling, the machine is much smarter than humans at this game. Even without the time limit, the winning rate of human players is only 12%. When there is a 10 s time limit, human players win 7% of the time, and when there is a 5 s time limit, human players only win 2% of the time. In addition, from the comparison of reaction times human players takes eight times as long to play as the machine does.

2. When reaction time was limited, human players perform much worse and had much less of a chance (less than one quarter) of winning compared with the machine, which indicates that the decision of human players is rushed. Interestingly, through repeated training certain people were able to make more careful judgements within the 5 s time limit.
3. Figure 5 shows the comparison of reaction times between a typical human player and the machine. From Figure 5, it can be seen that the threat matrix has a disadvantage compared with human player, namely, that the reaction time of the machine is proportional to the number of planes in play. Thus, the increase in the time spent with the increase in the number of steps is smooth, and there are no fast decisions. However, this is very different for humans. Although it generally takes human players longer, this does not increase in a strictly monotonic fashion, and at times human players make exceptionally fast decisions. Sometimes, the human player spends much less time than her/his average time, or even less than the machine (for example, at 29 steps, 46 steps and 54 steps). The reason for this is there are special positions, such as hit, dotfly, and jump, where human players can make quick decisions without calculation, whereas a machine must evaluate all planes and then make a choice by comparing evaluation functions. In this particular case, the algorithm can be improved. However, the special positions require extra judgments that take up additional memory space and CPU time. After balancing the potential improvements against the resources required, the company abandoned the optional improvements.

6.2. Comparison of One Move Ahead and Two Moves Ahead

Obviously, for different positions (S, R), the amount of computation is likely to be different, which leads to different levels of computational complexity. This indicates that the evaluations of one move and multiple moves are not comparable. To show the pros and cons of evaluating one move and two moves, we must compare (S, R) at the same position. Thus, we set up twenty different typical positions (S, R) in order to evaluate the difference. Figure 6a shows the comparison of the evaluation result, f, and Figure 6b shows the comparison of the evaluation times.

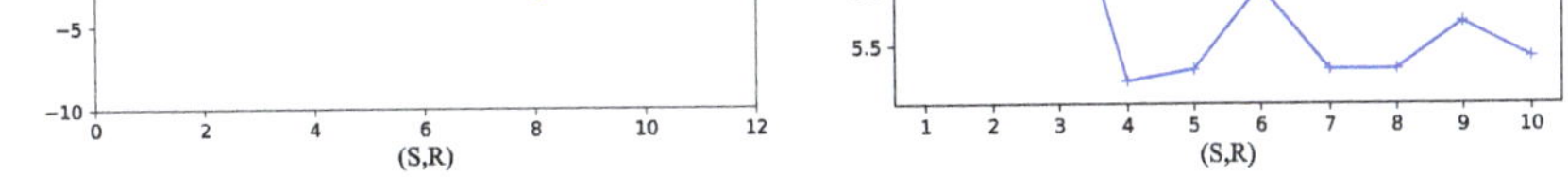

(**a**) Comparison of evaluation value (**b**) Comparison of evaluation time

Figure 6. Comparison of the evaluation of one move and two moves ahead. The vertical axis indicates the evaluation results, f, in (**a**) and indicates the evaluation time in (**b**).

From Figure 6a, two evaluations are consistent. In order to see their impact on the selection of planes, we created the following formula:

$$\mu(i,j) = \frac{\|f_i^1(S,R) - f_i^2(S,R)\|}{\|f_i^1(S,R) - f_j^1(S,R)\|}, \tag{19}$$

where $f_i^1(S,R)$ and $f_i^2(S,R)$ represent the evaluation of the number i plane at one move ahead and at two moves ahead, respectively. Similarly, $f_j^1(S,R)$ represents the evaluation of the number j plane at one move ahead.

In Equation (19), if $\mu(i,j) \geq 1$, it indicates that the difference between the evaluations of one move and two moves of the same plane is larger than those of one move of two different planes. Therefore, the choice of planes at the same position will be different. Otherwise, if $\mu(i,j) < 1$, it means that the evaluations of one move ahead and two moves ahead will not change the choice of planes. According to Equation (19), we calculate $\mu(i,j)$ for the twenty positions, as shown in Table 3.

Table 3. $\mu(i,j)$ for twenty different chess positions (S,R). The * in the table means blank, indicating that the opponent's plane is in the staging area.

(S, R)	1	2	3	4	5	6	7	8	9	10	11	12	13	14	15	16	17	18	19	20
$\mu(2,1)$	0.21	0.32	0.23	0.14	0.52	0.27	0.54	0.21	0.35	0.12	0.42	0.19	0.51	0.72	0.43	0.62	0.75	0.21	*	0.46
$\mu(2,3)$	0.17	0.13	0.03	0.41	0.12	0.07	0.27	0.24	0.15	0.02	0.12	0.29	0.31	0.25	0.03	0.02	0.15	0.41	0.65	0.07
$\mu(2,4)$	0.31	0.21	0.10	0.21	0.22	0.08	0.23	0.53	0.81	0.02	0.33	0.13	*	*	0.07	0.12	0.60	0.02	0.36	0.21

In addition to the above twenty positions, we compared another fifty positions which provided a consistent result, i.e., the evaluation of two moves ahead did not change the choice of planes based compared to the evaluation of one move ahead. In fact, by looking at Equation (16), cited below, we can find the theoretical reasons for this:

$$f_i(S^{n+1},R^{n+1}) = V_1 \times f_i(S^n,R^n) + V_2 \times f_i(S^n,R^n) + T_1 \times f_i(S^n,R^n) + T_2 \times f_i(S^n,R^n). \quad (20)$$

In the evaluation of two moves ahead, the two moves need two more rolls of the dice; however, these are hypothetical. Thus, the further evaluation of the second move ahead is evaluated under the weight of $1/6 \times 1/6 = 1/36$, which makes the impact on the further evaluation small. We can see that this is different from other forms of chess.

However, from Figure 6b, the time cost of evaluating two moves ahead is twenty times of that of evaluating one move ahead. In addition, the memory consumption is increased by a large amount.

6.3. Comparison of Real-Time Computing and the Threat Matrix

In order to compare real-time computing and the threat matrix, we conducted experiments using a 32-bit Windows XP platform.

According to the game rules, the results of dice rolling is random, which means that it is hard to acquire the same position (i.e., the initial conditions for the computation) to compare the two evaluations. Therefore, similar to the previous comparison in Section 5.2, we set typical chess positions in order to ensure the consistency of the initial conditions. The results are shown in Figure 7.

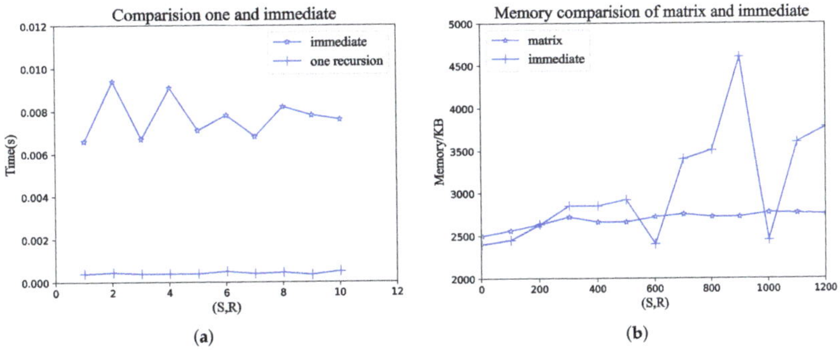

Figure 7. Comparison of real-time computing and the threat matrix in terms of time complexity and memory usage: (**a**) comparison of time complexity; (**b**) comparison of memory usage.

It can be seen from the results that the time consumption of evaluation using the threat matrix takes no more than 0.001 s, while the average time when using real-time calculation is 0.008 s, which is highly unstable compared to the threat matrix. The memory consumption of real-time computation is far greater than that of the threat matrix as well.

7. Conclusions

High-performance computers have significantly contributed to the development of artificial intelligence in the gaming industry, however, this should not be a reason to ignore the effort of pursuing a perfect evaluation method. This paper presents a Chinese Ludo program. Unlike most chess programs, which depend on high machine performance, the evaluation function in our program is only a linear sum of four factor values. The other contribution of this research is that we innovatively constructed a threat matrix that allows us to easily acquire the threat between any two dice on any two positions. The threat matrix approach can greatly reduce the amount of calculation, which allows the program to run on a 32-bit 80 $\times$ 86 SCM with a 100 MHz CPU while supporting a recursive algorithms to search plies. Our results show that, when compared with the real-time computing, our threat matrix approach can reduce computation costs by nearly 90%. Furthermore, the memory consumption is both reduced and relatively stable, which speeds up evaluation by 58%.

Author Contributions: F.H. designed the main methodology and the model; M.Z. developed the theoretical framework; All authors provided critical feedback and contributed to the research and analysis of the manuscript. All authors have read and agreed to the published version of the manuscript.

Funding: This work is supported by the National Natural Science Foundation of China (6217071437; 62072200).

Conflicts of Interest: The authors declare no conflict of interest.

References

1. Danilovic, S.; de Voogt, A. Making sense of abstract board games: Toward a cross-ludic theory. *Games Cult.* **2021**, *16*, 499–518. [CrossRef]
2. Abdalzaher, M.S.; Muta, O. A Game-Theoretic Approach for Enhancing Security and Data Trustworthiness in IoT Applications. *IEEE Internet Things J.* **2020**, *7*, 11250–11261. doi: [CrossRef]
3. Silver, D.; Huang, A.; Maddison, C.J.; Guez, A.; Sifre, L.; Van Den Driessche, G.; Schrittwieser, J.; Antonoglou, I.; Panneershelvam, V.; Lanctot, M.; others. Mastering the game of Go with deep neural networks and tree search. *Nature* **2016**, *529*, 484–489. [CrossRef] [PubMed]
4. Silver, D.; Hubert, T.; Schrittwieser, J.; Antonoglou, I.; Lai, M.; Guez, A.; Lanctot, M.; Sifre, L.; Kumaran, D.; Graepel, T.; others. A general reinforcement learning algorithm that masters chess, shogi, and Go through self-play. *Science* **2018**, *362*, 1140–1144. [CrossRef] [PubMed]

5. Polese, J.; Sant'Ana, L.; Moulaz, I.R.; Lara, I.C.; Bernardi, J.M.; Lima, M.D.d.; Turini, E.A.S.; Silveira, G.C.; Duarte, S.; Mill, J.G. Pulmonary function evaluation after hospital discharge of patients with severe COVID-19. *Clinics* **2021**, *76*. [CrossRef] [PubMed]
6. Zhang, Z.; Ning, H.; Shi, F.; Farha, F.; Xu, Y.; Xu, J.; Zhang, F.; Choo, K.K.R. Artificial intelligence in cyber security: research advances, challenges, and opportunities. *Artif. Intell. Rev.* **2021**, *55*, 1029–1053. [CrossRef]
7. Fu, Y.; Li, C.; Yu, F.R.; Luan, T.H.; Zhang, Y. A survey of driving safety with sensing, vehicular communications, and artificial intelligence-based collision avoidance. *IEEE Trans. Intell. Transp. Syst.* **2021**, 1–22. [CrossRef]
8. Morandin, F.; Amato, G.; Gini, R.; Metta, C.; Parton, M.; Pascutto, G.C. SAI a Sensible Artificial Intelligence that plays Go. In Proceedings of the 2019 International Joint Conference on Neural Networks (IJCNN), Budapest, Hungary, 14–19 July 2019; IEEE: Piscataway, NJ, USA, 2019; pp. 1–8.
9. Nilsson, A.; Smith, S.; Ulm, G.; Gustavsson, E.; Jirstrand, M. A performance evaluation of federated learning algorithms. In Proceedings of the Second Workshop on Distributed Infrastructures for Deep Learning, Rennes, France, 10 December 2018; Association for Computing Machinery: New York, NY, USA, 2018; pp. 1–8.
10. Nkambule, M.S.; Hasan, A.N.; Ali, A.; Hong, J.; Geem, Z.W. Comprehensive evaluation of machine learning MPPT algorithms for a PV system under different weather conditions. *J. Electr. Eng. Technol.* **2021**, *16*, 411–427. [CrossRef]
11. Shayesteh, S.; Nazari, M.; Salahshour, A.; Sandoughdaran, S.; Hajianfar, G.; Khateri, M.; Yaghobi Joybari, A.; Jozian, F.; Fatehi Feyzabad, S.H.; Arabi, H.; et al. Treatment response prediction using MRI-based pre-, post-, and delta-radiomic features and machine learning algorithms in colorectal cancer. *Med. Phys.* **2021**, *48*, 3691–3701. [CrossRef] [PubMed]
12. Paveglio, T.B. From Checkers to Chess: Using Social Science Lessons to Advance Wildfire Adaptation Processes. *J. For.* **2021**, *119*, 618–639. [CrossRef]
13. Torrado, R.R.; Bontrager, P.; Togelius, J.; Liu, J.; Perez-Liebana, D. Deep reinforcement learning for general video game ai. In Proceedings of the 2018 IEEE Conference on Computational Intelligence and Games (CIG), Maastricht, The Netherlands, 14–17 August 2018; IEEE: Piscataway, NJ, USA, 2018; pp. 1–8.
14. Zhu, J.; Villareale, J.; Javvaji, N.; Risi, S.; Löwe, M.; Weigelt, R.; Harteveld, C. Player-AI Interaction: What Neural Network Games Reveal About AI as Play. In Proceedings of the 2021 CHI Conference on Human Factors in Computing Systems, Yokohama, Japan, 8–13 May 2021; Association for Computing Machinery: New York, NY, USA, 2021; pp. 1–17.
15. Lee, C.S.; Tsai, Y.L.; Wang, M.H.; Kuan, W.K.; Ciou, Z.H.; Kubota, N. AI-FML Agent for Robotic Game of Go and AIoT Real-World Co-Learning Applications. In Proceedings of the 2020 IEEE International Conference on Fuzzy Systems (FUZZ-IEEE), Glasgow, UK, 19–24 July 2020; pp. 1–8.
16. Lüscher, T.F.; Lyon, A.; Amstein, R.; Maisel, A. Artificial intelligence: The pathway to the future of cardiovascular medicine. *Eur. Heart J.* **2022**, *43*, 556–558. [CrossRef] [PubMed]
17. Feng, X.; Slumbers, O.; Wan, Z.; Liu, B.; McAleer, S.; Wen, Y.; Wang, J.; Yang, Y. Neural auto-curricula in two-player zero-sum games. In Proceedings of the Thirty-Fifth Conference on Neural Information Processing Systems (NeurIPS 2021), Virtual, 6–14 December 2021.
18. Bošanský, B.; Lisý, V.; Lanctot, M.; Čermák, J.; Winands, M.H. Algorithms for computing strategies in two-player simultaneous move games. *Artif. Intell.* **2016**, *237*, 1–40. [CrossRef]
19. Nangrani, S.; Nangrani, I.S. Artificial Intelligence Based State of Charge Estimation of Electric Vehicle Battery. In *Smart Technologies for Energy, Environment and Sustainable Development*; Springer: Berlin/Heidelberg, Germany, 2022; Volume 2, pp. 679–686.
20. Hankey, A. Kasparov versus Deep Blue: An Illustration of the Lucas Gödelian Argument. *Cosm. Hist. J. Nat. Soc. Philos.* **2021**, *17*, 60–67.
21. Silver, D.; Schrittwieser, J.; Simonyan, K.; Antonoglou, I.; Huang, A.; Guez, A.; Hubert, T.; Baker, L.; Lai, M.; Bolton, A.; others. Mastering the game of go without human knowledge. *Nature* **2017**, *550*, 354–359. [CrossRef] [PubMed]
22. Bai, W.; Cao, Y.; Dong, J. Research on the application of low-polygon design to Chinese chess: Take the Chinese chess "knight" as an example. In Proceedings of the 2021 the 3rd World Symposium on Software Engineering, Xiamen, China, 24–26 September 2021; pp. 199–204.
23. Ma, X.; Shang, T.; others. Study on Traditional Ethnic Chess Games in Gansu Province: The Cultural Connotations and Values. *Adv. Phys. Educ.* **2021**, *11*, 276–283.
24. He, H. The Race Card: From Gaming Technologies to Model Minorities by Tara Fickle. *J. Asian Am. Stud.* **2021**, *24*, 515–517. [CrossRef]
25. Tomašev, N.; Paquet, U.; Hassabis, D.; Kramnik, V. Assessing game balance with AlphaZero: Exploring alternative rule sets in chess. *arXiv* **2020**, arXiv:2009.04374
26. Available online: https://en.wikipedia.org/wiki/AeroplaneChess (accessed on 1 April 2022).
27. Polk, A.W.S. Enhancing AI-based game playing using Adaptive Data Structures. Ph.D. Thesis, Carleton University, Ottawa, ON, Canada, 2016.

electronics

Article

Simulation of Authentication in Information-Processing Electronic Devices Based on Poisson Pulse Sequence Generators

Volodymyr Maksymovych [1], Elena Nyemkova [1,*], Connie Justice [2], Mariia Shabatura [1], Oleh Harasymchuk [1], Yuriy Lakh [1] and Morika Rusynko [1]

1 Department of Information Technology Security, Lviv Polytechnic National University, 79013 Lviv, Ukraine; volodymyr.m.maksymovych@lpnu.ua (V.M.); mariia.m.mandrona@lpnu.ua (M.S.); oleh.i.harasymchuk@lpnu.ua (O.H.); yurii.v.lakh@lpnu.ua (Y.L.); morika.k.rusynko@lpnu.ua (M.R.)

2 Purdue School of Engineering and Technology, Indiana University—Purdue University Indianapolis (IUPUI), Indianapolis, IN 46202, USA; cjustice@iupui.edu

* Correspondence: olena.a.niemkova@lpnu.ua; Tel.: +38-032-235-8323

Abstract: Poisson pulse sequence generators are quite well studied, have good statistical properties, are implemented both in software and hardware, but have not yet been used for the purpose of authentication. The work was devoted to modeling authenticators of information-processing electronic devices by creating a bit template simulator based on a Poisson pulse sequence generator (PPSG). The generated templates imitated an important property of real bit templates, which reflected the physical uniqueness of electronic devices, namely Hamming distances between arbitrary template pairs for the same device were much smaller than the distance between arbitrary template pairs for two different devices. The limits of the control code values were determined by setting the range of the average frequency values of the output pulse sequence with the Poisson distribution law. The specified parameters of the output pulse sequence were obtained due to the optimization of the parameters of the PPSG structural elements. A combination of pseudo-random sequences with the control code's different values formed the bit template. The comparison of the Hamming distance between the standard and real-time templates with a given threshold value was used as a validation mechanism. The simulation experiment results confirmed the unambiguous authentication of devices. The simulation results also showed similarities with the real data obtained for the bit templates of personal computers' own noise. The proposed model could be used for improving the cybersecurity of a corporate network as an additional factor in the authentication of information-processing electronic devices for which the measurement of noise with the required accuracy is not possible or significantly difficult.

Keywords: cybersecurity; authentication; bit template; information-processing electronic device; Poisson pulse sequences generators

Citation: Maksymovych, V.; Nyemkova, E.; Justice, C.; Shabatura, M.; Harasymchuk, O.; Lakh, Y.; Rusynko, M. Simulation of Authentication in Information-Processing Electronic Devices Based on Poisson Pulse Sequence Generators. *Electronics* **2022**, *11*, 2039. https://doi.org/10.3390/electronics11132039

Academic Editor: Krzysztof Szczypiorski

Received: 24 May 2022
Accepted: 26 June 2022
Published: 29 June 2022

1. Introduction

The ability to authenticate electronic information-processing devices based on their unique characteristics attracts the attention of researchers who work to ensure the cybersecurity of information systems. The uniqueness of electronic devices at the physical level makes it possible to carry out authentication using various methods, namely physically non-cloneable functions for the Internet of Things (IoT) [1,2], error vector trajectories of the Global System for Mobile Communications (GSM) mobile phone signals [3], the spectrum of noise in the signal of radioactivity sensors [4], spontaneous electromagnetic radiation from operating mobile phones, light-emitting diode (LED) screens, laptops [5], wireless fidelity transmitters [6], etc. Authentication accuracy depends on the chosen method and conditions of experiments, but usually does not reach 100 percent.

The implementation of the idea of authentication by individual characteristics is based on a preliminary measurement of a physical quantity; for example, electromagnetic interference from operating units of the device, the so-called internal electrical noise. The method of authentication for electronic information-processing devices (personal computers) by the individual forms of correlograms of their internal electrical noise is known [7]. The device authenticator—noise bit template—is calculated from the normalized autocorrelation function of noise. In the future, bit templates could be compared with each other using the selected metric; this is the Hamming distance in the simplest case. An important property of the bit template set, obtained from sequential measurements of the internal noise of a single device, is their closeness in the sense that the Hamming distances between possible pairs of bit templates are small, while the distances between pairs of templates for different devices are much larger. This makes it possible to reliably distinguish these devices, i.e., to authenticate them. The reference templates from each electronic device are pre-recorded on the server. During authentication, the device presents a real-time template, which is compared with a reference template. Authentication is confirmed if the distance between templates is less than a threshold value for the claimed device. Bit template variations provide dynamic authentication because the templates do not repeat exactly, and thus the authenticator reuse attack is eliminated. There are known experiments, a result of which made it possible to authenticate stationary personal computers of the same series with an accuracy of 98.6% [8].

An integrated sound card can be used to measure the internal noise of computers. Usually, noise bit templates of desktop computers are stable over time. For laptops the situation is slightly different. If the laptop gets into a location with a strong external electromagnetic field that significantly affects the internal noise of the laptop, then authentication errors occur [8]. In addition, not all electronic information-processing devices have integrated ADCs, for example, many microprocessors do not have an integrated ADC. For them, the use of internal noise as a sign of authentication is not possible. Therefore, in this study, the problem of modeling authentication features based on Poisson pulse sequence generators was formulated.

Generators of random or pseudo-random pulse sequences have been used for a long time to solve a wide range of problems in science and technology. Almost all standard program libraries have the embedded generators of pseudo-random sequences, which users could utilize. One of the most important generators is the Poisson pulse sequence generator (PPSG). These generators are widely used in different branches of techniques for simulating different processes that have a random temporal and spatial nature [9], for sociological and scientific research [10,11]. Such generators are effectively used to solve cybersecurity problems [12,13], to simulate the output signals of dosimetric detectors when designing them, and testing devices for measuring the parameters of ionizing radiation [14–19], because the number of radioactive decay particles detected by the detector over a period of time is subject to the Poisson distribution law.

In recently published works quite effective principles of realization of software and hardware PPSG are presented. Their structures, based on the use of pseudo-random number generators (PRNGs), were proposed [20–27] and methods for assessing the quality of their output signals were developed [28–31]. In this case the effectiveness of the possible application of the PPSG significantly depends on the quality of the designed generator and on the main characteristics of its output sequence.

The aim of this study was to model bit templates for the authentication of electronic information-processing devices based on a Poisson pulse sequence generator. The following tasks were solved to achieve this aim.

1. Optimization of the parameters of the PPSG's structural elements in order to obtain the specified parameters of the output pulse sequence. Definition of the limits of control code values, specification of the range of values of the average frequency of the output pulse sequence, which corresponds to Poisson's law of distribution.
2. The bit template simulator was proposed based on the Poisson pulse sequence generator. Bit sequences with given characteristics were the result of the simulator.
3. The simulation experiment was carried out to test the required properties of bit templates.

In this research, based on previously obtained results concerning Poisson pulse sequence generators and the development of the control code theory, the needed sequences were programmatically generated. Sample device bit templates were simulated based on these sequences and authentication was also performed programmatically. The examination consisted of calculating pairs of possible Hamming distances between the templates of the same device (intradistances) and for different devices (interdistances) and comparing them one with another.

Comparing the set of intradistances with the set of interdistances confirmed the main idea, that the generated templates could be unmistakably classified as being related to different devices. The threshold of distances was determined, according to which classification was made for specific parameters of the PPSG.

In this research the Poisson Pulse Sequence Generator was used for the first time to create device authentication templates based on the principle of biometrics. Compared with the best practices, which were using the measured values—electromagnetic radiation, internal electrical noise—the proposed method had several advantages. Benefits included 100% authentication, significantly more devices, and no time delay for measurements.

2. Materials and Methods

2.1. Structural Scheme PPSG and the Principle of Its Operation

The generator [16–18], whose structural scheme is illustrated in Figure 1, consisted of a modified additive Fibonacci generator (MAFG), which contained registers Rg1–Rg5, adders Ad1–Ad3, logical scheme LS, as well as a comparing scheme CS and logical element &. All the structural MAFG elements, except LS, worked in binary-decimal code.

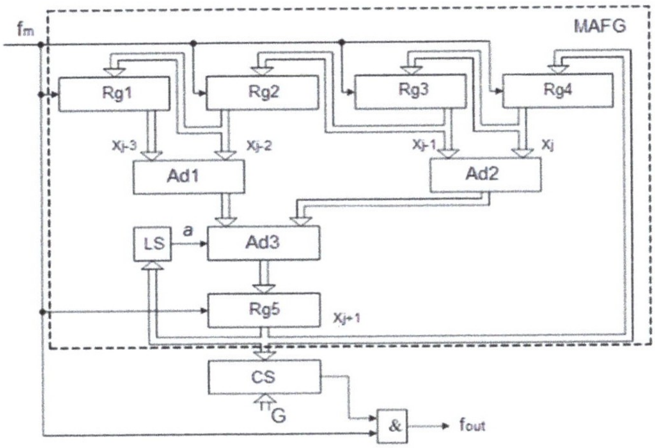

Figure 1. PPSG structural scheme based on MAFG.

On MAFG output, e.g., on Rg5 output, a sequence of pseudo-random numbers was formed in accordance with the following expression:

$$x_{j+1} = (x_j + x_{j-1} + x_{j-2} + x_{j-3} + a) mod m \tag{1}$$

where x_j, x_{j-1}, x_{j-2}, x_{j-3} are the numbers in registers Rg4, Rg3, Rg2, Rg1, correspondingly, $m = 10^q$, and q is the number of decades of the scheme's structural elements. The value of the variable a is determined by the logical equation

$$a = (a_{0_0} \oplus a_{0_1} \oplus a_{0_2} \oplus a_{0_3}) \oplus \ldots \oplus (a_{q-1_0} \oplus a_{q-1_1} \oplus a_{q-1_2} \oplus a_{q-1_3}) \tag{2}$$

where a_{i_j} $(i = 0,1,2,3;\ j = 0,1,\ldots,q-1)$ is the value of bits of the binary-decimal number in Rg5. The number of members of Equation (2) can be selected from the range $0 \ldots 4 \cdot q$.

The theoretical average value of the pulse frequency at the PPSG output is determining from the following Equation [16]:

$$f_{out} = \frac{G}{10^q} f_m \tag{3}$$

where G is the control code,f_m is the clock pulse frequency.

2.2. Output Signal Parameters and Internal Parameters of the Generator and Their Relationship

The main parameters of the output pulse sequence were as follows:

- average value of frequency f_{out};
- value range of f_{out};
- step of frequency f_{out} changing $-\Delta f_{out}$;
- the repetition period of the pulse sequence;
- compliance of the pulse sequence with the Poisson distribution law.

The parameters of the output pulse sequence were determined by the following internal parameters of the generator (Figure 1):

- the number of decades of the MAFG structural elements;
- initial settings of the registers Rg1–Rg5;
- number of members of Equation (2) involved in the implementation of the logic scheme LS.

The three internal parameters were clearly defined:

- the repetition period of pseudo-random numbers in the output of the MAFG register (Rg5 output);
- statistical characteristics of the number sequence of the MAFG output.

Based on the principle of PPSG construction, it could be argued that the repetition period of the output pulse sequence was equal to the repetition period of numbers in the MAFG output.

The repetition period and statistical characteristics of the sequence of numbers in the MAFG output determined the compliance of the output PPSG pulse sequence with the Poisson distribution law. However, that compliance significantly depended on the average frequency of an output sequence f_{out}, whose theoretical value was determined by Equation (3) and, therefore, depended on the correlation between control code value G and value 10^q. In fact, when value G was approaching the value 10^q, then f_{out} was approaching the clock frequency f_m and, under such conditions, the output sequence started losing its pseudo-random properties. From the other side, the lower the frequency of the output sequence f_{out}, the greater the time interval needed to be to determine its statistical characteristics. In this case such an interval should not exceed the repetition period of this sequence. Thus, in principle (theoretically), the original PPSG pulse sequence might conform to the Poisson distribution law for arbitrarily small average values f_{out}, however, the sequence repetition period should be of a sufficiently large value. As a limit, if the average value f_{out} went to zero, the repetition period should go to infinity.

These statements were practical in nature, satisfied most PPSG applications, and are confirmed below by specific calculations and simulation results. Theoretically, a more general approach to determine the correspondence of the output sequence to the Poisson distribution law could be considered, taking into account the value of the average repetition frequency, repetition period, observation time, and the chosen method of estimating statistical characteristics. However, such an approach needs to be refined to be applied in practice.

Taking into account the above, the average frequency f_{out}, the range of its values, and the step change could be calculated theoretically using Equation (3). The real values of these quantities were determined as a result of simulation and/or experimentally.

2.3. Estimation Method for Statistical Characteristics of the Output Signal

This research was carried out using a generalized method of studying the parameters of the output PPSG pulse sequence for compliance with the Poisson distribution law using Pearson's test [32].

In accordance with the proposed method, the flow of input pulses of the PPSG was divided into n equal groups, each of which consisted of i_{max} pulses. The maximum number of groups was n_{max}. The groups of input pulses corresponded to the groups of output pulses with the number of pulses $k_1, k_2, \ldots k_{n_{max}}$. The proposed method was based on the classical testing method of the hypothesis of the distribution of the general totality according to Poisson's law using Pearson's criterion (χ^2 criterion) [32–34]. In this case, taking into account the specifics of the PPSG construction, the following additions were proposed:

- we fixed nominal (theoretical) average value of numbers $k_1, k_2, \ldots k_{n_{max}} - k_c$, regardless of the control code value G;
- the value i_{max} was variable, depended on the value G, and was determined by the equation

$$i_{max} = \frac{10^q}{G} k_c \,. \tag{4}$$

As a result of the application of this method we obtained the value χ_c^2. According to the tables of critical distribution points of χ^2 [33,34], according to the selected level of significance α (usually α is assigned one of the three following values: 0.1; 0.05; 0.01), the number of degrees of freedom k could be obtained using the critical value χ_{cr}^2. If $\chi_c^2 < \chi_{cr}^2$ there was no reason not to accept the hypothesis that the pulse flux corresponded to the Poisson distribution law.

When determining the statistical characteristics of the PPSG output signal in the range of values of the control code G, it was useful to average the last (current) h values of χ_c^2. Obtained by such a way, variable χ_{cav}^2 was comparable with χ_{cr}^2. The averaging of the values χ_c^2 was necessary for a certain "smoothing" of the results. Based on the simulation experience, one could select value $h = 5$, which could be changed if needed for a clearer (more integrated) determination of the control code range G, in which the output pulse sequence corresponded to the Poisson distribution law.

When designing a PPSG, it is also useful to pre-determine the statistical characteristics of the number sequence, in this case at the MAFG output. This could be achieved using standard statistical tests, such as NIST statistical tests [22–27,32,35].

2.4. Defining the Limits of the Range of the Control Code Values

Lower G_1 and upper G_2 limits of the control code values G, in which the statistical characteristics of the output pulse sequence corresponded to the Poisson distribution law, could be determined based on the following.

The sequence evaluation time should not be longer than its repetition period T_n. That is, based on the above methodology, the following inequality must be satisfied:

$$i_{max} \cdot n_{max} \leq T_n \tag{5}$$

From Equation (4) and inequality (5) we obtain

$$G \geq \frac{10^q \cdot k_c \cdot n_{max}}{T_n} \tag{6}$$

This meant that the value G_1 was the smallest integer number satisfying Inequality (6). As a result of PPSG simulation, it was found that the value G_2 satisfied the following condition:

$$G \leq G_2 = s \cdot 10^q \tag{7}$$

In this case the value of the coefficient s was determined separately for a concrete number of MAFG decades q, and depended on the initial settings of the registers Rg1–Rg5, the number of involved members of Equation (2) and, under certain conditions, was close to 0.1.

3. Results

3.1. *Investigation of the PPSG Based on MAFG When $q = 3$*

3.1.1. Determining the Repetition Period of the MAFG

At a fixed number of decades, the MAFG repetition period of a pseudo-random sequence of numbers in its output T_n and, thus, the repetition period of the pulse sequence in the output of the PPSG, also depended on the number of involved members of Equation (2) and from the initial settings of the registers Rg1–Rg5.

The performed investigations showed that the initial settings of the registers affected the statistical characteristics of the output sequence. The values of these settings obtained as a simulation result, when the statistical characteristics were satisfactory, is shown below.

Dependence of the repetition period T_n on the used number of members from Equation (2) was significant. Some confirmed results are presented in Table 1, which were obtained for such initial states of registers Rg1–Rg5, correspondingly 1, 0, 0, 0, 0.

Table 1. Dependence T_n on output signal a of the logical scheme LS, $q = 3$.

a	T_n
$a = 0$	18,599
$a = a_{0_0}$	18,599
$a = a_{0_0} \oplus a_{0_1}$	103,404,839
$a = a_{0_0} \oplus a_{0_1} \oplus a_{0_2}$	4,348,679
$a = a_{0_0} \oplus a_{0_1} \oplus a_{0_2} \oplus a_{0_3}$	20,121,479
$a = a_{0_0} \oplus a_{0_1} \oplus a_{0_2} \oplus a_{0_3} \oplus a_{1_0}$	$> 10^9$
$a = (a_{0_0} \oplus a_{0_1} \oplus a_{0_2} \oplus a_{0_3}) \oplus (a_{1_0} \oplus a_{1_1} \oplus a_{1_2} \oplus a_{1_3}) \oplus (a_{2_0} \oplus a_{2_1} \oplus a_{2_2} \oplus a_{2_3})$	$> 10^9$

Optimization of equation choosing for the output signal LS was a separate partial task requiring additional research. Its solution would also affect the speed of the generator.

3.1.2. Determination of Statistical Characteristics and the Range of Values of the Control Code

Figure 2 illustrates the investigation results of PPSG statistical characteristics based on the MAFG for $q = 3$.

Here the following notations were used:

- SS_n—value χ_c^2;
- SS_n_pot—the average value of the last five (current) values χ_c^2—χ_{cav}^2;
- Level—number of values χ_{cav}^2 greater than χ_{cr}^2.

The results were obtained at the following values of the method parameters for evaluating the quality of the pulse sequence: $n_{max} = 1000$, $k_c = 10$, $\chi_{cr}^2 = 25$.

The output signal of the logic circuit of the LS was formed by the following expression:

$$a = \left(a_{0_0} \oplus a_{0_1} \oplus a_{0_2} \oplus a_{0_3}\right) \oplus \left(a_{1_0} \oplus a_{1_1} \oplus a_{1_2} \oplus a_{1_3}\right) \oplus \left(a_{2_0} \oplus a_{2_1} \oplus a_{2_2} \oplus a_{2_3}\right) \quad (8)$$

as a result of the search for initial states of various variants of registers Rg1–Rg5, it was found that the value of these settings was satisfactory—G, 0, 0, 0, 0, correspondingly. That is, the option in which the initial settings depended on the control code. This was for such initial settings for which results are presented in Figure 2.

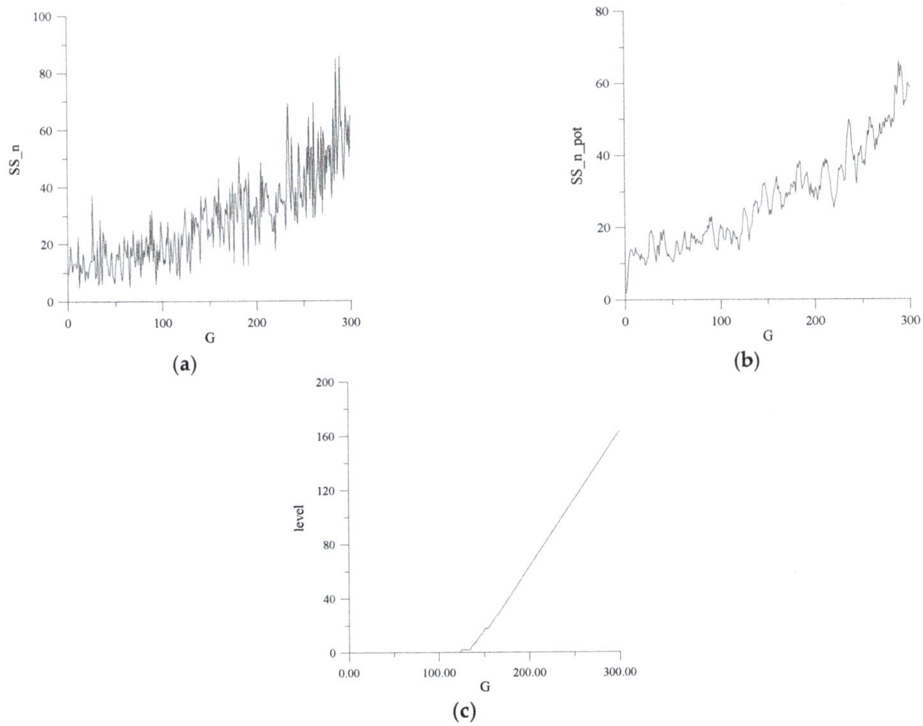

Figure 2. PPSG statistical characteristics based on MAFG ($q = 3$): (**a**) the value χ_c^2; (**b**) the average value of the last five (current) values $\chi_c^2 - \chi_{cav}^2$; (**c**) number of values χ_{cav}^2 greater than χ_{cr}^2. G—control code value.

Thus, the range of the control code values G $-(G_1 \div G_2)$, in which the original pulse sequence corresponded to the Poisson distribution law, in this case (when $q = 3$), was determined by the equation

$$G_1 = 1,\ G_2 = 124 \tag{9}$$

In this case, the value $G_1 = 1$, determined as a result of simulation, coincided with the value G_1, defined theoretically by the expression (6):

$$G \geq \frac{10^q \cdot k_c \cdot n_{max}}{T_n} = \frac{10^3 \cdot 10 \cdot 10^3}{10^9} = 10^{-2} \tag{10}$$

3.2. Dependence of the Average Value of the Output Signal Frequency on the Control Code

This section is divided by subheadings. It should provide a concise and precise description of the experimental results and their interpretation, as well as the experimental conclusions that can be drawn.

Figure 3a illustrates the dependence of the average frequency of the PPSG output pulse sequence on the control code G, while Figure 3b illustrates a fragment of that dependence.

Here solid lines show the dependences obtained by simulation, and dotted lines show theoretical values, calculated on the basis of Equation (3). Solid and dotted lines in Figure 3a almost coincide. To specify the calculations, it was accepted that $f_m = 1000$ Hz. All real dependences were obtained for the condition of formation of the LS output signal correspondingly with logical Equation (8) and explained initial states Rg1–Rg5: G, 0, 0, 0, 0, correspondingly.

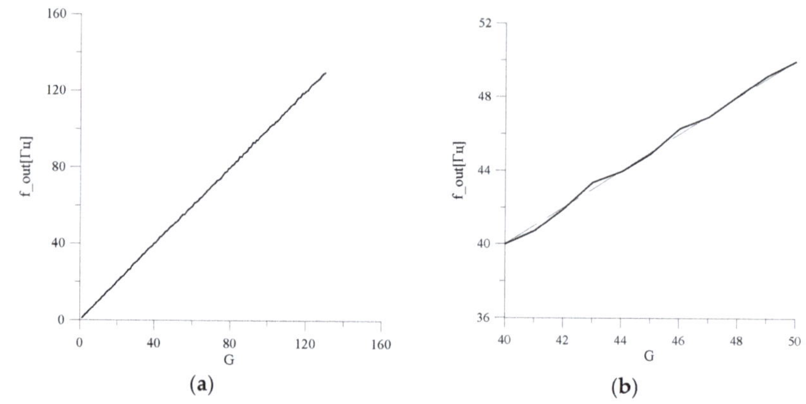

Figure 3. The value of the average frequency of the PPSG output signal based on MAFG ($q = 3$) when $\Delta G = 1$: (**a**) the dependence of the average frequency on the control code G; (**b**) the fragment of that dependence. G—control code value.

Thus, the dependences of the values of the average frequency of the output pulse sequence of the generator from the control code, obtained as a result of simulation, were close to theoretical. That practically allowed the use of Equation (3) while determining the average frequencies of the PPSG output signal.

3.3. *Investigation of the PPSG Based on MAFG When $q = 6$*

3.3.1. Determining the MAFG Repetition Period

Dependence of the repetition period T_n on the number of involved members of Equation (2) is presented in Table 2. The following initial states of the registers Rg1–Rg5, correspondingly 1, 0, 0, 0, 0, were obtained.

Table 2. Dependence of T_n on the output signal a of a logical scheme LS ($q = 6$).

a	T_n
$a = 0$	9,255,555
$a = a0_0$	4,649,999
$a = a0_0 \oplus a0_1$	$> 10^9$
$a = a0_0 \oplus a0_1 \oplus a0_2$	$> 10^9$
$a = a0_0 \oplus a0_1 \oplus a0_2 \oplus a0_3$	$> 10^9$
$a = a0_0 \oplus a0_1 \oplus a0_2 \oplus a0_3 \oplus a1_0$	$> 10^9$
$a = (a0_0 \oplus a0_1 \oplus a0_2 \oplus a0_3) \oplus (a1_0 \oplus a1_1 \oplus a1_2 \oplus a1_3) \oplus (a2_0 \oplus a2_1 \oplus a2_2 \oplus a2_3) \oplus (a3_0 \oplus a3_1 \oplus a3_2 \oplus a3_3) \oplus (a4_0 \oplus a4_1 \oplus a4_2 \oplus a4_3) \oplus (a5_0 \oplus a5_1 \oplus a5_2 \oplus a5_3)$	$> 10^{10}$

3.3.2. Determination of Statistical Characteristics and the Range of Values of the Control Code

Investigation results of the PPSG statistical characteristics based on MAFG for $q = 6$ are presented in Figure 4.

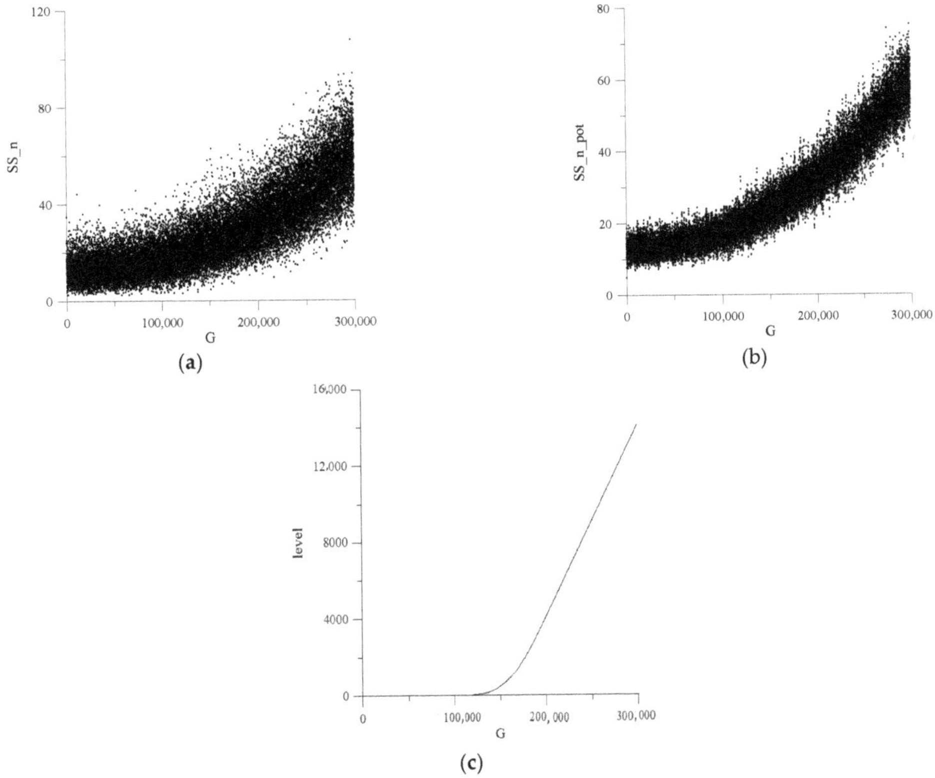

Figure 4. PPSG statistical characteristics based on MAFG ($q = 6$): (**a**) the value χ_c^2; (**b**) the average value of the last five (current) values $\chi_c^2 - \chi_{cav}^2$; (**c**) number of values χ_{cav}^2 greater than χ_{cr}^2. G—control code value.

The results were obtained for the same values of the parameters of the quality assessing method of the pulse sequence, as in the previous case (for $q = 3$): $n_{max} = 1000$, $k_c = 10$, $\chi_{cr}^2 = 25$.

Output signal of the logic scheme LS was formed according to the following expression:

$$a = \begin{array}{l} \left(a_{0_0} \oplus a_{0_1} \oplus a_{0_2} \oplus a_{0_3}\right) \oplus \left(a_{1_0} \oplus a_{1_1} \oplus a_{1_2} \oplus a_{1_3}\right) \oplus \left(a_{2_0} \oplus a_{2_1} \oplus a_{2_2} \oplus a_{2_3}\right) \oplus \\ \left(a_{3_0} \oplus a_{3_1} \oplus a_{3_2} \oplus a_{3_3}\right) \oplus \left(a_{4_0} \oplus a_{4_1} \oplus a_{4_2} \oplus a_{4_3}\right) \oplus \left(a_{5_0} \oplus a_{5_1} \oplus a_{5_2} \oplus a_{5_3}\right) \end{array} \tag{11}$$

where the initial settings of registers Rg1–Rg5 were correspondingly the following: G, 0, 0, 0, 0.

Control code range values $G - (G_1 \div G_2)$, in which the output pulse sequence corresponded to the Poisson distribution law, in this case (when $q = 6$), was determined by the following equation.

$$G_1 = 1, G_2 = 111010 \tag{12}$$

In this case the value $G_1 = 1$, determined as a result of simulation, coincided with the value G_1, determined theoretically by Expression (6):

$$G \geq \frac{10^q \cdot k_c \cdot n_{max}}{T_n} = \frac{10^6 \cdot 10 \cdot 10^3}{10^{10}} = 10^0. \tag{13}$$

3.4. Dependence of the OUTPUT Signal Frequency Average Value on the Control Code

Figure 5a illustrates the dependence of the output of the PPSG's pulse sequence average frequency on the control code G, while Figure 5b shows a fragment of this dependence.

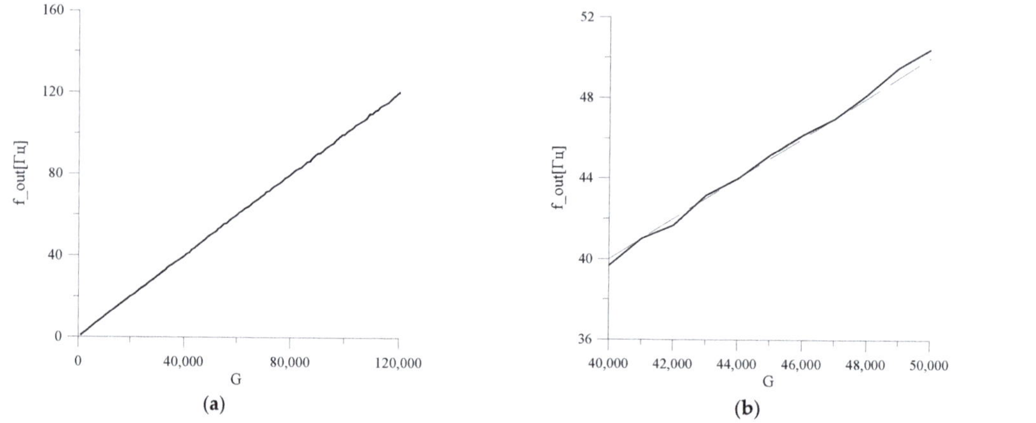

Figure 5. Output signal average frequency of the PPSG based on MAFG ($q = 6$), when $\Delta G = 1000$: (**a**) the dependence of the average frequency on the control code G; (**b**) the fragment of that dependence. G—control code value.

Here, similarly to Figure 3, solid lines illustrate the dependences obtained by simulation, and dotted lines illustrate theoretical values, calculated on the basis of Equation (3). Solid and dotted lines in Figure 5a almost coincide. It was accepted that. $f_m = 1000$ Hz. All the real dependences were obtained under the condition of LS output signal formation according to the logic of Equation (11) and the above-justified initial states Rg1–Rg5: G, 0, 0, 0, 0, correspondingly.

The fundamental difference between the dependencies presented in Figures 3 and 5 is that they were obtained using different values for the control code step changing $G - \Delta G$: in Figure 3 (for $q = 3$) when $\Delta G = 1$; while in Figure 5 (for $q = 6$) when $\Delta G = 1000$.

A decrease in value ΔG for $q = 6$, led to some ambiguity in establishing the average frequency value of the PPSG output sequence. This is illustrated in Figure 6 where the dependences were similar to the dependences in Figure 5, for $\Delta G = 100$.

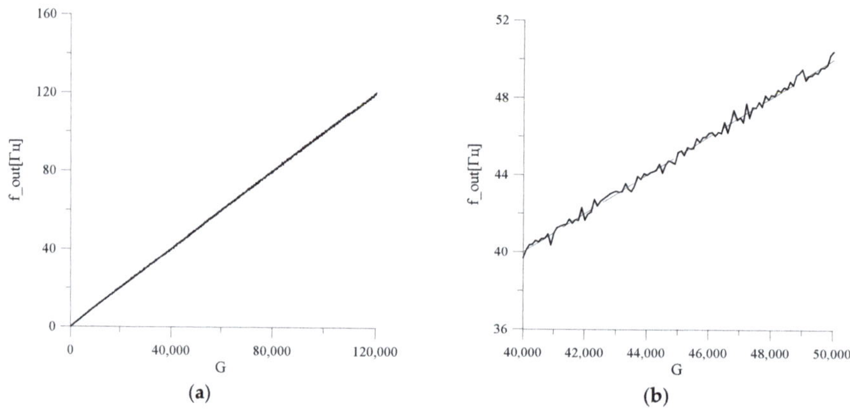

Figure 6. Output signal average frequency of the PPSG based on MAFG ($q = 6$), when $\Delta G = 100$: (**a**) the dependence of the average frequency on the control code G; (**b**) the fragment of that dependence. G—control code value.

3.5. Comparing PPSG Characteristics Based on MAFG for $q = 3$ and $q = 6$

Increasing the number of decades of the generator could significantly increase the repetition period of the sequence of numbers in the MAFG output and, thus, also the pulse sequence period of the PPSG output. However, this did not lead automatically to an increase in the generator's "distinguishing ability" concerning the established value of the output sequence average frequency f_{out}, which was actually setting the ability to specify the changing step $f_{out} - \Delta f_{out}$.

The performed research showed that "distinguishing ability", at a fixed value for the number of decades q, depended on the initial settings of the registers Rg1–Rg5 and on the involved members of Equation (2), which determined the logic of signal generation on the output of the LS scheme. In this case, the statistical characteristics of the original sequence depended on these parameters. Taking into account the above considerations, improving a generator's "distinguishing ability" could be the subject of a separate study.

As far as increasing the number of decades from $q = 3$ to $q = 6$, during the above-mentioned conditions, in fact did not lead to an increase in the PPSG's "distinguishing ability" (decreasing Δf_{out}) and expanded the range f_{out}. For future work, it would be worth considering the possibility of the practical use of this generator when $q = 3$.

3.6. Using the PPSG Based on the MAFG When $q = 3$

Figure 7 shows the structural scheme of the device, in which to expand the range of average values of the output frequency, an additional frequency divider FD was introduced.

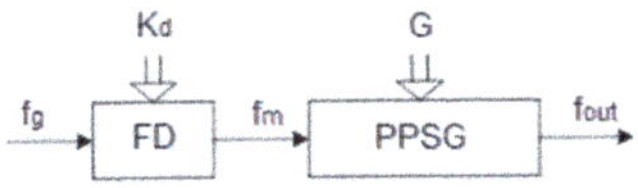

Figure 7. Structural scheme of the Poisson pulse sequence generator with an extended range of average values in the output frequency.

At the FD output the clock pulse sequence was formed for the PPSG, the frequency of which was determined by the equation

$$f_m = \frac{f_g}{K_d} \tag{14}$$

where K_d is the division factor FD and f_g is the reference generator frequency.

Some generator parameters are presented in the Table 3, of which one is presented in Figure 7, when f_g= 1 MHz. The PPSG was implemented based on the MAFG for $q = 3$, while its internal parameters corresponded to the above: the output signal of the logic scheme LS was formed by the expression (8), and the initial states of the registers Rg1–Rg5 were G, 0, 0, 0, 0, correspondingly. This allowed us to tell whether the statistical characteristics of the output pulse sequence in the given ranges of values f_{out}, corresponded to Poisson's law of distribution.

The construction of PPSGs based on MAFGs, all elements of which, except those of the LS, work in binary-decimal code, improves significantly the quality of the output sequence. This was confirmed by a generalized technique for studying parameters of the PPSG output pulse sequence for compliance with the Poisson distribution law using the Pearson test. Investigations of the proposed solutions illustrated that the dependences of the average frequency values of the generator's output pulse sequence from the control code, obtained as a result of simulation, were close to the theoretical ones. It was shown that the number of decades was enough to choose $q = 3$, because greater numbers of decades did not actually lead to an increased "distinguishing ability" for the PPSG; while scheme realization would be more complicated in that case. In order to expand the output frequency average values the introduction of a division factor into the PPSG structural scheme was proposed, which would be divided by the frequency of the reference generator. The question of the selecting number of the equation members to calculate the logical variable a, the value of which was obtained at the LS output, was rather significant. The number of data members of the equation and approaches to their choice significantly affected the size of the repetition period T_n. Further research is needed in this direction in order to improve the initial characteristics of the PPSG and increase its performance.

Table 3. PPSG parameters with additional FD.

K_d	f_m [Hz]	G	f_{out} [Hz]	Δf_{out} [Hz]
1	1,000,000	1 2 ... 100	1000 2000 ... 10,000	1000
10	100,000	1 2 ... 100	100 200 ... 10,000	100
100	10,000	1 2 ... 100	10 20 ... 1000	10
1000	1000	1 2 ... 100	1 2 ... 100	1
10,000	100	1 2 ... 100	0.1 0.2 ... 10	0.1
100,000	10	1 2 ... 100	0.01 0.02 ... 1	0.01
1,000,000	1	1 2 ... 100	0.001 0.002 ... 0.1	0.1

4. Discussion

4.1. Structural Scheme of the Simulator for the Authentication Bit Templates and the Principle of Its Operation

Real bit templates of the internal electrical noise of desktop computers (PC), which are calculated according to the normalized autocorrelation function, have a length of 1000 bits and contain approximately the same number of zero bits "0" and single bits "1". When comparing a pair of real-time templates of one PC, it turns out that they match for most positions. Only a few positions will have inverted bits. The positions of the inverted bits do not match for different pairs of templates. A comparison of the real-time bit noise templates of two different PCs showed much less similarity. The Hamming distances between the noise templates of different PCs were 5–10 times larger than the distances between the real-time noise templates of each PC. The developed Poisson pulse sequence generator made it possible to reproduce these properties.

The generator for $q = 6$ and G = 10,000 formed a bit sequence A, which contained mainly zero bits "0", and the number of single bits "1" for every thousand bits was an average of 10. The positions of the single bits in each fragment of 1000 bits did not match. Therefore, to simulate the real-time templates of the same device, it was advisable to choose the control code of the generator G = 10,000. On average, the Hamming distance between a pair of such fragments will be 20. If at $q = 6$ the value of the control code G = 100,000, for the generated sequence B, the number of single bits "1" per thousand bits was on average equal to 100. The Hamming distance between two 1000-bit fragments of sequence B will be on average 200. The formation of fragments of bit sequences A and B is shown in Figure 8.

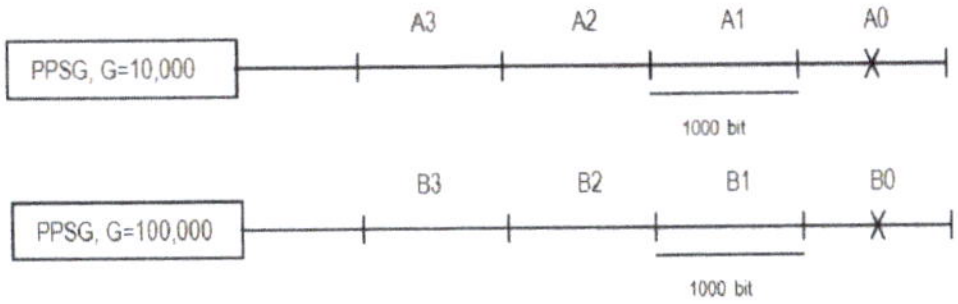

Figure 8. Derivation of groups A and B of bit sequences with a length of 1000 bits for the subsequent formation of bit templates.

From the beginning, the generation process was set, so the first 1000 bits were discarded for both sequence A ($A0$) and sequence B ($B0$).

A combination (direct sum) of fragments of sequences A and B was used to form bit templates. For each electronic device, a reference template was first created, and the real-time templates were compared with it. To form a reference bit template for electronic device N, there was a need to combine one 1000-bit fragment of sequence B, for example B_N with one 1000-bit fragment of sequence A, for example A_1, Figure 9.

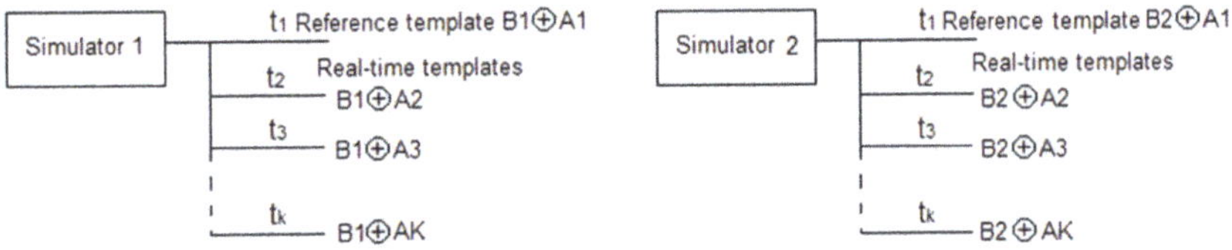

Figure 9. Formation of reference and real-time templates by two simulators, t_i is the time of template formation.

The fragment AM was used instead of A_1 to form the real-time template M of electronic device N. The bit templates of the electronic device N were calculated by the expression

$$BT_N^M = B_N \oplus A_M. \quad (15)$$

The structural scheme of the bit template simulator is shown in Figure 10.

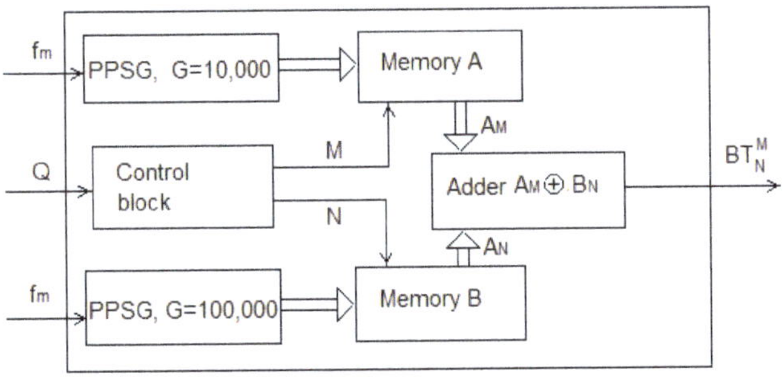

Figure 10. The structural scheme of the bit template simulator based on the PPSG.

The simulator functions were as follows. First, sequences A and B were generated and stored in the memory. To generate them, one could use one PPSG, which was started first with a control code G = 10,000, and then with a control code G = 100,000. Sequences were written to memory. Two PPSGs and two memory blocks were used to illustrate the process of forming and storing the necessary sequences in the scheme of Figure 10. The request Q for the template arrived at the control block, which sent a request to the memory for the required 1000-bit fragments of the A_M of sequence A and B_N of sequence B. Fragments A_M and B_N arrived at the adder, where the template BT_N^M was formed.

4.2. Results of the Simulation Experiment

Comparative analysis of the standard template set BT_J^1 of individual devices J was performed for ten devices. The Hamming distances $H\left(BT_J^1, BT_I^1\right)$ between pairs of standard template devices J and I were calculated. The calculated distances led to the following results in Table 4.

Table 4. The characteristics of the distance distribution between standard templates.

Average Value	Standard Deviation	Minimum Value	Maximum Value
184	10	165	205

The results in Table 4 were obtained for the 90 distances, I = 1..10, J = 1..10, I $\neq$ J. Each template was characterized by the template group—standard templates and real-time templates. For successful device authentication distances within each group needed to be significantly lower than the distances between standard templates. To check the adequacy of the proposed simulator model, two types of comparisons should be performed: the first type compares distances inside of the each group (for the each device), while the second one compares distances between different groups (between the different devices).

Simulation experiments were performed to generate templates for two devices, 10 templates for each. The distances between different templates $M \neq K$ of one device N (group $H1$, intradistances) and distances between different templates $M \neq K$ of two devices $N \neq L$ (group $H2$, interdistances) were found, only 90 distances for each group. The distances between bit templates were calculated by the following expressions.

$$\begin{aligned} H1(M,N) &= H\left(BT_N^M, BT_N^K\right), \\ H2(M,N) &= H\left(BT_N^M, BT_L^K\right). \end{aligned} \tag{16}$$

The distances for $M = K$ corresponded to the comparison of the reference template with itself for the same device and the comparison of the reference templates of two devices, and were not taken into account.

The results of calculations using Expression (16) are presented in Figure 11. The left side of the figure shows the distances between pairs of templates for the same device. (intradistances), whose numbers are indicated by columns and rows. The right side of the figure illustrates the distances between pairs of templates for different devices (interdistances).

The threshold value must be set in such a way to provide reliable authentication. For our calculations, as could be seen from Figure 12, the threshold value needed to exceed the maximum distance value of the intradistance group and be less than the minimum distance value of the interdistance group. In that case FRR and FAR were equal to zero.

The results of calculations of the distance distribution of group $H1$ for $N = 1$ and group $H2$ for $N = 1$, $L = 2$ are presented in Figure 12 as histograms.

	1	2	3	4	5	6	7	8	9	10
1	0	25	16	17	16	30	29	22	17	23
2	25	0	21	22	19	35	34	27	24	26
3	16	21	0	13	12	26	25	18	15	19
4	17	22	13	0	13	27	22	19	16	20
5	16	19	12	13	0	26	23	18	15	19
6	30	35	26	27	26	0	39	30	29	33
7	29	34	25	22	23	39	0	31	26	32
8	22	27	18	19	18	30	31	0	21	25
9	17	24	15	16	15	29	26	21	0	20
10	23	26	19	20	19	33	32	25	20	0

	1	2	3	4	5	6	7	8	9	10
1	189	203	202	199	200	208	207	206	199	203
2	203	189	205	202	201	211	210	209	204	204
3	202	205	189	201	202	210	209	208	203	205
4	199	202	201	189	199	207	206	205	200	202
5	200	201	202	199	189	208	209	206	201	203
6	208	211	210	207	208	189	215	212	209	211
7	207	210	209	206	209	215	189	213	2 6	210
8	206	209	208	205	206	212	213	189	207	209
9	199	204	203	200	201	209	206	207	189	202
10	203	204	205	202	203	211	210	209	202	189

Figure 11. Distances between pairs of templates of the same device (**top**) and two different devices (**bottom**).

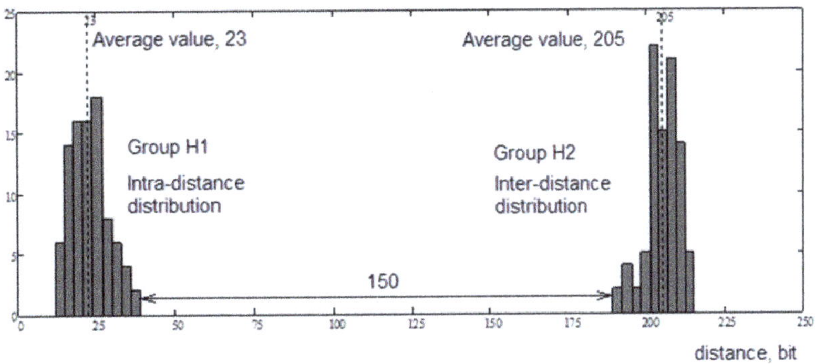

Figure 12. Histograms of the distance distribution between bit templates of one device (intradistances) and two devices (interdistances).

The characteristics of the distance distribution are shown in Table 5.

Table 5. The characteristics of the distance distribution.

Distribution	Average Value	Standard Deviation	Minimum Value	Maximum Value
Intradistances	23	6	12	39
Interdistances	205	5	189	215

The distance between the histograms was 150 bits, providing unambiguous authentication between the two devices. The average values were in good agreement with the theoretical estimates.

The repetition period for the developed generator under certain conditions was not less than 10^9 (Table 2). This allowed the estimation of the possible number of authenticated devices, which was determined by the repetition period and the length of the fragments of the sequence B, which was 10^6. The same estimate was valid for the number of authentication requests for each of the devices. These estimates determined the class of tasks for which the proposed model could be applied. For example, it could be a large enough network with up to a million devices. If you accept the service life of each device as 10 years, then such a device could be authenticated up to 250 times a day.

Let us compare the obtained simulation results with the existing practice, which uses authentication by internal electrical noise [8]. For comparison, the following parameters were selected: authentication reliability, the number of devices in the corporate network that could be simultaneously authenticated, the bit template calculation time. The results are presented in the Table 6.

Table 6. Comparison results by efficiency parameters.

Method	Reliability	Number of Devices	Measuring Time, s
Internal electric noises	98.6	175	2
Simulator based on a PPSG	100	1,000,000	-

As can be seen, the proposed method in the article provided better performance compared to the practice of authentication by internal electrical noise.

5. Conclusions

As a result of this research we executed the modeling of bit templates for information-processing electronic device authentication on the basis of the pulse Poisson sequences generator. For the purposes of the study, the Poisson pulse sequence generator was developed based on a modified additive Fibonacci generator. The developed generator had improved statistical characteristics for the output pulse sequence and expanded capabilities for solving specific practical problems.

The proposed simulator scheme contained two generators. The generator for the value of the control code $G = 10{,}000$ formed a bit sequence A, fragments of which had properties of the real-time templates of each device. The generator for the value of the control code $G = 100{,}000$ formed a bit sequence B, fragments of which reflected the difference between the series of real-time templates of different devices. In the bit template of the device, these properties were preserved by applying the action of the direct sum of fragments of sequences A and B.

An imitation experiment to generate templates for two devices confirmed the effectiveness of the proposed simulator. The properties of the generated bit templates allowed them to be used for the purpose of unambiguous authentication of information-processing electronic devices.

Further research will focus on protecting such bit templates from a variety of attacks. From the authors' point of view, the direction of detecting acoustic traps in speech recognition systems is also promising for the application of Poisson pulse sequence generators [36,37].

Author Contributions: Conceptualization, V.M. and E.N.; methodology, V.M. and C.J.; software, O.H.; validation, Y.L., M.R. and E.N.; formal analysis, C.J., M.S. and O.H.; investigation, V.M, E.N., C.J., M.S., O.H., Y.L. and M.R.; writing—original draft preparation, V.M., E.N. and M.S.; writing—review and editing, Y.L., C.J. and M.R.; supervision, E.N.; project administration, Y.L.; funding acquisition, E.N. and C.J. All authors have read and agreed to the published version of the manuscript.

Funding: This research was funded by CRDF Global, Grant Agreement G-202102-67366 "Developing software and hardware complex for dynamical authentication of information processing devices in a corporate network for cybersecurity purposes", supported by the U.S. Department of State, the Bureau of European and Eurasian Affairs.

Institutional Review Board Statement: Not applicable.

Informed Consent Statement: Not applicable.

Data Availability Statement: Not applicable.

Conflicts of Interest: The authors declare no conflict of interest.

References

1. Qureshi, M.; Munir, A. PUF-IPA: A PUF-based Identity Preserving Protocol for Internet of Things Authentication. In Proceedings of the 2020 IEEE 17th Annual Consumer Communications & Networking Conference (CCNC), Las Vegas, NV, USA, 10–13 January 2020. [CrossRef]
2. Scholz, A.; Zimmermann, L.; Sikora, A.; Tahoori, M.B.; Aghassi-Hagmann, J. Embedded Analog Physical Unclonable Function System to Extract Reliable and Unique Security Keys. *Appl. Sci.* **2020**, *10*, 759. [CrossRef]
3. Hasse, J.; Gloe, T.; Beck, M. Forensic identification of GSM mobile phones. In Proceedings of the first ACM Workshop on Information Hiding and Multimedia Security, Montpellier, France, 17–19 June 2013. [CrossRef]
4. Svoboda, J.; Schanfein, M. Transducer Signal Noise Analysis for Sensor. In Proceedings of the 53rd Annual INMM Meeting, Idaho Falls, ID, USA, 15–19 July 2012.
5. Chouchang, Y.; Alanson, P. Sample EM–ID: Tag–less Identification of Electrical Devices via Electromagnetic Emissions. In Proceedings of the 2016 IEEE International Conference on RFID (RFID), Orlando, FL, USA, 3–5 May 2016. [CrossRef]
6. Wang, X.; Zhang, Y.; Zhang, H. Identification and authentication for wireless transmission security based on RF-DNA fingerprint. *J. Wirel. Com. Netw.* **2019**, 230. [CrossRef]
7. Nyemkova, E. Authentication of Personal Computers with Unstable Internal Noise. *Int. J. Comput.* **2020**, *19*, 569–574. [CrossRef]
8. Sikora, A.; Nyemkova, E.; Lakh, Y. Accuracy Improvements of Identification and Authentication of Devices by EM-Measurements. In Proceedings of the 2020 IEEE 5th International Symposium on Smart and Wireless Systems within the Conferences on Intelligent Data Acquisition and Advanced Computing Systems (IDAACS-SWS), Dortmund, Germany, 17–18 September 2020. [CrossRef]
9. Kuźmiński, Ł. Using the Poisson Distribution to Estimate the Risk of Hydrological Danger. *Studia Ekon. Uniw. Ekon. W Katowicach* **2014**, *206*, 7–19. (In Polish)
10. Deon, A.; Menyaev, Y. Poisson Twister Generator by Cumulative Frequency Technology. *Algorithms* **2019**, *12*, 114. [CrossRef]
11. Kim, D.; Kim, J.; Cho, Y.S. A Poisson Cluster Stochastic Rainfall Generator that Accounts for the Interannual Variability of Rainfall Statistics: Validation at Various Geographic Locations across the United States. *J. Appl. Math.* **2014**, 560390. [CrossRef]
12. Bentley, M.; Stephenson, A.; Toscas, P.; Zhu, Z. A Multivariate Model to Quantify and Mitigate Cybersecurity Risk. *Risks* **2020**, *8*, 61. [CrossRef]
13. Leslie, N.O.; Harang, R.E.; Knachel, L.P.; Kott, A. Statistical Models for the Number of Successful Cyber Intrusions. *J. Def. Modeling Simul.* **2018**, *15*, 49–63. [CrossRef]
14. Veiga, A.; Spinelli, E. A Pulse Generator with Poisson-Exponential Distribution for Emulation of Radioactive Decay Events. In Proceedings of the IEEE 7th Latin American Symposium on Circuits & Systems (LASCAS), Florianopolis, Brazil, 28 February–2 March 2016. [CrossRef]
15. Arkani, M.; Khalafi, H.; Vosoughi, N. A Flexible Multichannel Digital Random Pulse Generator Based on FPGA. *World J. Nucl. Sci. Technol.* **2013**, *3*, 109–116. [CrossRef]
16. Maksymovych, V.; Mandrona, M.; Harasymchuk, O. Dosimetric Detector Hardware Simulation Model Based on Modified Additive Fibonacci Generator. In *Advances in Intelligent Systems and Computing*; Hu, Z., Petoukhov, S., Dychka, I., He, M., Eds.; Springer: Cham, Switzerland, 2020; Volume 938, pp. 162–171.
17. Maksymovych, V.N.; Harasymchuk, O.I.; Mandrona, M.N. Designing Generators of Poisson Pulse Sequences Based on the Additive Fibonacci Generators. *J. Autom. Inf. Sci.* **2017**, *49*, 1–13. [CrossRef]
18. Maksymovych, V.; Harasymchuk, O.; Opirskyy, I. The Designing and Research of Generators of Poisson Pulse Sequences on Base of Fibonacci Modified Additive Generator. In *Advances in Intelligent Systems and Computing*; Hu, Z., Petoukhov, S., Dychka, I., He, M., Eds.; Springer: Cham, Switzerland, 2018; Volume 754, pp. 43–53.
19. Pomme, S.; Keightley, J.; Fitzgerald, R. Uncertainty of Nuclear Counting. *Metrologia* **2015**, *53*. [CrossRef]
20. Takami, K.; Shin-ichi, N.; Shigeru, Y. A Generation of Random-Time Pulses Having a Poisson Distribution. *Keisoku Jido Seigyo Gakkai Ronbunshu* **1981**, *17*, 409–414.
21. Linares-Barranco, A.; Cascado, D.; Jimenez, G.; Civit, A.; Oster, M.; Linares-Barranco, B. Poisson AER Generator: Inter-Spike-Intervals Analysis. In Proceedings of the 2006 IEEE International Symposium on Circuits and Systems, Kos, Greece, 21–24 May 2006. [CrossRef]
22. Maksymovych, V.; Shabatura, M.; Harasymchuk, O.; Karpinski, M.; Jancarczyk, D.; Sawicki, P. Development of Additive Fibonacci Generators with Improved Characteristics for Cybersecurity Needs. *Appl. Sci.* **2022**, *12*, 1519. [CrossRef]

23. Mandrona, M.M.; Maksymovych, V.M.; Harasymchuk, O.I.; Kostiv, Y.M. Generator of Pseudorandom Bit Sequence with Increased Cryptographic Immunity. *Metall. Min. Ind.* **2014**, *5*, 25–29.
24. Maksymovych, V.; Harasymchuk, O.; Karpinski, M.; Shabatura, M.; Jancarczyk, D.; Kajstura, K. A New Approach to the Development of Additive Fibonacci Generators Based on Prime Numbers. *Electronics* **2021**, *10*, 2912. [CrossRef]
25. Mandrona, M.N.; Maksymovych, V.N. Comparative Analysis of Pseudorandom Bit Sequence Generators. *J. Autom. Inf. Sci.* **2017**, *49*, 78–86. [CrossRef]
26. Maksymovych, V.M.; Mandrona, M.M.; Garasimchuk, O.I.; Kostiv, Y.M. A Study of the Characteristics of the Fibonacci Modified Additive Generator with a Delay. *J. Autom. Inf. Sci.* **2016**, *48*, 76–82. [CrossRef]
27. Maksymovych, V.N.; Mandrona, M.N.; Kostiv, Y.M.; Harasymchuk, O.I. Investigating the Statistical Characteristics of Poisson Pulse Sequences Generators Constructed in Different Ways. *J. Autom. Inf. Sci.* **2017**, *49*, 11–19. [CrossRef]
28. Blanco, A.; Orúe, A.B.; López, A.; Martín, A. On-the-Fly Testing an Implementation of Arrow Lightweight PRNG Using a LabVIEW Framework. In *Advances in Intelligent Systems and Computing*; Kacprzyk, J., Ed.; Springer: Cham, Switzerland, 2019; Volume 951, pp. 175–184.
29. Jakobsson, K.S. Theory, Methods and Tools for Statistical Testing of Pseudo and Quantum Random Number Generators. Master's Thesis, Linköpings Universitet, Linköping, Sweden, 2014; p. 143.
30. Faster Randomness Testing with the NIST Statistical Test Suite. Available online: https://crocs.fi.muni.cz/_media/public/crocs/sys_space_2014.pdf (accessed on 20 December 2021).
31. Gorbenko, I.D.; Gorbenko, Y.I. *Applied Cryptology: Theory Practice Application*; Fort Publishing House: Kharkiv, Ukraine, 2012; p. 880.
32. Holland, R.; St. John, R. Chi square variants: The lehman distribution. In *Statistical Electromagnetics Book*; CRC Press: Boca Raton, FL, USA, 1999; p. 48. [CrossRef]
33. Almeida, F.M.L., Jr.; Barbi, M.; do Vale, M.A.B. A Proposal for a Different Chi-Square Function for Poisson Distributions. *Nucl. Instrum. Methods Phys. Res. Sect. A Accel. Spectrometers Detect. Assoc. Equip.* **2000**, *449*, 383–395. [CrossRef]
34. Horoneskul, M. Tables of Functions and Critical Distribution Points. Sections: Probability Theory. Mathematical Statistics, Mathematical Methods in Psychology. 2009. (in Ukrainian). Available online: http://repositsc.nuczu.edu.ua/bitstream/123456789/1530/1/Tablici.pdf (accessed on 20 December 2021).
35. NIST SP 800-22. A Statistical Test Suite for Random and Pseudorandom Number Generators for Cryptographic Applications. Available online: http://csrc.nist.gov/publications/nistpubs/800-22-rev1a/SP800-22rev1a.pdf (accessed on 20 December 2021).
36. Kwon, H.; Yoon, H.; Park, K.-W. Acoustic-decoy: Detection of adversarial examples through audio modification on speech recognition system. *Neurocomputing* **2020**, *417*, 357–370. [CrossRef]
37. Kwon, H.; Kim, Y.; Yoon, H.; Choi, D. Selective Audio Adversarial Example in Evasion Attack on Speech Recognition System. *IEEE Trans. Inf. Forensics Secur.* **2019**, *15*, 526–538. [CrossRef]

MDPI
St. Alban-Anlage 66
4052 Basel
Switzerland
Tel. +41 61 683 77 34
Fax +41 61 302 89 18
www.mdpi.com

Electronics Editorial Office
E-mail: electronics@mdpi.com
www.mdpi.com/journal/electronics

www.ingramcontent.com/pod-product-compliance
Lightning Source LLC
Chambersburg PA
CBHW061621220926
51850CB00005B/562
9783036569062